REVERSE OSMOSIS

A Guide for the Nonengineering Professional

REVERSE OSMOSIS

A Guide for the
Nonengineering Professional

Frank R. Spellman

CRC Press
Taylor & Francis Group
Boca Raton London New York

CRC Press is an imprint of the
Taylor & Francis Group, an **informa** business

CRC Press
Taylor & Francis Group
6000 Broken Sound Parkway NW, Suite 300
Boca Raton, FL 33487-2742

First issued in paperback 2019

© 2016 by Taylor & Francis Group, LLC
CRC Press is an imprint of Taylor & Francis Group, an Informa business

No claim to original U.S. Government works

ISBN-13: 978-1-4987-2750-1 (hbk)
ISBN-13: 978-0-367-37751-9 (pbk)

Library of Congress Cataloging-in-Publication Data

Spellman, Frank R.
  Reverse osmosis : a guide for the nonengineering professional / author, Frank R. Spellman.
      pages cm
  Includes bibliographical references and index.
  ISBN 978-1-4987-2750-1
  1. Water--Purification--Reverse osmosis process--Popular works. I. Title.

TD442.5.S64 2015
628.1'6744--dc23   2015004538

Visit the Taylor & Francis Web site at
http://www.taylorandfrancis.com

and the CRC Press Web site at
http://www.crcpress.com

Contents

Preface

Reverse osmosis (RO) is a membrane treatment process primarily used to separate dissolved solutes from water. To explain RO in its most basic form, *Reverse Osmosis: A Guide for the Nonengineering Professional* is written in an engaging, highly readable style. It is ideal for municipal managers, departmental and administrative assistants, equipment sales and marketing personnel, customer services representatives, and members of utility municipality boards, as well as professionals and general readers with little or no science or engineering background.

Reverse osmosis is most commonly known for its use in drinking water purification, particularly with regard to removing salt and other effluent materials from water molecules. In addition to explaining the role of RO in desalination and other freshwater applications, this text also discusses RO applications in the food industry, maple syrup production, hydrogen production, reef aquariums, and window cleaning.

Many people have heard about RO or have studied osmosis in biology classes. Unfortunately, though, there exists a wide gap in knowledge about RO, and personnel who should have a basic understanding of membrane technology instead know little about the process or have not kept pace with advancements in RO technology and applications. This is especially the case for those without science or engineering backgrounds. The concept of ion exchange technology may be familiar to some, but they may not understand how such systems actually operate. *Reverse Osmosis: A Guide for the Nonengineering Professional* is designed to fill this gap.

The book begins with a comprehensive discussion of the nuts and bolts that make up RO systems and then describes the concepts involved in RO system operation. It also provides a description of contaminants found in water today, with particular emphasis on pharmaceuticals and personal care products (PPCPs). PPCP contaminants are not typically removed from wastewater by conventional treatment processes; however, they can be removed by sophisticated membrane filtration processes. The text then further clarifies the process, the function the process serves in water or wastewater treatment, and the basic equipment required. It details how the processes fit together within a drinking water or wastewater treatment system and surveys the fundamental concepts that make up water and wastewater treatment processes as a whole.

By design, this text does not include in-depth surveys of the associated mathematics, engineering, chemistry, or biology; the math operations presented here are included simply to aid in understanding what might otherwise be complex explanations. The numerous illustrations, as well as an extensive glossary of terms, further contribute to comprehending the concepts and processes. The book is presented in conversational style to ensure that there is no failure to communicate with the reader. Failure to communicate is not an option.

Author

Frank R. Spellman, PhD, is a retired assistant professor of environmental health at Old Dominion University, Norfolk, Virginia, and is the author of more than 96 books covering topics ranging from concentrated animal feeding operations (CAFOs) to all areas of environmental science and occupational health. Many of his texts are readily available online, and several have been adopted for classroom use at major universities throughout the United States, Canada, Europe, and Russia; two have been translated into Spanish for South American markets. Dr. Spellman's works have been cited in more than 850 publications. He serves as a professional expert witness for three law groups and as an incident/accident investigator for the U.S. Department of Justice and a northern Virginia law firm. In addition, he consults nationwide on homeland security vulnerability assessments for critical infrastructures, including water/wastewater facilities, and conducts pre-Occupational Safety and Health Administration and U.S. Environmental Protection Agency audits throughout the country. Dr. Spellman receives frequent requests to co-author with well-recognized experts in several scientific fields; for example, he is a contributing author of the prestigious text *The Engineering Handbook*, 2nd ed. (CRC Press). He lectures throughout the country on wastewater treatment, water treatment, homeland security, and safety topics and teaches water/wastewater operator short courses at Virginia Tech, Blacksburg. Recently, Dr. Spellman traced and documented the ancient water distribution system at Machu Picchu, Peru, and surveyed several drinking water resources in Amazonia Coco, Ecuador. He also studied and surveyed two separate potable water supplies in the Galapagos Islands; while there, he also researched Darwin's finches. He earned a BA in public administration, a BS in business management, an MBA, and both an MS and a PhD in environmental engineering.

Acronyms

AI	Aggressive index
BAT	Best available technology
BNR	Biological nitrogen removal
BOD	Biochemical oxygen demand
CAFOs	Concentrated animal feeding operations
CBOD	Carbonaceous biochemical oxygen demand
CCL	Contaminant candidate list
CNS	Central nervous system
COD	Chemical oxygen demand
CWA	Clean Water Act
DMR	Discharge Monitoring Report
DO	Dissolved oxygen
DPR	Direct potable reuse
EBNR	Enhanced biological nutrient removal
ESWTR	Enhanced Surface Water Treatment Rule
GAC	Granular activated carbon
IMS	Integrated membrane system
IPR	Indirect potable reuse
LSI	Langelier saturation index
MBR	Membrane bioreactor
MCL	Maximum contaminant level
MCLG	Maximum contaminant level goal
MCRT	Mean cell residence time
MLE	Modified Ludzack–Ettinger
MLSS	Mixed liquor suspended solids
MLVSS	Mixed liquor volatile suspended solids
NOD	Nitrogenous oxygen demand
NOM	Natural organic matter (humic and fulvic acids)
NPDES	National Pollutant Discharge Elimination System
NSAIDs	Nonsteroidal antiinflammatory drugs
PCBs	Polychlorinated biphenyls
PMCL	Proposed maximum contaminant level
PMF	Pharmaceutical manufacturing facility
POTW	Publicly owned treatment works
PPCPs	Pharmaceuticals and personal care products
ppm	Parts per million
RAS	Return activated sludge
RASS	Return activated sludge solids
RBC	Rotating biological contactor
RO	Reverse osmosis
SC	Standard conditions

SDI	Silt density index
SDWA	Safe Drinking Water Act
SRT	Sludge retention time
SSV	Settled sludge volume
STP	Standard temperature and pressure
SVI	Sludge volume index
SWTR	Surface Water Treatment Rule
TDS	Total dissolved solids
TKN	Total Kjeldahl nitrogen
TMP	Transmembrane pressure
TSCA	Toxic Substances Control Act
TSS	Total suspended solids
TTHMs	Total trihalomethanes
USDA	U.S. Department of Agriculture
USDW	Underground source of drinking water
USEPA	U.S. Environmental Protection Agency
UV	Ultraviolet
WAS	Waste activated sludge
WASS	Waste activated sludge solids
WWTP	Wastewater treatment plant

1 Introduction

Thousands have lived without love, not one without water.

—W.H. Auden, poet

WHY A BOOK ABOUT REVERSE OSMOSIS?

Reverse osmosis (RO) is a common buzzword that most people living outside the Third World have encountered, particularly on drinking water bottle labels, as shown in Figure 1.1. It is beyond the scope of this book to discuss the pluses or minuses of bottled water, but it is important to note that the bottled water industry is growing. So, whether the reader is a bottoms-up or kick-the-habit type with regard to bottled water is not the point here. The point is that more and more people are concerned about the quality of their tap water, so they are choosing to purchase and consume bottled water instead. Why do people read the label on a bottle of water? Many read the label to find out the source of the water, what (if any) chemicals were added to the water, and how it was processed or treated. Also, as shown in Figure 1.1, labels commonly include claims about how pure the water is, pointing out that it came from a stream in some high alpine meadow and was filtered and treated by reverse osmosis, all with the intention of impressing potential consumers. Although the term reverse osmosis can often be found printed on bottled water labels and comes up in various discussions about water purification and other industrial processes, it is one of those terms not fully or even remotely understood by the average person. Thus, reverse osmosis joins the growing ranks of other buzzwords (for example, algorithms, benchmarking, fuzzy logic, real-time, podcasting, viral, tagging, and cloud computing and all of its derivatives ... private cloud, community cloud, public cloud, hybrid cloud) that we often hear or even use but only vaguely understand or do not understand at all. Sometimes it seems cool or even appropriate to use such terms. Many times our purpose, of course, is to use such terms to make others think we actually understand their real meaning; sometimes we want others to think we are intelligent and well informed, even when we are not.

For most people, reverse osmosis is just another buzzword on a label of bottled water that they know little about, unless they happen to be engineers or engineering specialists such as drinking water purification technicians; desalination specialists; technical managers or operators; water treatment plant managers or operators; wastewater water plant managers or operators; food production specialists; maple syrup producers; hydrogen producers; reef aquarium developers; window cleaning application designers; producers of nonalcoholic beverages; or professionals directly involved in several other industrial processes that use reverse osmosis.

1

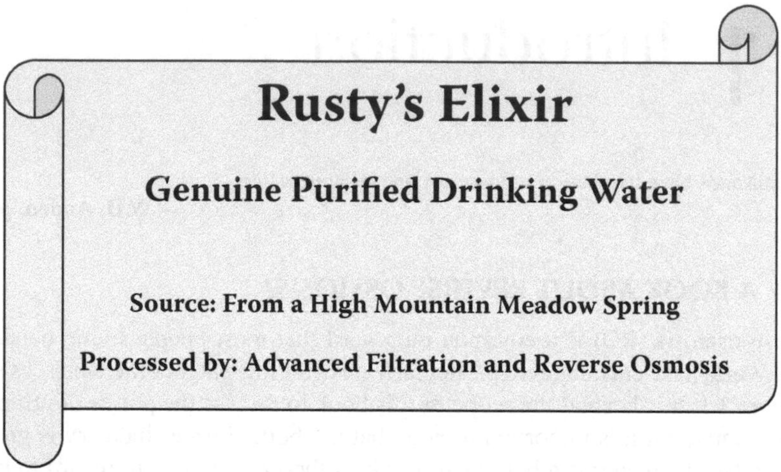

Rusty's Elixir

Genuine Purified Drinking Water

Source: From a High Mountain Meadow Spring

Processed by: Advanced Filtration and Reverse Osmosis

FIGURE 1.1   Bottled water label.

Is a lack of understanding of what RO is or what it entails or how it operates really a problem? It depends. For individuals who just want to add a technical-sounding term to their ever-expanding vocabulary without having an actual understanding of the technology itself it is not a problem. But, what about owners or managers of companies that utilize RO technology within their production or industrial processes? Certainly, such individuals would want and need to know more about the concept of reverse osmosis and should have at least a basic understanding of the process. Others who should be aware of what reverse osmosis is all about include professional salespeople who sell technical equipment directly or indirectly related to some process where RO is used.

Moreover, when RO is used in the public sector (for example, to purify drinking water for public use), general managers and other administrative executives or personnel should have knowledge of RO, how it works, and how it is used within their area of responsibility. Then there are elected officials and commissioners who have oversight responsibility for public operations in which RO is a major component. If such officials are to function as effective representatives of the people, they need to learn the nuts and bolts of reverse osmosis, at least to the point where they can think through issues concerning the technology. Also, it is always helpful when decision makers, especially those who control and allocate public funds, are able to understand presentations made to them by experts in their respective fields.

The personnel just mentioned may be professionals, but they are not necessarily engineers or scientists or technicians. They may hold positions in public administration, management, accounting, finance, sales, marketing, procurement, or any of hundreds of other areas but typically are not trained in the engineering, science, and design of reverse osmosis, even when an understanding of RO would be helpful in carrying out their job responsibilities.

So, how does a nonengineering professional become more familiar with RO, with its operation, applications, advantages and disadvantages, and limitations? An earlier publication by this author, *Water and Wastewater Treatment: A Guide for the Nonengineering Professional*, became an industrywide bestseller and was specifically lauded for its engaging, highly readable style. It became the premier guide for nonengineering professionals involved with water and wastewater operations, and a similar approach to the topic of reverse osmosis seemed called for. *Reverse Osmosis: A Guide for the Nonengineering Professional* uses step-by-step, jargon-free language to present all of the basic processes involved in reverse osmosis operations. Each process is described in basic terms and illustrated for easier understanding, including how RO units are used and function in various applications. Most importantly, the fundamental concepts of RO operations are discussed, without relying too much on mathematics, chemistry, and biology. When it is impossible to ignore mathematics, chemistry, and biology, they are presented in a simple, user-friendly form. The illustrations allow for easy comprehension of the more technical concepts and processes, and an extensive glossary of terms is provided for quick reference.

The bottom line is that whether the reader is a nonengineering professional, a nonprofessional of any type, or simply a bottled-water label reader who wants more information about reverse osmosis, this book is for you.

RECOMMENDED READING

Drinan, J. and Spellman, F.R. (2013). *Water and Wastewater Treatment: A Guide for the Nonengineering Professional*. Boca Raton, FL: CRC Press.

2 Reverse Osmosis
The Nuts and Bolts

Water is the driving force of all nature.

—**Leonardo de Vinci**

ESSENTIAL AND PRACTICAL DETAILS*

The foundation upon which reverse osmosis stands depends on the nuts and bolts of the principles behind it. Voltaire wisely stated, "If you wish to converse with me, please define your terms." To present the principles and operations of reverse osmosis (RO) in an understandable form, the terms associated with the technology are presented here in plain English. Moreover, to make sure that the essential and practical details and basic aspects of RO are easily understood, the concepts, units of expression, and pertinent nomenclature are also presented.

CONCEPTS

MISCIBILITY AND SOLUBILITY

Miscible refers to being capable of being mixed in all proportions. Simply, when two or more substances disperse themselves uniformly in all proportions when brought into contact they are said to be completely soluble in one another, or completely miscible. More precisely, a miscible solution can be defined as a "homogeneous molecular dispersion of two or more substances" (Jost, 1992). Some examples include the following:

- All gases are completely miscible.
- Water and alcohol are completely miscible.
- Water and mercury (in its liquid form) are immiscible.

Between the two extremes of miscibility lies a range of *solubility*; that is, various substances mix with one another up to a certain proportion. In many environmental situations, a rather small amount of contaminant is soluble in water in contrast to the complete miscibility of water and alcohol. The amounts are measured in parts per million (ppm).

---

* Based on Spellman, F.R., *The Science of Water: Concepts and Applications*, 3rd ed., CRC Press, Boca Raton, FL, 2015.

SUSPENSION, SEDIMENT, BIOSOLIDS, AND PARTICLES

Often water carries solids or particles in *suspension*. These dispersed particles are much larger than molecules and may be comprised of millions of molecules. The particles may be suspended in flowing conditions and initially under quiescent conditions, but eventually gravity causes settling of the particles. The resultant accumulation by settling is known as *sediment* or *biosolids* (sludge) or as residual solids in wastewater treatment vessels. Between the extremes of readily falling out due to gravity and permanent dispersal as a solution at the molecular level, there are intermediate types of dispersions or suspensions. *Particles* can be so finely milled or of such small intrinsic size as to remain in suspension almost indefinitely and in some respects similarly to solutions.

EMULSIONS

Emulsions represent a special case of a suspension. As the reader knows, oil and water do not mix. Oil and other hydrocarbons derived from petroleum generally float on water and have negligible solubility in water. In many instances, oils may be dispersed as fine oil droplets (an emulsion) in water and not readily separated by floating because of size and/or the addition of dispersal-promoting additives. Oil and, in particular, emulsions can prove detrimental to many treatment technologies and must be treated in the early steps of a multi-step treatment train.

IONS

An ion is an electrically charged particle. For example, sodium chloride (table salt) forms charged particles upon dissolution in water. Sodium is positively charged (a cation), and chloride is negatively charged (an anion). Many salts similarly form cations and anions upon dissolution in water.

CONCENTRATION

The concentration of an ion or substance in water is often expressed in terms of parts per million (ppm) or mg/L. Sometimes parts per thousand or parts per trillion (ppt) or parts per billion (ppb) are also used. These are known as units of expression. A ppm is analogous to a full shot glass of swimming pool water as compared to the entire contents of a standard swimming pool full of water. A ppb is analogous to one drop of water from an eye dropper added to the total amount of water in a standard swimming pool full of water.

$$\text{Parts per million (ppm)} = \text{Mass of substance} \div \text{Mass of solutions} \qquad (2.1)$$

Because 1 kg of a solution with water as the solvent has a volume of approximately 1 liter,

$$1 \text{ ppm} \approx 1 \text{ mg/L}$$

PERMEATE

The portion of the feed stream that passes through a reverse osmosis membrane is the permeate.

CONCENTRATE, REJECT, RETENTATE, BRINE, OR RESIDUAL STREAM

The concentrate, reject, retentate, brine, or residual stream is the membrane output stream that contains water that has not passed through the membrane barrier and concentrated feedwater constituents that are rejected by the membrane.

TONICITY

Tonicity is a measure of the effective osmotic pressure gradient (as defined by the water potential of the two solutions) of two solutions separated by a semipermeable membrane. It is important to point out that, unlike osmotic pressure, tonicity is only influenced by solutes that cannot cross this semipermeable membrane, as only these exert an effective osmotic pressure. Solutes able to freely cross do not affect tonicity because they will always be in equal concentrations on both sides of the membrane. There are three classifications of tonicity that one solution can have relative to another (Sperelakis, 2011):

- *Hypertonic* refers to a greater concentration. In biology, a hypertonic solution is one with a higher concentration of solutes outside the cell than inside the cell; the cell will lose water by osmosis.
- *Hypotonic* refers to a lesser concentration. In biology, a hypotonic solution has a lower concentration of solutes outsid e the cell than inside the cell; the cell will gain water through osmosis.
- *Isotonic* refers to a solution in which the solute and solvent are equally distributed. In biology, a cell normally wants to remain in an isotonic solution, where the concentration of the liquid inside it equals the concentration of liquid outside it; there will be no net movement of water across the cell membrane.

OSMOSIS

Osmosis is the naturally occurring transport of water through a membrane from a solution of low salt content to a solution of high salt content in order to equalize salt concentrations.

OSMOTIC PRESSURE

Osmotic pressure is a measurement of the potential energy difference between solutions on either side of a semipermeable membrane due to osmosis. Osmotic pressure is a colligative property, meaning that the property depends on the concentration of the solute, but not on its identity.

OSMOTIC GRADIENT

The osmotic gradient is the difference in concentration between two solutions on either side of a semipermeable membrane. It is used to indicate the difference in percentages of the concentration of a specific particle dissolved in a solution. Usually, the osmotic gradient is used when comparing solutions that have a semipermeable membrane between them, allowing water to diffuse between the two solutions, toward the hypertonic solution. Eventually, the force of the column of water on the hypertonic side of the semipermeable membrane will equal the force of diffusion on the hypotonic side, creating equilibrium. When equilibrium is reached, water continues to flow, but it flows both ways in equal amounts as well as force, thus stabilizing the solution.

MEMBRANE

A membrane is thin layer of material capable of separating materials as a function of their chemical or physical properties when a driving force is applied.

SEMIPERMEABLE MEMBRANE

A semipermeable membrane is a membrane permeable only by certain molecules or ions.

RO SYSTEM FLOW RATING

Although the influent and reject flows are usually not indicated, the product flow rate is used to derive an RO system flow rating. A 600-gpm RO system, for example, yields 600 gpm of permeate.

RECOVERY CONVERSION

The recovery conversion is the ratio of the permeate flow to the feed flow, which is fixed by the designer and is generally expressed as a percentage. It is used to describe what volume percentage of influent water is recovered. Exceeding the design recovery can result in accelerated and increased fouling and scaling of the membranes.

$$\% \text{ Recovery} = (\text{Recovery flow/Feed flow}) \times 100 \qquad (2.2)$$

CONCENTRATION FACTOR

The concentration factor is the ratio of solute contamination in the concentrate stream to solute concentration in the feed system. The concentration factor is related to recovery in that, at 40% recovery, for example, the concentrate would be 2/5 that of the influent water.

REJECTION

The term *rejection* is used to describe what percentage of an influent species a membrane retains. For example, 97% rejection of salt means that the membrane will retain 97% of the influent salt. It also means that 3% of influent salt will pass through the membrane into the permeate; this is known as *salt passage*. Equation 2.3 is used to calculate the rejection of a given species.

$$\% \text{ Rejection} = [(C_i - C_p)/C_i] \times 100 \tag{2.3}$$

where
C_i = Influent concentration of a specific component.
C_p = Permeate concentration of a specific component.

The RO system uses a semipermeable membrane to reject a wide variety of impurities. Table 2.1 is a partial list of the general rejection ability of the most commonly used thin-film composite (TFC) RO membranes. Note that these percentages are averaged based on experience and are generally accepted within the industry. They are not a guarantee of performance. Actual rejection can vary according to the chemistry of the water, temperature, pressure, pH, and other factors (Pure Water Products, 2014).

TABLE 2.1

Estimated Reverse Osmosis Rejection Percentages of Selected Impurities for Thin-Film Composite Membranes

Impurity	Rejection Percentage	Impurity	Rejection Percentage
Aluminum	97–98%	Lead	96–98%
Ammonium	85–95%	Magnesium	96–98%
Arsenic	95–96%	Manganese	96–98%
Bacteria	99+%	Mercury	96–98%
Bicarbonate	95–96%	Nickel	97–99%
Boron	50–70%	Nitrate	93–96%
Bromide	93–96%	Phosphate	99+%
Cadmium	96–98%	Radioactivity	95–98%
Calcium	96–98%	Radium	97%
Chloride	94–95%	Selenium	97%
Chromium	96–98%	Silica	85–90%
Copper	97–99%	Silicate	95–97%
Cyanide	90–95%	Silver	95–97%
Detergents	97%	Sodium	92.98%
Fluoride	94–96%	Sulfate	99+%
Herbicides	97%	Sulfite	96–98%
Insecticides	97%	Virus	99+%
Iron	98–99%	Zinc	98–99%

Source: Adapted from Pure Water Products, *Reverse Osmosis Rejection Percentages*, Pure Water Products, LLC, Denton, TX, 2014.

Flux

The word *flux* comes from the Latin *fluxus* ("flow") or *fluere* ("to flow") (Weekly, 1967). This term was introduced into differential calculus as *fluxion* by Sir Isaac Newton. With regard to RO systems, flux is the rate of water flow (volumetric flow rate) across a unit surface area (membrane); it is expressed as gallons of water per square foot of membrane area per day (gfd) or liters per hour per square meter (liters per square meter per hour (LMH). In general, flux is proportional to the density of flow; it varies by how the boundary faces the direction of flow and is proportional within the area of the boundary.

Specific Flux (Permeability)

Specific flux, or *permeability*, refers to the membrane flux normalized for temperature and pressure, expressed as gallons per square foot per day per pound per square inch (gfd/psi) or liters per square meter per hour per bar (LMH/bar). Specific flux is sometimes discussed when comparing the performance of one type of membrane with another. In comparing membranes, the higher the specific flux the lower the driving pressure required to operate the RO system (Kucera, 2010).

Concentration Polarization

Similar to the flow of water through a pipe (see Figure 2.1A,B), concentration polarization is the phenomenon of increased solute (e.g., salt) concentration relative to the bulk solution that occurs in a thin boundary layer at the membrane surface on the feed side (Figure 2.1C). Let's look first at Figure 2.1A, which shows that flow may be *laminar* (streamline), and then look at Figure 2.1B, where the flow may be *turbulent*. Laminar flow occurs at extremely low velocities. The water moves in straight parallel lines, called *streamlines* or *laminae*, which slide upon each other as they travel rather than mixing up. Normal pipe flow is turbulent flow, which occurs because of friction encountered on the inside of the pipe. The outside layers of flow are thrown into the inner layers, and the result is that all of the layers mix and are moving in different directions and at different velocities, although the direction of flow is forward. Figure 2.1C shows the hydraulic boundary layer formed by fluid flow through a pipe. Concentration polarization has a negative effect on the performance of an RO membrane; specifically, it reduces the throughput of the membrane (Kucera, 2010). Flow may be steady or unsteady. For our purposes, we consider steady-state flow only; that is, most of the hydraulic calculations in this text assume steady-state flow.

Membrane Fouling

Membrane fouling is a process where a loss of membrane performance occurs due to the deposition of suspended or dissolved substances on its external surfaces, at its pore openings, or within its pores, forming a fouling layer. It can also be caused by internal changes in the membrane material. Both forms of fouling can cause membrane permeability to decline.

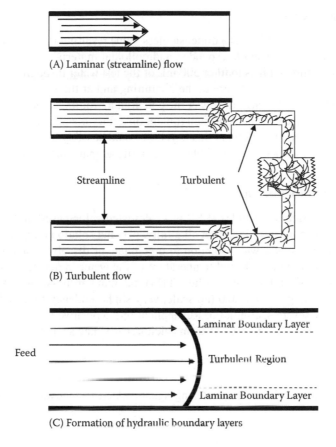

(A) Laminar (streamline) flow

Streamline   Turbulent

(B) Turbulent flow

Feed

Laminar Boundary Layer

Turbulent Region

Laminar Boundary Layer

(C) Formation of hydraulic boundary layers

FIGURE 2.1 (A) Laminar (streamline) flow. (B) Turbulent flow. (C) Hydraulic boundary layer formed when fluid flows through a pipe.

MEMBRANE SCALING

Membrane scaling is a form of fouling on the feed-concentrate side of the membrane that occurs when dissolved species are concentrated in excess of their solubility limit. Scaling is exacerbated by low cross-flow velocity and high membrane flux (Kucera, 2010).

DID YOU KNOW?

The fouling of a reverse osmosis membrane is almost inevitable. Particulate matter will be retained and is an ideal nutrient for biomass, resulting in biofouling.

SILT DENSITY INDEX

The silt density index (SDI) is a dimensionless value resulting from an empirical test used to measure the level of suspended and colloidal material in water. It is calculated from the time it takes to filter 500 mL of the test water through a 0.45-µm pore diameter filter at 30 psi pressure at the beginning and at the end of a specified test duration. The lower the SDI, the lower the potential for fouling a membrane with suspended solids. Visually, the deposited foulant on a filter membrane can be identified by its color. For example, foulant that is yellow could possibly indicate iron or organics, red foulant indicates iron, and black may indicate manganese (Kucera, 2010).

LANGELIER SATURATION INDEX

The Langelier Saturation Index (LSI) is a calculated value based on total dissolved solids, calcium concentration, total alkalinity, pH, and solution temperature. It indicates the tendency of a water solution to precipitate or dissolve calcium carbonate. The LSI is based on the pH and temperature of the water in question as well as the concentrations of total dissolved solids (TDS), calcium hardness, and alkalinity. The LSI generally ranges from <0.0 (no scale, very slight tendency to dissolve scale) to 3.0 (extremely severe scaling). For RO applications, a positive LSI indicates that the influent water has the tendency to form calcium carbonate scale (Kucera, 2010).

ANTISCALANTS

Antiscalants are chemical sequestering agents added to feedwater to inhibit scale formation.

BASIC TERMS AND DEFINITIONS

Chemistry—The science that deals with the composition and changes in composition of substances. Water is an example of this composition; it is composed of two gases: hydrogen and oxygen. Water also changes form from liquid to solid to gas but does not necessarily change composition.

Matter—Anything that has weight (mass) and occupies space. Kinds of matter include elements, compounds, and mixtures.

Solids—Substances that maintain definite size and shape. Solids in water fall into one of the following categories:

Dissolved solids—Solids in a single-phase (homogeneous) solution consisting of dissolved components and water. Dissolved solids are the material in water that will pass through a glass fiber filter and remain in an evaporating dish after evaporation of the water.

Colloidal solids (sols)—Solids that are uniformly dispersed in solution. They form a solid phase that is distinct from the water phase.

Suspended solids—Solids in a separate phase from the solution. Some suspended solids are classified as settleable solids. Settleable solids are determined by placing a sample in a cylinder and measuring the amount of solids that have settled after a set amount of time. The size of solids increases moving from dissolved solids to suspended solids. Suspended solids are the material deposited when a quantity of water, sewage, or other liquid is filtered through a glass fiber filter.

Total solids—The solids in water, sewage, or other liquids; include suspended solids (largely removable by a filter) and filterable solids (those that pass through the filter).

Liquids—Having a definite volume but not shape, liquids will fill containers to certain levels and form free level surfaces.

Gases—Having neither definite volume nor shape, gases completely fill any container in which they are placed.

Mixture—A physical, not chemical, intermingling of two of more substances. Sand and salt stirred together form a mixture.

Element—The simplest form of chemical matter. Each element has chemical and physical characteristics different from all other kinds of matter.

Compound—A substance of two or more chemical elements chemically combined. Examples include water (H_2O), which is a compound formed by hydrogen and oxygen. Carbon dioxide (CO_2) is composed of carbon and oxygen.

Atom—The smallest particle of an element that can unite chemically with other elements. All of the atoms of an element have the same chemical behavior, although they may differ slightly in weight. Most atoms can combine chemically with other atoms to form molecules.

Molecule—The smallest particle of matter or a compound that possesses the same composition and characteristics as the rest of the substance. A molecule may consist of a single atom, two or more atoms of the same kind, or two or more atoms of different kinds.

Radical—Two or more atoms that unite in a solution and behave chemically as if a single atom.

Ion—An atom or group of atoms that carries a positive or negative electric charge as a result of having lost or gained one or more electrons.

Cation—A positively charged ion.

Anion—A negatively charged ion.

Ionization—The formation of ions by the splitting of molecules or electrolytes in solution. Water molecules are in continuous motion, even at lower temperatures. When two water molecules collide, a hydrogen ion is transferred from one molecule to the other. The water molecule that loses the hydrogen ion becomes a negatively charged hydroxide ion. The water molecule that gains the hydrogen ion becomes a positively charged hydronium ion. This process is commonly referred to as the *self-ionization of water.*

Organic—Chemical substances of animal or vegetable origin made of carbon structure.

Inorganic—Chemical substances of mineral origin.

Solvent—The component of a solution that does the dissolving.

Solute—The component of a solution that is dissolved by the solvent.

Saturated solution—The physical state in which a solution will no longer dissolve more of the dissolving substance (solute).

Colloidal—Any substance in a certain state of fine division in which the particles are less than 1 micron in diameter.

Turbidity—A condition in water caused by the presence of suspended matter. Turbidity results in the scattering and absorption of light rays.

Precipitate—A solid substance that can be dissolved but is separated from the solution because of a chemical reaction or change in conditions such as pH or temperature.

GAS LAWS

Because gases can be pollutants as well as the conveyors of pollutants into various water bodies used as sources of drinking water and other types of water usage, it is important to have a fundamental understanding of the gas laws. Air (which is mainly nitrogen) is usually the main gas stream. Gas conditions are usually described in two ways: *standard temperature and pressure* (STP) and *standard conditions* (SC). STP represents 0°C (32°F) and 1 atm. The more commonly used SC value represents typical room conditions of 20°C (70°F) and 1 atm; SC is usually measured in cubic meters (m^3), normal cubic meters (Nm^3), or standard cubic feet (scf).

To understand the physics of air it is imperative to have an understanding of the various physical laws that govern the behavior of pressurized gases. One of the more well-known physical laws is *Pascal's law*, which states that a confined gas (fluid) transmits externally applied pressure uniformly in all directions, without a change in magnitude. This parameter can be seen in a container that is flexible, as it will assume a spherical (balloon) shape. The reader has probably noticed that most compressed-gas tanks are cylindrical in shape; the spherical ends contain the pressure more effectively and allow the use of thinner sheets of steel without sacrificing safety.

BOYLE'S LAW

Though gases are compressible, note that, for a given mass flow rate, the actual volume of gas passing through the system is not constant within the system due to changes in pressure. This physical property (the basic relationship between the pressure of a gas and its volume) is described by Boyle's law, named for the Irish physicist and chemist Robert Boyle, who discovered this property in 1662. It states: "The absolute pressure of a confined quantity of gas varies inversely with its volume, if its temperature does not change." For example, if the pressure of a gas doubles, its volume will be reduced by a half, and *vice versa*; that is, as pressure goes up, volume goes down, and the reverse is true. This means, for example, that if 12 ft^3 of air at 14.7 psia (pounds per square inch absolute) is compressed to 1 ft^3, air pressure will rise to 176.4 psia, as long as the air temperature remains the same. This relationship can be calculated as follows:

$$P_1 \times V_1 = P_2 \times V_2 \tag{2.4}$$

where

P_1 = Original pressure (units for pressure must be absolute).

P_2 = New pressure (units for pressure must be absolute).

V_1 = Original gas volume at pressure P_1.
V_2 = New gas volume at pressure P_2.

This equation can be rewritten as

$$P_2/P_1 = V_1/V_2 \quad \text{or} \quad P_1/P_2 = V_2/V_1 \tag{2.5}$$

To allow for the effects of atmospheric pressure, always remember to convert from gauge pressure (psig, or pounds per square inch gauge) *before* solving the problem, then convert back to gauge pressure *after* solving it.

Pounds per square inch absolute (psia) = psig + 14.7 psi

and

Pounds per square inch gauge (psig) = psia − 14.7 psi

Note that, in a pressurized gas system where gas is caused to move through the system by the fact that gases will flow from an area of high pressure to that of low pressure, we will always have a greater actual volume of gas at the end of the system than at the beginning (assuming the temperature remains constant).

CHARLES'S LAW

Another physical law dealing with temperature is Charles's law, discovered by French physicist Jacques Charles in 1787. It states: "The volume of a given mass of gas at constant pressure is directly proportional to its absolute temperature." The absolute temperature is the temperature in Kelvin (273 + °C); absolute zero = −460°F, or 0°R on the Rankine scale. This is calculated by using the following equation:

$$P_2 = P_1 \times (T_2/T_1) \tag{2.6}$$

Charles's law also states: "If the pressure of a confined quantity of gas remains the same, the change in the volume (V) of the gas varies directly with a change in the temperature of the gas," as given below:

$$V_2 = V_1 \times (T_2/T_1) \tag{2.7}$$

IDEAL GAS LAW

The ideal gas law combines Boyle's and Charles's laws because air cannot be compressed without its temperature changing. The ideal gas law can be expressed as

$$(P_1 \times V_1)/T_1 = (P_2 \times V_2)/T_2 \tag{2.8}$$

Note that the ideal gas law is still used as a design equation even though the equation shows that the pressure, volume, and temperature of the second state of a gas are equal to the pressure, volume, and temperature of the first state. In actual practice, however, other factors such as humidity, heat of friction, and efficiency losses all affect the gas. Also, this equation uses absolute pressure (psia) and absolute temperatures (°R) in its calculations.

In air science practice, the importance of the ideal gas law cannot be overstated. It is one of the fundamental principles used in calculations involving gas flow in air-pollution-related work. This law is used to calculate actual gas flow rates based on the quantity of gas present at standard pressures and temperatures. It is also used to determine the total quantity of that contaminant in a gas that can participate in a chemical reaction. The ideal gas law has three important variables:

- Number of moles of gas
- Absolute temperature
- Absolute pressure

In practical applications, practitioners generally use the following standard ideal gas law equation:

$$V = nRT/P \quad \text{or} \quad PV = nRT \tag{2.9}$$

where
V = Volume.
n = Number of moles.
R = Universal gas constant.
T = Absolute temperature.
P = Absolute pressure.

SOLUTIONS

A solution is a condition in which one or more substances are uniformly and evenly mixed or dissolved. In other words, a solution is a homogeneous mixture of two or more substances. Solutions can be solids, liquids, or gases, such as drinking water, seawater, or air. Here, we are focusing primarily on liquid solutions. A solution has two components: a solvent and a solute (see Figure 2.2). The *solvent* is the component that does the dissolving; typically, the solvent is the species present in the greater

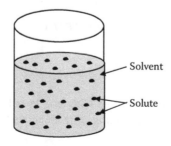

Solvent

Solute

FIGURE 2.2   Solution with two components: solvent and solute.

quantity. The *solute* is the component that is dissolved. When water dissolves substances, it creates solutions with many impurities. Generally, a solution is usually transparent and not cloudy and visible to longer wavelength ultraviolet light. Because water is colorless, the light necessary for photosynthesis can travel to considerable depths. However, a solution may be colored when the solute remains uniformly distributed throughout the solution and does not settle with time.

Basic Solution Calculations

Remember, in chemical solutions, the substance being dissolved is called the *solute*, and the liquid present in the greatest amount in a solution (and that does the dissolving) is called the *solvent*. We should also be familiar with another term, *concentration*—the amount of solute dissolved in a given amount of solvent. Concentration is measured as

$$
\begin{aligned}
\% \text{ Strength} &= \frac{\text{Weight of solute}}{\text{Weight of solution}} \times 100 \\
&= \frac{\text{Weight of solute}}{\text{Weight of solute} + \text{Weight of solvent}} \times 100
\end{aligned}
\tag{2.10}
$$

■ EXAMPLE 2.1

Problem: If 30 lb of chemical is added to 400 lb of water, what is the percent strength (by weight) of the solution?

Solution:

$$
\% \text{ Strength} = \frac{\text{Wt. of solute}}{\text{Wt. of solute} + \text{Wt. of solvent}} \times 100 = \frac{30 \text{ lb}}{30 \text{ lb} + 400 \text{ lb}} \times 100 = 7\%
$$

Important to making accurate computations of chemical strength is a complete understanding of the dimensional units involved; for example, it is important to understand exactly what *milligrams per liter* (mg/L) signifies:

$$
\text{Milligrams per liter (mg/L)} = \text{Milligrams of solute/Liters of solution} \tag{2.11}
$$

Another important dimensional unit commonly used when dealing with chemical solutions is *parts per million* (ppm):

$$
\text{Parts per million (ppm)} = \text{Parts of solute/Million parts of solution} \tag{2.12}
$$

Note: "Parts" is usually a weight measurement.

For example:

$$
9 \text{ ppm} = (9 \text{ lb solids})/(1,000,000 \text{ lb solution})
$$

This leads to two important parameters that water practitioners should commit to memory:

- 1 mg/L = 1 ppm
- 1% = 10,000 mg/L

When working with chemical solutions, it is also necessary to be familiar with the two chemical properties of *density* and *specific gravity*. Density is the weight of a substance per a unit of its volume—for example, pounds per cubic foot or pounds per gallon:

$$\text{Density} = \text{Mass of substance/Volume of substance} \qquad (2.13)$$

Here are a few key facts about density:

- Density is measured in units of lb/cf, lb/gal, or mg/L.
- Density of water = 62.5 lb/cf = 8.34 lb/gal.
- Density of concrete = 130 lb/cf.
- Density of liquid alum at 60°F = 1.33.
- Density of hydrogen peroxide (35%) = 1.132.

Specific gravity is the ratio of the density of a substance to a standard density:

$$\text{Specific gravity} = \text{Density of substance/Density of water} \qquad (2.14)$$

Here are a few facts about specific gravity:

- Specific gravity has no units.
- Specific gravity of water = 1.0.
- Specific gravity of concrete = 2.08.
- Specific gravity of liquid alum at 60°F = 1.33.
- Specific gravity of hydrogen peroxide (35%) = 1.132.

When molecules dissolve in water, the atoms making up the molecules come apart (dissociate) in the water. This dissociation in water is called *ionization*. When the atoms in the molecules come apart, they do so as charged atoms (both negatively and positively charged), which, as described earlier, are called *ions*. The positively charged ions are called *cations* and the negatively charged ions are called *anions*.

The ionization that occurs when calcium carbonate ionizes is a good example:

$$CaCO_3 \quad \leftrightarrow \quad Ca^{2+} \quad + \quad CO_3^{2-}$$

calcium carbonate	calcium ion	carbonate ion
	(cation)	(anion)

Another good example is the ionization that occurs when table salt (sodium chloride) dissolves in water:

$$NaCl \quad \leftrightarrow \quad Na^+ \quad + \quad Cl^-$$

sodium chloride	sodium ion	chloride ion
	(cation)	(anion)

Some of the common ions found in water are listed below:

Hydrogen	H^+
Sodium	Na^+
Potassium	K^+
Chloride	Cl^-
Bromide	Br
Iodide	I^-
Bicarbonate	HCO_3^-

Solutions serve as a vehicle to (1) allow chemical species to come into close proximity so they can react; (2) provide a uniform matrix for solid materials, such as paints, inks, and other coatings so they can be applied to surfaces; and (3) dissolve oil and grease so they can be rinsed away.

Water dissolves *polar substances* better than *nonpolar substances*. For example, polar substances such as mineral acids, bases, and salts are easily dissolved in water. Nonpolar substances such as oils and fats and many organic compounds do not dissolve as easily in water.

CONCENTRATIONS

Because the properties of a solution depend largely on the relative amounts of solvent and solute, the concentrations of each must be specified.

Note: Chemists use both relative terms, such as saturated and unsaturated, as well as more exact concentration terms, such as weight percentages, molarity, and normality.

Although polar substances dissolve better than nonpolar substances in water, polar substances dissolve in water only to a point; that is, only so much solute will dissolve at a given temperature. When that limit is reached, the resulting solution is saturated. At this point, the solution is in equilibrium—no more solute can be dissolved. A liquid/solids solution is supersaturated when the solvent actually dissolves more than an equilibrium concentration of solute (usually when heated).

Specifying the relative amounts of solvent and solute, or specifying the amount of one component relative to the whole, usually gives the exact concentrations of solution. Solution concentrations are sometimes specified as weight percentages.

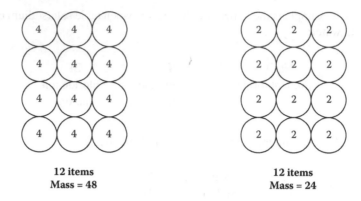

12 items
Mass = 48

12 items
Mass = 24

FIGURE 2.3 The example on the left is twice as heavy, so twice the mass is required to equal the same number of items.

MOLES

To understand the concepts of molarity, molality, and normality, we must first understand the concept of a mole. The mole is defined as the amount of a substance that contains exactly the same number of items (i.e., atoms, molecules, or ions) as 12 g of carbon-12. By experiment, Avogadro determined this number to be 6.02×10^{23} (to three significant figures). If 1 mole of carbon atoms equals 12 g, for example, what is the mass of 1 mole of hydrogen atoms? Note that carbon is 12 times heavier than hydrogen; therefore, we need only 1/12 the weight of hydrogen to equal the same number of atoms of carbon (see Figure 2.3 for an example).

Note: One mole of hydrogen equals 1 gram.

By the same principle:

- One mole of $CO_2 = 12 + 2(16) = 44$ g.
- One mole of $Cl^- = 35.5$ g.
- One mole of Ra = 226 g.

In other words, we can calculate the mass of a mole if we know the formula of the substance.

Molarity (*M*) is defined as the number of moles of solute per liter of solution. The volume of a solution is easier to measure in the lab than its mass:

$$M = \text{(No. of moles of solute)/(No. of liters of solution)}$$

Molality (*m*) is defined as the number of moles of solute per kilogram of solvent:

$$m = \text{(No. of moles of solute)/(No. of kilograms of solutions)}$$

Note: Molality is not used as frequently as molarity, except in theoretical calculations.

Especially for acids and bases, the *normality* (N) rather than the molarity of a solution is often reported. Normality is the number of equivalents of solute per liter of solution (1 equivalent of a substance reacts with 1 equivalent of another substance):

$$N = \text{(No. of equivalents of solute)/(No. of liters of solution)}$$

In acid/base terms, an *equivalent* (or gram equivalent weight) is the amount that will react with 1 mole of H^+ or OH^-; for example,

- One mole of HCl will generate 1 mole of H^+; therefore, 1 mole HCl = 1 equivalent.
- One mole of $Mg(OH)_2$ will generate 2 moles of OH^-; therefore, 1 mole of $Mg(OH)_2$ = 2 equivalents.

$$HCl \Rightarrow H^+ + Cl^-$$

$$Mg(OH)^{2+} \Rightarrow Mg^{2+} + 2OH^-$$

By the same principle:

- A 1-M solution of H_3PO_4 is 3 N.
- A 2-N solution of H_2SO_4 is 1 M.
- A 0.5-N solution of NaOH is 0.5 M.
- A 2-M solution of HNO_3 is 2 N.

Chemists titrate acid/base solutions to determine their normality. An endpoint indicator is used to identify the point at which the titrated solution is neutralized.

Note: If it takes 100 mL of 1-N HCl to neutralize 100 mL of NaOH, then the NaOH solution must also be 1 N.

PREDICTING SOLUBILITY

Predicting solubility is difficult, but there are a few general rules of thumb, such as "like dissolves like."

- *Liquid–liquid solubility*—Liquids with similar structure and hence similar intermolecular forces will be completely miscible. For example, we would correctly predict that methanol and water are completely soluble in any proportion.
- *Liquid–solid solubility*—Solids *always* have limited solubilities in liquids, in general because of the difference in magnitude of their intermolecular forces. Therefore, the closer the temperature is to its melting point, the better the match between a solid and a liquid.

Note: At a given temperature, lower melting solids are more soluble than higher melting solids. Structure is also important; for example, nonpolar solids are more soluble in nonpolar solvents.

- *Liquid–gas solubility*—As with solids, the more similar the intermolecular forces, the higher the solubility; therefore, the closer the match between the temperature of the solvent and the boiling point of the gas, the higher the solubility. When water is the solvent, an additional *hydration* factor promotes solubility of charged species. Other factors that can significantly affect solubility are temperature and pressure. In general, raising the temperature typically increases the solubility of solids in liquids.

Note: Dissolving a solid in a liquid is usually an endothermic process (i.e., heat is absorbed), so raising the temperature will fuel this process. In contrast, dissolving a gas in a liquid is usually an exothermic process (i.e., it evolves heat), so lowering the temperature generally increases the solubility of gases in liquids.

Note: Thermal pollution is a problem because of the decreased solubility of O_2 in water at higher temperatures.

Pressure has an appreciable effect on the solubility of gases in liquids. For example, carbonated beverages such as soda water are typically bottled at significantly higher atmospheres. When the beverage is opened, the decrease in the pressure above the liquid causes the gas to bubble out of solution. When shaving cream is used, dissolved gas comes out of solution, bringing the liquid with it as foam.

COLLIGATIVE PROPERTIES

Properties of a solution that depend on the concentrations of the solute species rather than their identity include the following:

- Lowering vapor pressure
- Raising boiling point
- Decreasing freezing point
- Osmotic pressure

True colligative properties are directly proportional to the concentration of the solute but entirely independent of its identity.

REFERENCES AND RECOMMENDED READING

AWWA. (2007). *Reverse Osmosis and Nanofiltration*, 2nd ed. Denver, CO: American Water Works Association.

Bird, R.B, Stewart, W.E., and Lightfoot, E.N. (1960). *Transport Phenomena*. New York: Wiley

Jost, N.J. (1992). Surface and ground water pollution control technology. In: *Fundamentals of Environmental Science and Technology* (Knowles, P.C., Ed.). Rockville, MD: Government Institutes.

Kucera, J. (2010). *Reverse Osmosis: Industrial Applications and Processes*. New York: John Wiley & Sons.

Pure Water Products. (2014). *Reverse Osmosis Rejection Percentages*. Denton, TX: Pure Water Products, LLC, http://www.purewaterproducts.com/articles/ro-rejection-rates.

Spellman, F.R. (2015). *The Science of Water: Concepts and Applications*, 3rd ed. Boca Raton, FL: CRC Press.

Sperelakis, N. (2011). *Cell Physiology Source Book: Essentials of Membrane Biophysics*. New
    York: Academic Press.
Weekley, E. (1967). *An Etymological Dictionary of Modern English*. New York: Dover.

3 Reverse Osmosis and Filtration Spectrum

No water, no life. No blue, no green.

—Sylvia Earle, marine biologist

OSMOSIS

It is difficult for most people to gain an understanding of reverse osmosis unless they first understand the principles of natural biological osmosis. In the simplest terms, osmosis can be defined as the naturally occurring process whereby water is transported through a membrane from a solution with a low salt content to a solution with a high salt content in order to equalize the salt concentration (Figure 3.1).

OSMOTIC PRESSURE

In a well-practiced experimental demonstration of osmotic pressure as applied to osmosis, water moves spontaneously from an area of high vapor pressure to an area of low vapor pressure (Figure 3.2). If this experiment were allowed to continue, in the end all of the water would move to the solution (Figure 3.3). A similar process will occur when pure water is separated from a concentrated solution by a *semipermeable membrane* (i.e., a membrane that only allows the passage of water molecules) (Figure 3.4). The osmotic pressure is the pressure that is just adequate to prevent osmosis. In dilute solutions, the osmotic pressure is directly proportional to the solute concentration and is independent of its identity (Figure 3.5).

REVERSE OSMOSIS PROCESS*

Reverse osmosis is a separation or purification process (not properly a filtration process) that uses pressure to force a solvent through a semipermeable membrane that retains the solute on one side and allows the pure solvent to pass to the other side, forcing it from a region of high solute concentration through a membrane to a region of low solute concentration by applying a pressure in excess of osmotic pressure. The difference between normal osmosis and reverse osmosis is shown in Figure 3.6. Although many solvents (liquids) may be used and many applications are described in this book, the primary application of RO discussed in this book is water-based

---

* Adapted from USEPA, *Capsule Report: Reverse Osmosis Process*, EPA/625/R-96/009, U.S. Environmental Protection Agency, Washington, DC, 1996; Spellman, F.R., *Physics for Nonphysicists*, Government Institutes, Lanham, MD, 2009.

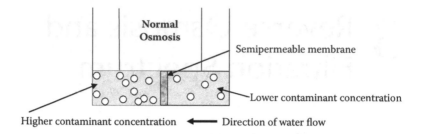

FIGURE 3.1   Osmosis process.

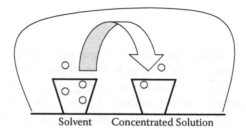

FIGURE 3.2   Osmotic pressure.

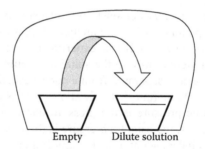

FIGURE 3.3   Osmotic pressure.

systems. Therefore, after an explanation of the RO process and its many different applications, the major emphasis of the discussion will focus on water as the liquid solvent; that is, drinking water purification, wastewater reuse, and desalination processes will be discussed in detail.

DID YOU KNOW?

Reverse osmosis (RO) is not properly a filtration method. In RO, an applied pressure is used to overcome osmotic pressure, a colligative property, that is driven by chemical potential (Pure Water Solutions, 2013).

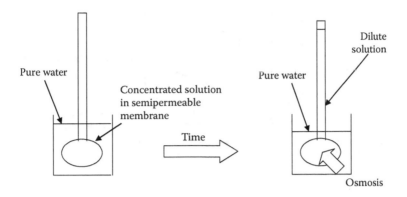

FIGURE 3.4 Passage of water molecules only, a colligative property.

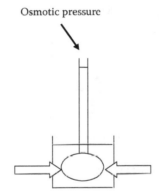

FIGURE 3.5 Osmotic pressure is the pressure just adequate to prevent osmosis.

In the RO process (Figure 3.6), water passes through a membrane, leaving behind a solution with a smaller volume and a higher concentration of solutes. The solutes can be contaminants or useful chemicals or reagents, such as copper, nickel, and chromium compounds, which can be recycled for further use in metals plating or other metal finishing processes. The recovered water can be recycled or treated downstream, depending on the quality of the water and the needs of the plant. As shown in Figure 3.7, the water that passes through the membrane is the *permeate*, and the concentrated solution left behind is the *retentate* (or concentrate).

The RO process does not require thermal energy, only an electrically driven feed pump. RO processes have simple flow sheets and a high energy efficiency. However, RO membranes can be fouled or damage. This can result in holes in the membrane

DID YOU KNOW?

The energy consumption of reverse osmosis is directly related to the salt concentration, because a high salt concentration has a high osmotic pressure.

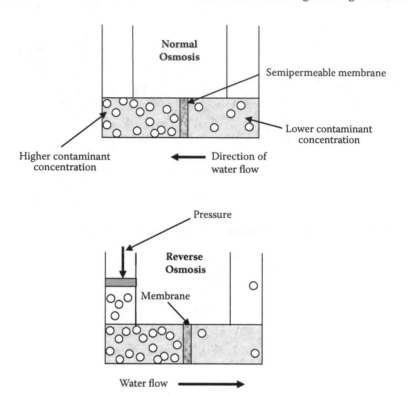

FIGURE 3.6   Normal osmosis and reverse osmosis.

and passage of the concentrated solution to clean water—and thus a release to the environment. In addition, some membrane materials are susceptible to attack by oxidizing agents, such as free chlorine.

The flux of a component *A* (recall that RO flux is the rate of water flow across a unit surface area) through an RO membrane is given by Equation 3.1:

$$N_A = P_A(\Delta\Phi/L) \tag{3.1}$$

where

N_A = Flux of component *A* through the membrane (mass/time length^2).
P_A = Permeability of *A* (mass-length/time-force).
$\Delta\Phi$ = Driving force (DF) of *A* across the membrane, either pressure difference or concentration difference (force/length^2 or mass/length^2).
L = Membrane thickness (length).

At equilibrium, the pressure difference between the two sides of the RO membrane equals the osmotic pressure difference. At low solute concentration, the osmotic pressure (π) of a solution is given by Equation (3.2):

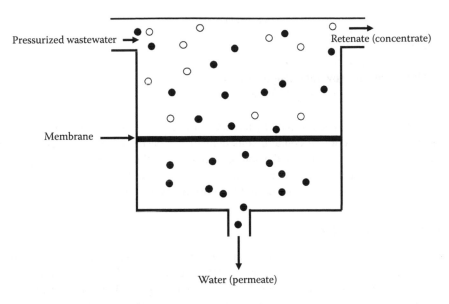

FIGURE 3.7 Reverse osmosis process.

$$\pi = C_S RT \tag{3.2}$$

where
    π = Osmotic pressure (force/length^2).
    C_S = Concentration of solutes in solution (moles/length^3).
    R = Ideal gas constant (force-length/mass-temperature).
    T = Absolute temperature (K or $^\circ$R).

As a mixture is concentrated by passing water through the membrane, osmotic pressure of the solution increases, thereby reducing the driving force for further water passage. An accurate characterization of the pressure to drive the RO process must be based on an osmotic pressure computed from the average of the feed and retentate stream compositions. The water recovery of an RO process may be expressed by Equation 3.3:

DID YOU KNOW?

It is not uncommon to confuse reverse osmosis with filtration; however, there are key differences between the two. The predominant removal mechanism in membrane filtration is straining, or size exclusion, so the process can theoretically achieve perfect exclusion of particles regardless of operational parameters such as influent pressure and concentration. On the other hand, reverse osmosis involves a diffusive mechanism so that separation efficiency is dependent on solute concentration, pressure, and water flux rate (Pure Water Solutions, 2013).

$$REC = (Q_p/Q_F) \times 100 \qquad\qquad (3.3)$$

where

REC = Water recovery (%).
Q_p = Permeate flow rate (length^2/time).
Q_F = Feed flow rate (length^2/time).

Water recovery is determined by temperature, operating pressure, and membrane surface area. Rejection of contaminants determines permeate purity, whereas water recovery primarily determines the volume reduction of the feed or amount of permeate produced. Generally, for concentrations of waters from the metal finishing industry, greater water recoveries are desirable to obtain overall greater volume reduction.

FILTRATION SPECTRUM

Again, reverse osmosis is not filtration *per se*; however, when RO is compared to other membrane processes, the comparison is generally labeled as a comparison of membrane filtration techniques. When constructing and labeling an illustrative filtration spectrum chart, comparisons are made according to the various "filtration" techniques. The relative particle sizes shown in Figure 3.8 and the filtration spectrum chart provided in Figure 3.9 compare the sizes of typical water contaminants and the various filtration technologies that can address each contaminant (a detailed and thorough discussion of water contaminants is presented later). Note that the figures highlight the importance of pretreatment before a RO unit to avoid premature pluggage by larger suspended solids.

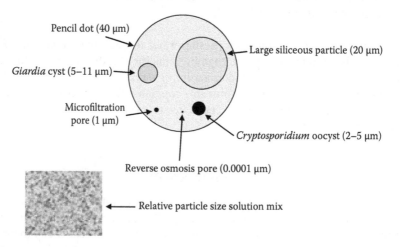

FIGURE 3.8 Relative particle sizes. Note the 0.0001 pore size characteristic of reverse osmosis which allows the RO membrane to prevent passage of larger contaminants. (Adapted from Wachinski, A.M., *Membrane Processes for Water Reuse*, McGraw-Hill, New York, 2013.)

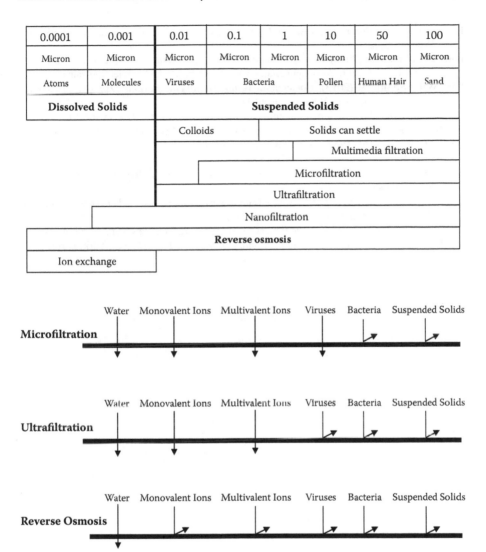

0.0001	0.001	0.01	0.1	1	10	50	100
Micron	Micron	Micron	Micron	Micron	Micron	Micron	Micron

FIGURE 3.9 Reverse osmosis is able to reject almost all of the contaminants of concern to public health officials. (Adapted from Puretec, *Filtration Spectrum*, puretecwater.com/resources/filtration-spectrum.pdf, 2014.)

REFERENCES AND RECOMMENDED READING

AWWA. (2007). *Reverse Osmosis and Nanofiltration*, 2nd ed. Denver, CO: American Water Works Association.

Bird, R.B, Stewart, W.E., and Lightfoot, E.N. (1960). *Transport Phenomena*. New York: Wiley.

Jost, N.J. (1992). Surface and ground water pollution control technology. In: *Fundamentals of Environmental Science and Technology* (Knowles, P.C., Ed.). Rockville, MD: Government Institutes.

Kucera, J. (2010). *Reverse Osmosis: Industrial Applications and Processes*. New York: John Wiley & Sons.

Pure Water Products. (2014). *Reverse Osmosis Rejection Percentages*. Denton, TX: Pure Water Products, LLC, http://www.purewaterproducts.com/articles/ro-rejection-rates.

Pure Water Solutions. (2013). *Commercial Water Treatment: Reverse Osmosis*, http://www.purewatersolutions.us/Reverse_Osmosis.html.

Puretec. (2014). *Filtration Spectrum*, puretecwater.com/resources/filtration-spectrum.pdf.

Spellman, F.R. (2009). *Physics for Nonphysicists*. Lanham, MD: Government Institutes.

Spellman, F.R. (2015). *The Science of Water: Concepts and Applications*, 3rd ed. Boca Raton, FL: CRC Press.

Sperelakis, N. (2011). *Cell Physiology Source Book: Essentials of Membrane Biophysics*. New York: Academic Press.

USEPA. (1996). *Capsule Report: Reverse Osmosis Process*, EPA/625/R-96/009. Washington, DC: U.S. Environmental Protection Agency.

Wachinski, A.M. (2013). *Membrane Processes for Water Reuse*. New York: McGraw-Hill.

Weekley, E. (1967). *An Etymological Dictionary of Modern English*. New York: Dover.

4 Reverse Osmosis Equipment and System Configuration

> Water is the only drink for a wise man.
>
> —Henry David Thoreau

MEMBRANE MATERIALS

The membrane material refers to the substance from which the membrane itself is made. Normally, the membrane material is manufactured from a synthetic polymer, although other forms, including ceramic and metallic "membranes," may be available. Currently, almost all membranes manufactured for drinking water production are made of polymeric material, because they are significantly less expensive than membranes constructed of other materials.

The material properties of the membrane may significantly impact the design and operation of the filtration system. For example, membranes constructed of polymers that react with oxidants commonly used in drinking water treatment should not be used with chlorinated feed water. Mechanical strength is another consideration, as a membrane with greater strength can withstand larger transmembrane pressure (TMP) levels, allowing for greater operational flexibility and the use of higher pressures with pressure-based direct integrity testing. Similarly, a membrane with bidirectional strength may allow cleaning operations or integrity testing to be performed from either the feed or the filtrate side of the membrane. Material properties influence the exclusion characteristics of a membrane as well. A membrane with a particular surface charge may achieve enhanced removal of particulate or microbial contaminants of the opposite surface charge due to electrostatic attraction. In addition, a membrane can be characterized as being hydrophilic (i.e., water attracting or, as the author defines it, water loving) or hydrophobic (i.e., water repelling or water hating). These terms describe the ease with which membranes can be wetted, as well as the propensity of the material to resist fouling to some degree.

Reverse osmosis (RO) membranes are generally manufactured from cellulose acetate or polyamide materials (and their respective derivatives), and there are various advantages and disadvantages associated with each. While cellulose membranes are susceptible to biodegradation and must be operated within a relatively narrow pH range of about 4 to 8, they do have some resistance to continuous low-level oxidant exposure. In general, for example, chlorine doses of 0.5 mg/L or less may control biodegradation as well as biological fouling without damaging the membrane.

Polyamide (PA) membranes, by contrast, can be used under a wide range of pH conditions and are not subject to biodegradation. Although PA membranes have very limited tolerance for the presence of strong oxidants, they are compatible with weaker oxidants such as chloramines. PA membranes require significantly less pressure to operate and have become the predominant material used for RO applications.

In a symmetric membrane, the membrane is uniform in density or pore structure throughout the cross-section, whereas in an asymmetric membrane the density of the membrane material changes across the cross-sectional area. Some asymmetric membranes have a graded construction, in which the porous structure gradually decreases in density from the free to the filtrate side of the membrane. In other asymmetric membranes, there may be a distinct transition between the dense filtration layer (i.e., the skin) and the support structure. The more densely skinned layer is exposed to the feed water and acts as the primary filtration barrier, while the thicker and more porous understructure serves primarily as mechanical support. Some hollow fibers may be manufactured as single- or double-skinned membranes, with the double skin providing filtration at both the outer and inner walls of the fibers. Like the asymmetric skinned membranes, composite membranes also have a thin, dense layer that serves as the filtration barrier. However, in composite membranes the skin is a different material than the porous substructure onto which it is cast. This surface layer is designed to be thin so as to limit the resistance of the membrane to the flow of water, which passes more freely through the porous substructure. RO membrane construction is typically either asymmetric or composite.

MEMBRANE MODULES

The module is the housing that contains the membrane. Membrane modules are commercially available in four configurations:

- Plate-and-frame
- Spiral-wound
- Hollow-fiber
- Tubular

PLATE-AND-FRAME MODULES

The plate-and-frame configuration is one of the earliest membrane models developed. As shown in Figure 4.1, plate-and-frame modules use flat sheet membranes that are layered between spacers and supports. The supports also form a flow channel

---

DID YOU KNOW?

Plate-and-frame modules are relatively easy to clean, which makes them ideal for use in high suspended solids applications. The best cleaning technique involves removing the plates and hand-cleaning each individual sheet of the membrane.

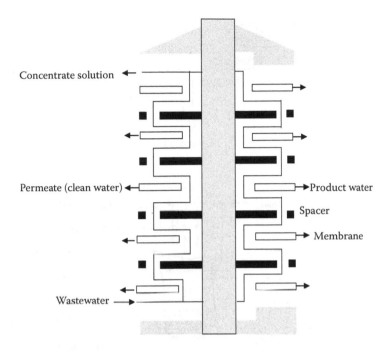

FIGURE 4.1   Plate-and-frame reverse osmosis module.

for the permeate water. The feed water flows across the flat sheets and from one layer to the next. Because of the very low surface area-to-volume ratio, the plate-and-frame configuration is considered inefficient and is therefore seldom used in drinking water applications. Recent innovations have increased the packing densities for new designs of plate-and-frame modules. Maintenance on plate-and-frame modules is possible due to the nature of their assembly. They offer high recoveries with their long feed channels and are used to treat feed streams that often cause fouling problems. Advanced designs of plate-and-frame modules capable of operating up to 25% dissolved solids and operating pressures up to 4500 psia (pounds per square in absolute) have been placed in operation in Germany. This development opens new opportunities for the use of reverse osmosis for concentration of metal finishing wastewaters.

SPIRAL-WOUND MODULES

Spiral-wound modules were developed as an efficient configuration for the use of semipermeable membranes to remove dissolved solids, and thus are most often associated with RO processes. The basic unit of a spiral-wound module is a sandwich arrangement of flat membrane sheets, called a *leaf*, which is wound around a central perforated tube (Figure 4.2). One leaf consists of two membrane sheets placed back to back and separated by a fabric spacer called a *permeate carrier*. The layers of the leaf are glued along three edges, and the unglued edge is sealed around the perforated central tub. A single spiral-wound module 8 inches in diameter may contain

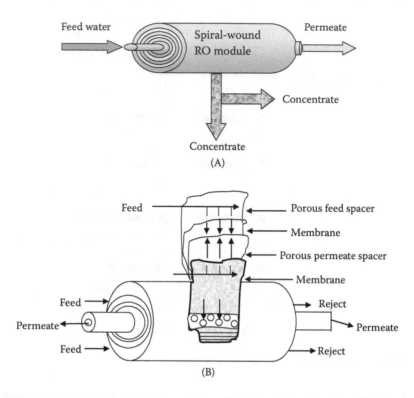

FIGURE 4.2 (A) Cutaway view of a spiral-wound RO module consisting of internal-wound product spacers, RO membranes, feed spacers, and RO membranes. (B) Internal construction of spiral-wound module.

up to approximately 20 leaves, each separated by a layer of plastic mesh, a *spacer*, that serves as the feed water channel. Feed water enters the spacer channels at the end of the spiral-wound element in a path parallel to the central tube. As the feed water flows across the membrane surface through the spaces, a portion permeates through either of the two surrounding membrane layers and into the permeate carrier, leaving behind any dissolved and particulate contaminants that are rejected by the semipermeable membrane. The filtered water in the permeate carrier travels spirally inward around the element toward the central collector tube, while the water in the feed spacer that does not permeate through the membrane layer continues to flow across the membrane surface, becoming increasingly concentrated in rejected contaminants. This concentrated stream exits the element parallel to the central tube through the opposite end from which the feed water entered.

DID YOU KNOW?

The spiral-bound membrane is the most commonly used module in RO systems.

HOLLOW-FIBER MODULES

Most hollow-fiber modules used in drinking water treatment applications are manufactured to accommodate porous membranes and designed to filter particulate matter. As the name suggests, these modules are comprised of hollow-fiber membranes (see Figure 4.3), which are long and very narrow tubes that may be constructed of

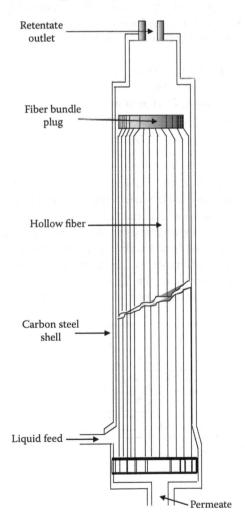

FIGURE 4.3 Hollow-fiber module.

any of the various membrane materials described earlier. The fibers may be bundled in one of several different arrangements. In one common configuration used by many manufacturers, the fibers are bundled together longitudinally, potted in a resin on both ends, and encased in a pressure vessel that is included as a part of the hollow-fiber module. These modules are typically mounted vertically, although horizontal mounting may also be utilized. One alternative configuration is similar to spiral-wound modules in that both are inserted into pressure vessels that are independent of the module itself. These modules (and the associated pressure vessels) are mounted horizontally. Another configuration in which the bundled hollow fibers are mounted vertically and submerged in a basin does not utilize a pressure vessel. A typical commercially available hollow-fiber module may consist of several hundred to over 10,000 fibers. Hollow-fiber modules offer the greatest packing densities of the other module configurations described in this section. Figure 4.3 shows a hollow-fiber module. Although specific dimensions vary by manufacturer, approximate ranges for hollow-fiber construction are as follows:

- Outside diameter—0.5 to 2.0 mm
- Inside diameter—0.3 to 1.0 mm
- Fiber wall thickness—0.1 to 0.6 mm
- Fiber length—1 to 2 m

TUBULAR MODULES

Tubular membranes are essentially a larger, more rigid version of hollow-fiber membranes. Tubular module have membranes supported within the inner part of tubes. The operator can easily service feed and permeate channels to remove fouling layers. Tubular modules are somewhat resistant to fouling when operated with a turbulent feed flow. This is accomplished with larger flow channels than those used with hollow-fiber and spiral-wound modules. The drawbacks of tubular modules are their high energy requirements for pumping large volumes of water, high capital costs, and low membrane surface area per unit volume of module (see Figure 4.4).

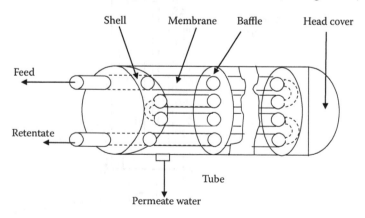

FIGURE 4.4   Tubular module.

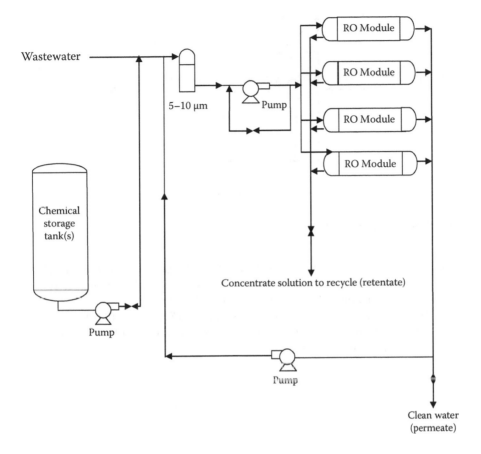

FIGURE 4.5 Reverse osmosis system.

SYSTEM CONFIGURATION

Figure 4.5 illustrates a schematic of an early 1990s RO system with four modules in parallel, chemical pretreatment, and an upfront filtration step. Figure 4.6 illustrates the typical symbol used in membrane schematics. Figure 4.7 illustrates a typical RO membrane system with one influent stream (i.e., feed) and two effluent streams (i.e., permeate and concentration) that is commonly used today. As shown in the figure, a typical RO membrane systems consists of three separate subsystems: pretreatment, the membrane process, and posttreatment.

All sources of input to an RO system must undergo some type and level of pretreatment. Pretreatment is necessary because RO thin-film composite membranes are subject to fouling by many substances:

- *Biological fouling*—Bacteria, microorganisms, viruses, and protozoans; pretreatment accomplished by chlorination
- *Particle fouling*—Suspended solids, sand, clay, and turbidity ingredients; pretreatment accomplished by filtration

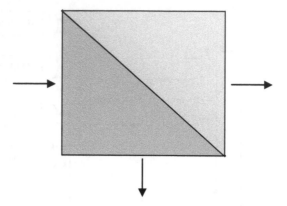

FIGURE 4.6 Typical membrane symbol.

- *Colloidal fouling*—Organic and inorganic complexes, colloidal particles, and microalgae; pretreatment accomplished by coagulation and filtration, with flocculation and sedimentation typically included
- *Organic fouling*—Natural organic matter (NOM), including humic and fulvic acids; pretreatment accomplished by coagulation, filtration, and activated carbon adsorption

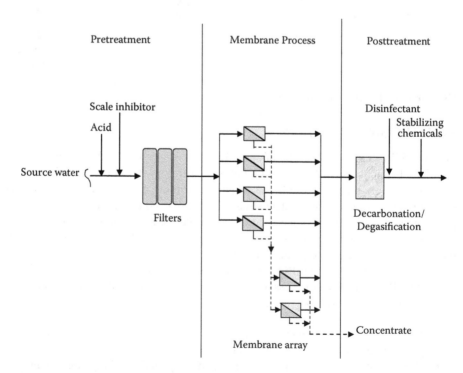

FIGURE 4.7 Typical RO membrane system.

- *Mineral fouling*—Calcium, magnesium, barium, or strontium sulfates and carbonates; pretreatment accomplished by acidification and antiscalant dosing
- *Oxidant fouling*—Chlorine, ozone, $KMNO_4$ (potassium permanganate); pretreatment accomplished by oxidant scavenger dosing with sodium (metabisulfite and granulated activated carbon)

Pretreatment processes usually involve adding acid, scale inhibitor, or both to prevent precipitation of sparingly soluble salts as the reject ions become more concentrated, followed by cartridge filtration (5-micron) as the last step to protect the RO membranes from damage and particulate fouling (i.e., debris, sand, and piping materials). In some cases, additional pretreatment is required upstream of the filter cartridges for those input waters with higher fouling potential. Posttreatment usually includes units processes common to conventional drinking water treatment, such as aeration, degasification pH adjustment, addition of corrosion control chemicals, fluoridation, and disinfection and other unit processes that are discussed later.

REFERENCES AND RECOMMENDED READING

AWWA. (2007). *Reverse Osmosis and Nanofiltration*, 2nd ed. Denver, CO: American Water Works Association.
Bailer, H.H., Lykins, Jr., B.W., Fronk, C.A., and Kramer, S.H. (1987). Using reverse osmosis to remove agricultural chemicals from groundwater. *J. AWWA*, 79(8): 55–60.
Bird, R.B, Stewart, W.E., and Lightfoot, E.N. (1960). *Transport Phenomena*. New York: Wiley.
Kucera, J. (2010). *Reverse Osmosis: Industrial Applications and Processes*. New York: John Wiley & Sons.
Pure Water Products. (2014). *Reverse Osmosis Rejection Percentages*. Denton, TX: Pure Water Products, LLC, http://www.purewaterproducts.com/articles/ro-rejection-rates.
Spellman, F.R. (2015). *The Science of Water: Concepts and Applications*, 3rd ed. Boca Raton, FL: CRC Press.
Sperelakis, N. (2011). *Cell Physiology Source Book: Essentials of Membrane Biophysics*. New York: Academic Press.
Wachinski, A.M. (2013). *Membrane Processes for Water Reuse*. New York: McGraw-Hill.
Weekley, E. (1967). *An Etymological Dictionary of Modern English*. New York: Dover.

5 Water Contaminants

ARE WE TO WAIT UNTIL ALL FROGS "CROAK"?*

Are we to wait until all frogs "croak"?

The earliest chorus of frogs—those high-pitched rhapsodies of spring peepers, those "jug-o-rum" calls of bullfrogs, those banjo-like bass harmonies of green frogs, those long and guttural cadences of leopard frogs, their singing—a prelude to the splendid song of birds beside an otherwise still pond on an early spring evening—heralds one of nature's most dramatic events: metamorphosis. This metamorphosis begins with masses of eggs that soon hatch into gill-breathing, herbivorous, fishlike tadpole larvae. As they feed and grow, warmed by the spring sun, almost imperceptibly a remarkable transformation begins. Hind legs appear and gradually lengthen. Tails shorten. Larval teeth vanish, and lungs replace gills. Eyes develop lids. Forelegs emerge. In a matter of weeks, the aquatic, vegetarian tadpole (should it escape the many perils of the pond) will complete its metamorphosis into an adult, carnivorous frog.

This springtime metamorphosis is special. This anticipated event (especially for the frog) marks the end of winter, the rebirth of life, a rekindling of hope (especially for mankind). This yearly miracle of change sums up in a few months each spring what occurred over 3000 million years ago, when the frog evolved from its ancient predecessor. Today, however, something is different, strange, and wrong with this striking and miraculous event.

In the first place, where are all the frogs? Where have they gone? Why has their population decreased so dramatically in recent years? The second problem is that this natural metamorphosis process (perhaps a reenactment of some Paleozoic drama whereby, over countless generations, the first amphibian types equipped themselves for life on land) now demonstrates aberrations of the worst kind, of monstrous proportions and with dire results for frog populations in certain areas. The U.S. Environmental Protection Agency has received reports about deformed frogs in certain sections of the United States, particularly Minnesota, as well as in Canada and parts of Europe.

Most of the deformities have been in the rear legs and appear to be developmental. The question is why? Researchers have noted that neurological abnormalities have also been found. Again, the question is why?

Researchers have pointed the finger of blame at parasites, pesticides and other chemicals, ultraviolet radiation, acid rain, and metals. Something is going on. What is it? We do not know!

---

* Adapted from Spellman, F.R., *The Science of Water*, 3rd ed., CRC Press, Boca Raton, FL, 2015.

The next question, then, is what are we going to do about it? Are we to wait until all the frogs croak before we act—before we find the source, the cause, the polluter—before we see similar results in other species … maybe our own?

The final question is obvious: When frogs are forced by mutation into becoming something else, is this evolution by gunpoint?

Are humans holding the gun?

WATER CONSTITUENTS

To prevent the croaking of frogs and the rest of us, it is extremely important (to all of us) to ensure that we are ingesting water and other fluids and foods that will do us no harm. With regard to our drinking water supplies and other water or fluid usage, there are a number of steps we can take to ensure our health and well-being, and many of them are discussed in this book, including pretreatment, treatment, and posttreatment technology that works (when properly applied and operated) to ensure our safety. Before we discuss reverse osmosis (RO) system applications, how the technology is used it clean up our drinking water, and its other important uses, we must have some basic appreciation for and knowledge of the constituents and contaminants contained or possibly contained in our water.

On a personal note, in the not too distant past, while teaching a college classroom full of environmental health, science, and engineering undergraduate and graduate students, I was lecturing on water quality. Specifically, I was discussing the major water constituents and contaminants that can be present in many sources of freshwater selected to be used for drinking water. In my mind, the only thing worse than presenting a monotone lecture for three hours is having to *listen* to the same. So, to capture students' attention and to ensure that something (anything) I present to them registers on their brain cells and keeps them off their iPads and awake, I have found that storytelling (with silly embellishments that they can relate to) is a great attention getter; it actually works.

In the beginning minutes of this particular lecture session, when I mentioned that there is no such thing as naturally pure water at ground level on Earth, I was instantly flagged down (as I knew I would be) by one of my super-smart (really—all my students are smart) undergrad students who was stirring up that butterfly effect of flapping wings we all feel even when flapped at a great distance from us. However, instead of flapping filamentous wings in South America, she was flapping both arms to get my total attention. This she did because I usually notice what a student who sits directly in front of me during a lecture is doing, or not doing.

So, I paused and acknowledged that the over-eager student had the floor. I asked if she had a question (don't they all)?

"No, no, not a question … but I completely disagree [I love when this happens] with your statement that there is no such thing as perfectly clear or pure rainwater."

"Oh," I said. "Please explain what you mean."

I noticed that she had the full attention of all 78 students, all of whom were eagerly awaiting her explanation. They were hooked; it does not get any better than this.

The student stated, "Well, my family and I collect rainwater in barrels that we have placed below our house gutters and downspouts."

"Yes," I said.

"Yes," she replied. "You see, my family and I do not like to drink tapwater or that phony bottled water, so we collect and drink pure rainwater."

"What makes you think rainwater is pure?" I asked.

"Well, professor, any fool [giggles and laughter from the students; a deadpan look from me], make that *anyone*, knows that water from the heavens above has not had the opportunity to be contaminated from exposure to human pollution."

"Hmmmm. I see. Well, have you ever tested your collected rainwater to determine it purity?"

"Nope, no need. When it falls from the sky, well, it just has to be clean and pure and wholesome—the purest form of water available on the Earth. The fact is that evaporation of the water due to the sun allows any contaminants to be left behind. Rainwater is much better than any of that water that is treated at the local waterworks or bottled water that is supposedly taken from some pure mountain stream."

"I see," I said. To myself I thought: Let her dig the hole deeper so that only I can bale her out with the facts ... and some common sense.

She continued, "We also use the rainwater for our garden vegetables so that they are not contaminated by that horrible tapwater stuff. You know, they add disinfectant to it. It just about gags a maggot to drink it, and it stinks terrible."

I have to admit that after this particular comment I stood in front of that classroom full of bright-eyed and bushy-tailed students almost in total shock. Almost. Now I am certain that there are those out there in la-la land reading this account and wondering how I was going to counter-argue the student's assertion that rainwater is the purest form of water available on Earth. Well, in the first place, at my advanced age and having been the recipient of countless amounts of real-world experience, both good and bad, I had learned not to argue with anyone. Why should I? Experience has taught me that calmly showing and telling can be a hundred times more effective as a training tool than any form or amount of argument. Thus, after regaining my cool, I turned to my usual friends to help me show the student and her fellow students the error of her statements. The friends? Well, for me that could only be the Rabbit and the Grasshopper, as well as, in this particular instance, the addition of Mr. Squirrel, of course.

As a learned rebuttal, I related the following to that classroom of some of America's finest.

A RECENT RABBIT AND GRASSHOPPER CONVERSATION

Grasshopper stated to his friend Rabbit: "You know, dear friend, I am careful, downright cautious, about the water I drink and wallow in at times, including rainwater."

Rabbit replied, "I've always preferred rainwater myself because it is so pure."

While they continued their discussion, Mr. Squirrel scampered from a copse of vine-maple below them, and Grasshopper and Rabbit watched him move off to the left. Mr. Squirrel, seemingly in a hurry, was constantly jerking his head to the right, over his shoulder. He dashed toward the marshy bank and stopped to sniff the ground. Some noise or an odor carried on the wind suddenly seemed to startle Mr. Squirrel, and he caught sight of Grasshopper and Rabbit. He scurried toward them and away from the tall marsh grass.

Mr. Squirrel said, "Hi, friends! What you up to?"

Grasshopper and Rabbit replied in unison that, "We were just discussing the purity of rainfall for drinking and just generally wallowing around in."

Mr. Squirrel looked at his two friends in wonder while scratching at some itch near his bushy tail. "The purity of rainwater? Gee, the only concern I ever have about rainwater is will there be enough of it? Will it rain enough for me to quench my thirst? I'm not worried about purity. If it's wet, that's all that matters to me."

"Well," Rabbit stated, "I just think that the water we take from humans' water bottles and sinks and hoses and puddles is not as good as basic rainwater. Rainwater is the purest form of water available on Earth, or so they say. Except for Grasshopper, that is. He thinks I might be mistaken about the purity of rainwater. What do you think, Mr. Squirrel?"

Looking all bright-eyed and bushy-tailed (remember this saying originated with the accepted description of a squirrel) and full of energy, vigor, and more energy, Mr. Squirrel replied: "Hey, wet is wet, and I love to drink up rainwater, any and all rainwater. Who cares about purity?"

After deliberate and well-practiced thumping of his foot, Rabbit replied to his friend, Mr. Squirrel: "Grasshopper has told me that, although rainwater is considered the purest form of water, it is often much less than pure when it reaches the surface of the Earth. As rain moves through the atmosphere, it picks up particles and impurities that are in the air. Chemicals and pollutants are among the impurities that can be picked up by the raindrops. These chemicals can drastically change the purity of water as it falls toward the ground."

"Wow," said Mr. Squirrel, while looking for the nearest puddle to quench his thirst.

Rabbit asked, "So, what do you think now, Mr. Squirrel?"

Pausing a moment to scratch again and having spotted a puddle a few feet away, Mr. Squirrel said, "Right on, Grasshopper!" and scampered away to the puddle.

Classroom stories like this one actually work and can deliver the intended message. For some reason, the students understand and get the message better. I have found out that one good Grasshopper, Rabbit, and Mr. Squirrel story is better than any lecture I have ever given or ever could give. There has never been a failure to communicate with my students.

Let's fast forward to our discussion of water constituents. Natural water, whether rainwater or not, can contain a number of substances, or what we call impurities or *constituents*. When a particular constituent can affect the quality or health of the water user, it is called a *contaminant* or *pollutant*. It is these contaminants that the environmental specialist or waterworks operator tries to prevent from entering the drinking water supply.

SOLIDS

Other than gases, all contaminants of water contribute to the solids content. Natural waters carry a lot of dissolved solids and non-dissolved solids. The non-dissolved solids are nonpolar substances and relatively large particles of materials, such as *silt*, that will not dissolve. Classified by their size and state, by their chemical characteristics, and by their size distribution, solids can be dispersed in water in both suspended and dissolved forms. With regard to size, solids in water can be classified as

- Suspended solids
- Settleable solids
- Colloidal solids
- Dissolved solids

Total solids are those solids, both suspended and dissolved, that remain behind when the water is removed by evaporation. Solids are also characterized as being *volatile* or *nonvolatile*. Colloidal solids are extremely fine suspended solids less than 1 micron in diameter that can still make water cloudy; they are so small they will not settle even if allowed to sit quietly for days or weeks.

> **Note:** Though not technically accurate from a chemical point of view because some finely suspended material can actually pass through the filter, *suspended solids* are defined as those that can be filtered out in the suspended solids laboratory test. The material that passes through the filter is defined as *dissolved solids*.

Turbidity

One of the first characteristics people notice about water is its *clarity*. Turbidity is a condition in water caused by the presence of suspended matter, resulting in the scattering and absorption of light rays. In plain English, turbidity is a measure of the light-transmitting properties of water. Natural water that is very clear (low turbidity) allows one to see images at considerable depths. High-turbidity water appears cloudy. Even water with low turbidity, however, can still contain dissolved solids. Dissolved solids do not cause light to be scattered or absorbed; thus, the water looks clear. High turbidity causes problems in water treatment because the components that cause high turbidity can cause taste and odor problems and will reduce the effectiveness of disinfection.

Color

Water can be colored, but often the color of water can be deceiving. In fact, color is considered an aesthetic quality of water and has no direct health impact. Many of the colors associated with water are not "true" colors but the result of colloidal suspension. This *apparent color* can be attributed to dissolved tannin extracted from decaying plant material. *True color* is the result of dissolved chemicals, most often organics that cannot be seen.

DISSOLVED OXYGEN

Gases can be dissolved in water; for example, gases such as oxygen, carbon dioxide, hydrogen sulfide, and nitrogen dissolve in water. Gases dissolved in water are important. Carbon dioxide, for example, is important because of the role it plays in pH and alkalinity. Carbon dioxide is released into the water by microorganisms and

TABLE 5.1

Common Metals Found in Water

Metal	Health Hazard
Barium	Circulatory system effects and increased blood pressure
Cadmium	Concentration in the liver, kidneys, pancreas, and thyroid
Copper	Nervous system damage and kidney effects; toxic to humans
Lead	Same as copper
Mercury	Central nervous system (CNS) disorders
Nickel	CNS disorders
Selenium	CNS disorders
Silver	Gray skin
Zinc	Taste effects; not a health hazard

consumed by aquatic plants. Dissolved oxygen (DO) in water is of most importance to waterworks operators because it is an indicator of water quality. Just as solutions can become saturated with solute, this is also the case with water and oxygen. The amount of oxygen that can be dissolved at saturation depends on the temperature of the water. In the case of oxygen, however, the effect is just the opposite of other solutes. The higher the temperature, the lower the saturation level; the lower the temperature, the higher the saturation level.

METALS

Metals are constituents or impurities often carried by water. At normal levels, most metals are not harmful, but a few metals can cause taste and odor problems in drinking water. Some metals may be toxic to humans, animals, and microorganisms. Most metals enter water as part of compounds that ionize to release the metal as positive ions. Table 5.1 lists some metals commonly found in water and their potential health hazards.

ORGANIC MATTER

Organic compounds contain the element carbon and are derived from material that was once alive (i.e., plants and animals). Organic compounds include fats, dyes, soaps, rubber products, plastics, wood, fuels, cotton, proteins, and carbohydrates. Organic compounds in water are usually large, nonpolar molecules that do not dissolve well in water. They often provide large amounts of energy to animals and microorganisms.

INORGANIC MATTER

Inorganic matter or inorganic compounds are carbon free, not derived from living matter, and easily dissolved in water. Inorganic matter is of mineral origin and includes acids, bases, salts, etc. Several inorganic components are important in establishing and controlling water quality.

TABLE 5.2
Relative Strengths of Acids in Water

Acid	Formula
Perchloric acid	$HClO_4$
Sulfuric acid	H_2SO_4
Hydrochloric acid	HCl
Nitric acid	HNO_3
Phosphoric acid	H_3PO_4
Nitrous acid	HNO_2
Hydrofluoric acid	HF
Acetic acid	CH_3COOH
Carbonic acid	H_2CO_3
Hydrocyanic acid	HCN
Boric acid	H_3BO_3

Acids

An acid is a substance that produces hydrogen ions (H^+) when dissolved in water. Hydrogen ions are hydrogen atoms that have been stripped of their electrons. A single hydrogen ion is nothing more than the nucleus of a hydrogen atom. Lemon juice, vinegar, and sour milk are acidic or contain acid. The common acids used in treating water are hydrochloric acid (HCl), sulfuric acid (H_2SO_4), nitric acid (HNO_3), and carbonic acid (H_2CO_3). Note that in each of these acids, hydrogen (H) is one of the elements. The relative strengths of acids in water, listed in descending order of strength, are classified in Table 5.2.

Bases

A base is a substance that produces hydroxide ions (OH^-) when dissolved in water. Bitter things, such as lye or common soap, contain bases. The bases used in water-works operations are calcium hydroxide, $Ca(OH)_2$; sodium hydroxide, $NaOH$; and potassium hydroxide, KOH. Note that the hydroxyl group (OH) is found in all bases. In addition, note that bases contain metallic substances, such as sodium (Na), calcium (Ca), magnesium (Mg), and potassium (K). These bases contain the elements that produce the alkalinity in water.

Salts

When acids and bases chemically interact, they neutralize each other. The compounds (other than water) that form from the neutralization of acids and bases are salts. Salts constitute, by far, the largest group of inorganic compounds. A common salt used in waterworks operations, copper sulfate, is utilized to kill algae in water.

pH

pH is a measure of the hydrogen ion (H^+) concentration. Solutions range from very acidic (having a high concentration of H^+ ions) to very basic (having a high concentration of OH^- ions). The pH scale ranges from 0 to 14, with 7 being the neutral value. The pH of water is important to the chemical reactions that take place within water, and pH values that are too high or low can inhibit the growth of microorganisms. High pH values are considered basic, and low pH values are considered acidic. Stated another way, low pH values indicate a high H^+ concentration, and high pH values indicate a low H^+ concentration. Because of this inverse logarithmic relationship, there is a tenfold difference in H^+ concentration.

Natural water varies in pH depending on its source. Pure water has a neutral pH, with an equal number of H^+ and OH^-. Adding an acid to water causes additional positive ions to be released, so the H^+ ion concentration goes up and the pH value goes down. Changing the hydrogen ion activity in solutions can shift the chemical equilibrium of water. Thus, pH adjustment is used to optimize coagulation, softening, and disinfection reactions, as well as for corrosion control. To control water coagulation and corrosion, it is necessary for the waterworks operator to test for the hydrogen ion concentration of the water to get the pH. In coagulation tests, as more alum (acid) is added, the pH value is lowered. If more lime (alkali, or base) is added, the pH value is raised. This relationship is important, and if good floc is formed the pH should then be determined and maintained at that pH value until there is a change in the new water.

ALKALINITY

Alkalinity is defined as the capacity of water to accept protons (positively charged particles); it can also be defined as a measure of the ability of the water to neutralize an acid. Stated in even simpler terms, alkalinity is a measure of the capacity of the water to absorb hydrogen ions without a significant pH change (i.e., to neutralize acids). Bicarbonates, carbonates, and hydrogen cause alkalinity compounds in a raw or treated water supply. Bicarbonates are the major components, because of the action of carbon dioxide on the basic materials of soil; borates, silicates, and phosphates may be minor components. Alkalinity of raw water may also contain salts formed from organic acids, such as humic acid. Alkalinity in water acts as a *buffer* that tends to stabilize and prevent fluctuations in pH. It is usually beneficial to have significant alkalinity in water because it would tend to prevent quick changes in pH. Quick changes in pH interfere with the effectiveness of the common water treatment processes. Low alkalinity also contributes to the corrosive tendencies of water. When alkalinity is below 80 mg/L, it is considered to be low.

HARDNESS

Hardness may be considered a physical or chemical parameter of water. It represents the total concentration of calcium and magnesium ions, reported as calcium carbonate. Hardness causes soaps and detergents to be less effective and contributes to scale formation in pipes and boilers. Hardness is not considered a health hazard; however, water that contains hardness must often be softened by lime precipitation or ion exchange. Low hardness contributes to the corrosive tendencies of water. Hardness and alkalinity often occur together because some compounds can contribute both alkalinity and hardness ions. Hardness is generally classified as shown in Table 5.3.

TABLE 5.3
Water Hardness Classification

Classification	mg/L CaCo$_3$
Soft	0–75
Moderately hard	75–150
Hard	150–00
Very hard	Over 300

IMPORTANT PROPERTIES OF WATER

SOLUBILITY

Compounds that can form hydrogen bonds with water tend to be far more soluble in water than compounds that cannot form hydrogen bonds.

SURFACE TENSION

Water has a high surface tension. Surface tension governs surface phenomena and is an important factor in physiology.

DENSITY

Density is mass per unit volume. Water has its maximum liquid density at 4°C. When water freezes, the resulting ice floats.

BOILING POINT

In general, the boiling point increases with molecular weight, but hydrogen bonding increases the boiling point of water above that predicted based on molecular weight alone.

HEAT CAPACITY

Heat capacity is the amount of energy required to raise the temperature of a substance 1°C. Water has a higher heat capacity than any other liquid except for ammonia. This attribute allows organisms and geographical regions to stabilize temperature.

HEAT OF VAPORIZATION

Heat of vaporization is the energy required to change a liquid to a vapor. Water has a higher heat of vaporization than any other material. This attribute affects the transfer of water molecules between surface water and the atmosphere.

Latent Heat of Fusion

The heat of fusion is the energy required to change a substance from a solid (ice) to a liquid (water).

Phase Transitions of Water

A phase transition is the spontaneous conversion of one phase to another that occurs at a characteristic temperature for a given pressure. For example, at 1 atm, ice is the stable phase of water below 0°C, but above this temperature the liquid is more stable. A phase diagram of water, for example, is a map of the ranges of pressure and temperature at which each phase of the water is the most stable.

WATER MICROBIOLOGY

Microorganisms are significant in water and wastewater because of their roles in disease transmission and they are the primary agents of biological treatment.

> **Note:** To have microbiological activity, the body of water or wastewater must have the appropriate environmental conditions. The majority of wastewater treatment processes, for example, are designed to operate using an aerobic process. The conditions required for aerobic operation include (1) sufficient free, elemental oxygen; (2) sufficient organic matter (food); (3) sufficient water; (4) enough nitrogen and phosphorus (nutrients) to permit oxidation of the available carbon materials; (5) proper pH (6.5 to 9.0); and (6) lack of toxic materials.

Microbiology: What Is It?

Biology is generally defined as the study of living organisms (i.e., the study of life). Microbiology is a branch of biology that deals with the study of microorganisms so small in size that they must be studied under a microscope. Microorganisms of interest to water and wastewater practitioners and those involved in the operation of RO systems include bacteria, protozoa, viruses, algae, and others. With regard to microbes, the primary concern is how to control microorganisms that cause waterborne diseases (waterborne pathogens) to protect the consumer (human and animal).

WATER AND WASTEWATER MICROORGANISMS

Microorganisms of interest to water and wastewater operators include bacteria, protozoa, viruses, algae, rotifers, fungi, and nematodes. These organisms are the most diverse group of living organisms on Earth and occupy important niches in the ecosystem. Their simplicity and minimal survival requirements allow them to exist in diverse situations. Because microorganisms are a major health issue, water treatment specialists are concerned about controlling the waterborne pathogens (e.g., bacteria, virus, protozoa) that cause waterborne diseases. The focus of wastewater operators, on the other hand, is on the millions of organisms that arrive at the plant with the influent. The majority of these organisms are nonpathogenic and beneficial to plant operations. From a microbiological standpoint, the mix of microorganism species depends on the characteristics of the influent, environmental conditions, process design, and mode of plant operation. This mix may also include pathogenic organisms responsible for diseases such as typhoid, tetanus, hepatitis, dysentery, and gastroenteritis, among others.

To understand how to minimize or maximize the growth of microorganisms and control pathogens, one must study the structure and characteristics of the microorganisms. The sections that follow will look at each of the major groups of microorganisms (those important to water/wastewater operators) with regard to their size, shape, types, nutritional needs, and control.

Note: In a water environment, water is not a medium for the growth of microorganisms but is instead a means of transmission (that is, it serves as a conduit; hence, the name *waterborne*) of the pathogen to the place where an individual is able to consume it and thus begin an outbreak of disease (Koren, 1991). This is contrary to the view taken by the average person. That is, when the topic of waterborne disease is brought up, we might mistakenly assume that waterborne diseases are at home in water. Nothing could be further from the truth. A water-filled ambience is not the environment in which the pathogenic organism would choose to live, if it had such a choice. The point is that microorganisms do not normally grow, reproduce, languish, and thrive in watery surroundings. Pathogenic microorganisms temporarily residing in water are simply biding their time, going with the flow, waiting for their opportunity to meet up with their unsuspecting host or hosts. To a degree, when the pathogenic microorganism finds its host or hosts, it is finally home or may have found its final resting place. The good news is that when we incorporate RO membranes in the source feedwater system that supplies our potable water systems, because the membranes are semipermeable and not porous, they are able to screen pathogenic microorganisms and particulate matter from our water supplies. This is the case whether the pathogen is right at home or not; the bottom line is we do not want them in our homes (Spellman, 1997).

KEY TERMS

Algae, simple—Plants, many microscopic, containing chlorophyll. Freshwater algae are diverse in shape, color, size, and habitat. They are the basic link in the conversion of inorganic constituents in water into organic constituents.

Algal bloom—Sudden spurts of algal growth which can affect water quality adversely and indicate potentially hazardous changes in local water chemistry.

Anaerobic—Able to live and grow in the absence of free oxygen.

Autotrophic organisms—Produce food from inorganic substances.

Bacteria—Single-celled microorganisms that possess rigid cell walls. They may be aerobic, anaerobic, or facultative. They can cause disease, but some are important in pollution control.

Biogeochemical cycle—The chemical interactions among the atmosphere, hydrosphere, and biosphere.

Coliform organisms—Microorganisms found in the intestinal tract of humans and animals. Their presence in water indicates fecal pollution and potentially adverse contamination by pathogens.

DID YOU KNOW?

Reverse osmosis membranes provide between 4- and 5-log (i.e., 99.99 to 99.999%) removal of viruses normally associated with waterborne disease (Lozier et al., 1994).

Denitrification—The anaerobic biological reduction of nitrate to nitrogen gas.
Fungi—Simple plants lacking the ability to produce energy through photosynthesis.
Heterotrophic organism—Organisms that are dependent on organic matter for foods.
Prokaryotic cell—The simple cell type, characterized by a lack of nuclear membrane and the absence of mitochondria.
Virus—The smallest form of microorganisms capable of causing disease.

MICROORGANISM CLASSIFICATION AND DIFFERENTIATION

The microorganisms we are concerned with are tiny organisms that make up a large and diverse group of free-living forms; they exist as single cells, cell bunches, or clusters. Found in abundance almost anywhere on Earth, the vast majority of microorganisms are not harmful. Many microorganisms, or microbes, occur as single cells (unicellular), others are multicellular, and still others (viruses) do not have a true cellular appearance. A single microbial cell, for the most part, exhibits the characteristic features common to other biological systems, such as metabolism, reproduction, and growth.

CLASSIFICATION

Greek scholar and philosopher Aristotle classified animals based on fly, swim, and walk/crawl/run. For centuries thereafter, scientists simply classified the forms of life visible to the naked eye as either animal or plant. We began to have trouble differentiating microorganisms, though, so this system of classification had to be revised. The Swedish naturalist Carolus Linnaeus organized much of the current knowledge about living things in 1735. The importance of organizing or classifying organisms cannot be overstated, for without a classification scheme it would be difficult to establish a criteria for identifying organisms and to arrange similar organisms into groups. Probably the most important reason for classifying organisms is to make things less confusing (Wistriech and Lechtman, 1980). Linnaeus was quite innovative in his classification of organisms. One of his innovations still with us today is the *binomial system of nomenclature*. Under the binomial system, all organisms are generally described by a two-word scientific name: *genus* and *species*. Genus and species are groups that are part of a hierarchy of groups of increasing size, based on their taxonomy:

Kingdom
Phylum
Class
Order
Family
Genus
Species

Using this system, a fruit fly might be classified as

Animalia
Arthropoda
Insecta
Diptera
Drosophilidae
Drosophila
melanogaster

This means that this organism is the species *melanogaster* in the genus *Drosophila* in the family Drosophilidae in the order Diptera in the class Insecta in the phylum Arthropoda in the kingdom Animalia.

To further illustrate how the hierarchical system is exemplified by the classification system, the standard classification of the mayfly is provided below:

Kingdom	Animalia
Phylum	Arthropoda
Class	Insecta
Order	Ephermeroptera
Family	Ephemeridae
Genus	*Hexagenia*
Species	*limbata*

Utilizing this hierarchy and Linnaeus' binomial system of nomenclature, the scientific name of any organism includes both the generic and specific names. To uniquely name a species, it is necessary to supply both the genus and the species; for our examples, those would be *Drosophila melanogaster* for the fruit fly and *Hexagenia limbota* for the mayfly. The first letter of the generic name is usually capitalized; for example, *E. coli* indicates that *coli* is the species and *Escherichia* (abbreviated to *E.*) is the genus. The largest, most inclusive category is the kingdom. The genus and species names are always in Latin, so they are usually printed in italics. Some organisms also have English common names. Microbe names of particular interest in water/wastewater treatment include

- *Escherichia coli* (a coliform bacterium)
- *Salmonella typhi* (the typhoid bacillus)
- *Giardia lamblia* (a protozoan)

TABLE 5.4

Simplified Classification of Microorganisms

Kingdom	Members	Cell Classification
Animal	Rotifers	
	Crustaceans	
	Worms and larvae	
Plant	Ferns	Eucaryotic
	Mosses	
Protist	Protozoa	
	Algae	
	Fungi	
	Bacteria	Prokaryotic
	Lower algae forms	

- *Shigella* spp.
- *Vibrio cholerae*
- *Campylobacter*
- *Leptospira* spp.
- *Entamoeba histolytica*
- *Crytosporidia*

Note: *Escherichia coli* is commonly referred to as simply *E. coli*, and *Giardia lamblia* is usually referred to by only its genus name, *Giardia*.

Generally, we use a simplified system of microorganism classification in water science by breaking down the classification into the kingdoms of Animalia, Plantae, and Protista. As a general rule, the animal and plant kingdoms contain all of the multicell organisms, and the protist kingdom includes all single-cell organisms. Along with a microorganism classification based on the animal, plant, and protist kingdoms, microorganisms can be further classified as being *eucaryotic* or *prokaryotic* (see Table 5.4).

Note: A eucaryotic organism is characterized by a cellular organization that includes a well-defined nuclear membrane. The prokaryotes have a structural organization that sets them off from all other organisms. They are simple cells characterized by a nucleus *lacking* a limiting membrane, an endoplasmic reticulum, chloroplasts, and mitochondria. They are remarkably adaptable and exist abundantly in the soil, sea, and freshwater.

DIFFERENTIATION

Differentiation among the higher forms of life is based almost entirely on morphological (form or structure) differences; however, differentiation (even among the higher forms) is not as easily accomplished as we might expect, because normal variations among individuals of the same species occur frequently. Because of this variation, even within a species, securing accurate classifications when dealing with

TABLE 5.5
Forms of Bacteria

Form	Technical Name		Example
	Singular	**Plural**	
Sphere	Coccus	Cocci	*Streptococcus*
Rod	Bacillus	Bacilli	*Bacillus typhosis*
Curved or spiral	Spirillum	Spirilla	*Spirillum cholera*

single-celled microscopic forms that present virtually no visible structural differences becomes extremely difficult. Under these circumstances, it is necessary to consider physiological, cultural, and chemical differences, as well as structure and form. Differentiation among the smaller groups of bacteria is based almost entirely on chemical differences.

BACTERIA

The simplest wholly contained life systems are bacteria or prokaryotes, which are the most diverse group of microorganisms. They are among the most common microorganisms in water. They are primitive, unicellular (single-celled) organisms possessing no well-defined nucleus and presenting a variety of shapes and nutritional needs. Bacteria contain about 85% water and 15% ash or mineral matter. The ash is largely composed of sulfur, potassium, sodium, calcium, and chlorides, with small amounts of iron, silicon, and magnesium. Bacteria reproduce by binary fission.

Note: Binary fission occurs when one organism splits or divides into two or more new organisms.

Bacteria, once considered the smallest living organism (although now it is known that smaller forms of matter exhibit many of the characteristics of life), range in size from 0.5 to 2 µm in diameter and about 1 to 10 µm long.

Note: A *micron* is a metric unit of measurement equal to 1/1000 of a millimeter. To visualize the size of bacteria, consider that about 1000 average bacteria lying side by side would reach across the head of a straight pin.

Bacteria are categorized into three general groups based on their physical form or shape (although almost every variation has been found; see Table 5.5). The simplest form is the sphere. Spherical-shaped bacteria are called *cocci* (meaning "berries"). They are not necessarily perfectly round but may be somewhat elongated, flattened on one side, or oval. Rod-shaped bacteria are called *bacilli*. Spiral-shaped bacteria, called *spirilla*, have one or more twists and are never straight. Such formations are usually characteristic of a particular genus or species. Within these three groups are many different arrangements. Some exist as single cells; others as pairs, as packets of four or eight, as chains, or as clumps.

Most bacteria require organic food to survive and multiply. Plant and animal material that gets into the water provides the food source for bacteria. Bacteria convert the food to energy and use the energy to make new cells. Some bacteria can use inorganics (e.g., minerals such as iron) as an energy source and can multiply even when organics (pollution) are not available.

Bacterial Growth Factors

Several factors affect the rate at which bacteria grow, including temperature, pH, and oxygen levels. The warmer the environment, the faster the rate of growth. Generally, for each increase of 10°C, the growth rate doubles. Heat can also be used to kill bacteria. Most bacteria grow best at neutral pH. Extreme acidic or basic conditions generally inhibit growth, although some bacteria may require acidic conditions and some alkaline conditions for growth.

Bacteria are aerobic, anaerobic, or facultative. If *aerobic*, they require free oxygen in the aquatic environment. *Anaerobic* bacteria exist and multiply in environments that lack dissolved oxygen. *Facultative* bacteria (e.g., iron bacteria) can switch from aerobic to anaerobic growth or grow in an anaerobic or aerobic environment.

Under optimum conditions, bacteria grow and reproduce very rapidly. As noted previously, bacteria reproduce by *binary fission*. An important point to consider with regard to bacterial reproduction is the rate at which the process can take place. The total time required for an organism to reproduce and the offspring to reach maturity is the *generation time*. Bacteria growing under optimal conditions can double their number about every 20 to 30 minutes. Obviously, this generation time is very short compared with that of higher plants and animals. Bacteria continue to grow at this rapid rate as long as nutrients hold out—even the smallest contamination can result in a sizable growth in a very short time.

> **Note:** Even though wastewater can contain bacteria counts in the millions per milliliter, in wastewater treatment under controlled conditions bacteria can help to destroy and identify pollutants. In such a process, bacteria stabilize organic matter (e.g., activated sludge processes) and thereby assist the treatment process in producing effluent that does not impose an excessive oxygen demand on the receiving body. Coliform bacteria can be used as an indicator of pollution by human or animal wastes.

Destruction and Removal of Bacteria

In water and wastewater treatment, the destruction of bacteria is usually referred to as *disinfection*. Disinfection does not mean that all microbial forms are killed. That would be *sterilization*. Instead, disinfection reduces the number of disease-causing organisms to an acceptable number. Growing bacteria are generally easy to control by disinfection; however, some bacteria form survival structures known as *spores*, which are much more difficult to destroy. It is these survival structures that necessitate the use of RO membranes (and other membrane technologies) to remove them. RO systems are highly effective in removing bacterium such as *Campylobacter*, *Salmonella*, *Shigella*, and *E. coli*.

Note: Inhibiting the growth of microorganisms is termed *antisepsis,* whereas destroying them is called *disinfection.*

Waterborne Bacteria

All surface waters contain bacteria, and these waterborne bacteria are responsible for infectious epidemic diseases. Bacterial numbers increase significantly during storm events when streams are high. Heavy rainstorms increase stream contamination by washing material from the ground surface into the stream. After the initial washing occurs, few impurities are left to be washed into the stream, which may then carry relatively "clean" water. A river of good quality shows its highest bacterial numbers during rainy periods; however, a much-polluted stream may show the highest numbers during low flows because of the constant influx of pollutants. Water and wastewater operators are primarily concerned with bacterial pathogens responsible for disease. These pathogens enter potential drinking water supplies through fecal contamination and are ingested by humans if the water is not properly treated and disinfected.

Note: Regulations require that owners of all public water supplies collect water samples and deliver them to a certified laboratory for bacteriological examination at least monthly. The number of samples required is usually in accordance with federal standards, which generally require that one sample per month be collected for each 1000 persons served by the waterworks.

PROTOZOA

Reverse osmosis systems have a very high effectiveness in removing protozoa. Protozoans (or "first animals") are a large group of eucaryotic organisms of more than 50,000 known species belonging to the kingdom Protista. They have adapted a form of cell that serves as the entire body; in fact, protozoans are one-celled, animal-like organisms with complex cellular structures. In the microbial world, protozoans are giants, many times larger than bacteria. They range in size from 4 to 500 μm. The largest ones can almost be seen by the naked eye. They can exist as solitary or independent organisms, such as the stalked ciliates (e.g., *Vorticella* sp.) (Figure 5.1), or they can colonize (e.g., the sedentary *Carchesium* sp.). Protozoa get their name because they employ the same type of feeding strategy as animals; that is, they are *heterotrophic,* meaning that they obtain cellular energy from organic substances such as proteins. Most are harmless, but some are parasitic. Some forms have two life stages: *active trophozoites* (capable of feeding) and *dormant cysts.*

The major groups of protozoans are based on their method of locomotion (motility). The Mastigophora are motile by means of one or more *flagella,* the whip-like projection that propels the free-swimming organisms (*Giardia lamblia* is a flagellated protozoan). The Ciliophora move by means of shortened modified flagella called *cilia,* which are short hair-like structures that beat rapidly and propel them through the water. The Sarcodina rely on *amoeboid movement,* which is a streaming or gliding action; the shape of amoebae changes as they stretch and then contract to move from place to place. The Sporozoa, in contrast, are nonmotile; they are simply swept along, riding the current of the water.

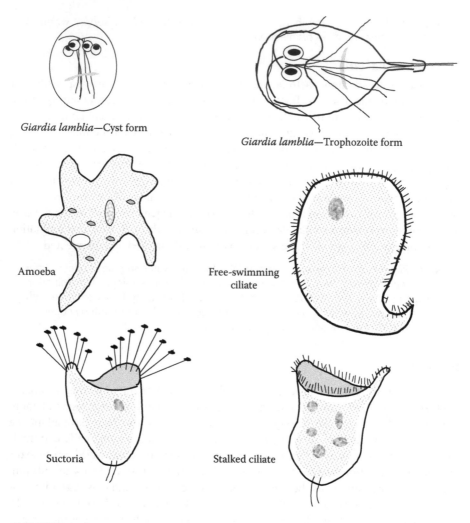

FIGURE 5.1 Protozoa.

Protozoa consume organics to survive; their favorite food is bacteria. Protozoa are mostly aerobic or facultative with regard to their oxygen requirements. Toxic materials, pH, and temperature affect protozoan rates of growth in the same way as they affect bacteria.

Most protozoan life cycles alternate between an active growth phase (*trophozoites*) and a resting stage (*cysts*). Cysts are extremely resistant structures that protect the organism from destruction when it encounters harsh environmental conditions—including chlorination.

> **Note:** Protozoans that are not completely resistant to chlorination require higher disinfectant concentrations and longer contact time for disinfection than normally used in water treatment.

The protozoans and associated waterborne diseases of most concern to water-works operators include

- *Entamoeba histolytica* (amoebic dysentery)
- *Giardia lamblia* (giardiasis)
- *Cryptosporidium* (cryptosporidiosis)

In wastewater treatment, protozoa are a critical part of the purification process and can be used to indicate the condition of treatment processes. Protozoa normally associated with wastewater include amoebae, flagellates, free-swimming ciliates, and stalked ciliates.

Amoebae are associated with poor wastewater treatment of a young biosolids mass (see Figure 5.1). They move through wastewater by a streaming or gliding motion. Moving the liquids stored within the cell wall effects this movement. They are normally associated with an effluent high in biochemical oxygen demand (BOD) and suspended solids.

Flagellates (flagellated protozoa) have a single, long, hair-like or whip-like projection (flagellum) that is used to propel the free-swimming organisms through wastewater and to attract food (see Figure 5.1). Flagellated protozoans are normally associated with poor treatment and a young biosolids. When the predominant organism is flagellated protozoa, the plant effluent will contain high levels of BOD and suspended solids.

The *free-swimming ciliated protozoan* uses its tiny, hair-like projections (cilia) to move itself through the wastewater and to attract food (see Figure 5.1). This type of protozoan is normally associated with a moderate biosolids age and effluent quality. When the free-swimming ciliated protozoan is the predominant organisms, the plant effluent will normally be turbid and contain a high amount of suspended solids.

The *stalked ciliated protozoan* attaches itself to the wastewater solids and uses its cilia to attract food (see Figure 5.1). The stalked ciliated protozoan is normally associated with a plant effluent that is very clear and contains low levels of both BOD and suspended solids.

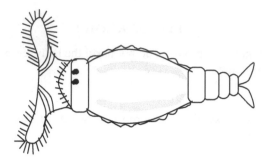

FIGURE 5.2   *Philodina*, a common rotifer.

Rotifers make up a well-defined group of the smallest, simplest multicellular microorganisms and are found in nearly all aquatic habitats (see Figure 5.2). Rotifers are a higher life form associated with cleaner waters. Normally found in well-operated wastewater treatment plants, they can be used to indicate the performance of certain types of treatment processes.

PATHOGENIC PROTOZOA

As mentioned, certain types of protozoans can cause disease. Of particular interest to the drinking water practitioner are *Entamoeba histolytica* (amoebic dysentery and amoebic hepatitis), *Giardia lamblia* (giardiasis), *Cryptosporidium* (cryptosporidiosis), and the emerging *Cyclospora* (cyclosporiasis). Sewage contamination transports eggs, cysts, and oocysts of parasitic protozoa and helminths (tapeworms, hookworms, etc.) into raw water supplies, leaving water treatment (in particular, filtration) and disinfection as the means by which to diminish the danger of contaminated water for the consumer.

Entamoeba histolytica is an amoeboid protozoan that lives in anaerobic environments. Like other pathogenic protozoa, *Entamoeba* is capable of forming cysts that can remain dormant for extended periods of time in the water. These cysts spread to new hosts when we ingest contaminated water.

To prevent the occurrence of *Entamoeba histolytica, Giardia*, and *Cryptosporidium* spp. in surface water supplies and to address increasing problems with waterborne diseases, the U.S. Environmental Protection Agency (USEPA) implemented the Surface Water Treatment Rule (SWTR) in 1989. The rule requires both filtration and disinfection of all surface water supplies as a means of primarily controlling *Giardia* and enteric viruses. Since implementation of the SWTR, the USEPA has also recognized that *Cryptosporidium* species are agents of waterborne disease. In its 1996 series of surface water regulations, the USEPA included *Cryptosporidium*.

To test the need for and the effectiveness of the USEPA's SWTR, LeChevallier et al. (1991) conducted a study on the occurrence and distribution of *Giardia* and *Cryptosporidium* organisms in raw water supplies to 66 surface water filter plants. These plants were located in 14 states and one Canadian province. A combined immunofluorescence test indicated that cysts and oocysts were widely dispersed

DID YOU KNOW?

Although most people who contract giardiasis recover naturally within a week or two, the illness sometimes lingers for up to a year.

in the aquatic environment. *Giardia* species were detected in more than 80% of the samples. *Cryptosporidium* species were found in 85% of the sample locations. Taking into account several variables, *Giardia* or *Cryptosporidium* species were detected in 97% of the raw water samples. After evaluating their data, the researchers concluded that the SWTR might have to be upgraded (subsequently, it has been) to require additional treatment.

Giardia

Giardia lamblia (also known as the hiker's/traveler's scourge or disease) is a microscopic parasite that can infect warm-blooded animals and humans. Although *Giardia* was discovered in the 19th century, not until 1981 did the World Health Organization (WHO) classify *Giardia* as a pathogen. An outer shell called a *cyst* allows *Giardia* to survive outside the body for long periods of time. If viable cysts are ingested, *Giardia* can cause the illness known as *giardiasis*, an intestinal illness that can cause nausea, anorexia, fever, and severe diarrhea. The symptoms last only for several days, and the body can naturally rid itself of the parasite in 1 to 2 months; however, for individuals with weakened immune systems, the body often cannot rid itself of the parasite without medical treatment.

In the United States, *Giardia* is the most commonly identified pathogen in waterborne disease outbreaks. Contamination of a water supply by *Giardia* can occur in two ways: (1) by the activity of animals in the watershed area of the water supply, or (2) by the introduction of sewage into the water supply. Wild and domestic animals are major contributors to the contamination of water supplies. Studies have also shown that, unlike many other pathogens, *Giardia* is not host specific. In short, *Giardia* cysts excreted by animals can infect and cause illness in humans. Additionally, in several major outbreaks of waterborne diseases, the *Giardia* cyst source was sewage-contaminated water supplies.

Treating the water supply, however, can effectively control waterborne *Giardia*. Chlorine and ozone are examples of two disinfectants known to effectively kill *Giardia* cysts. Filtration of the water can also effectively trap and remove the parasite from the water supply. The combination of disinfection and filtration is the most effective water treatment process available today for prevention of *Giardia* contamination.

In drinking water, *Giardia* is regulated under the SWTR. Although the SWTR does not establish a maximum contaminant level (MCL) for *Giardia,* it does specify treatment requirements to achieve at least 99.9% (3-log) removal or inactivation of *Giardia.* This regulation requires that all drinking water systems using surface water or groundwater under the influence of surface water must disinfect and filter the water. The Enhanced Surface Water Treatment Rule (ESWTR), which includes *Cryptosporidium* and further regulates *Giardia*, was established in December 1996.

Giardiasis

Giardiasis is recognized as one of the most frequently occurring waterborne diseases in the United States. *Giardia lamblia* cysts have been discovered in places as far apart as Estes Park, Colorado (near the Continental Divide); Missoula, Montana; Wilkes-Barre, Scranton, and Hazleton, Pennsylvania; and Pittsfield and Lawrence, Massachusetts, just to name a few (CDC, 1995).

Giardiasis is characterized by intestinal symptoms that can last a week or more and may be accompanied by one or more of the following: diarrhea, abdominal cramps, bloating, flatulence, fatigue, and weight loss. Although vomiting and fever are often listed as relatively frequent symptoms, people involved in waterborne outbreaks in the United States have not commonly reported them. Although most *Giardia* infections persist for only 1 or 2 months, some people experience a more chronic phase that can follow the acute phase or may become manifest without an antecedent acute illness. Loose stools and increased abdominal gassiness with cramping, flatulence, and burping characterize the chronic phase. Fever is not common, but malaise, fatigue, and depression may ensue; for a small number of people, the persistence of infection is associated with the development of marked malabsorption and weight loss (Weller, 1985). Similarly, lactose (milk) intolerance can be a problem for some people. This can develop coincidentally with the infection or be aggravated by it, causing an increase in intestinal symptoms after ingestion of milk products.

Some people may have several of these symptoms without evidence of diarrhea or have only sporadic episodes of diarrhea every three or four days. Still others may have no symptoms at all. The problem, then, may not be one of determining whether or not someone is infected with the parasite but how harmoniously the host and the parasite can live together. When such harmony does not exist or is lost, it then becomes a problem of how to get rid of the parasite, either spontaneously or by treatment.

> **Note:** Three prescription drugs are available in the United States to treat giardiasis: quinacrine, metronidazole, and furazolidone. In a recent review of drug trials in which the efficacies of these drugs were compared, quinacrine produced a cure in 93% of patients, metronidazole cured 92%, and furazolidone cured about 84% of patients.

Giardiasis occurs worldwide. In the United States, *Giardia* is the parasite most commonly identified in stool specimens submitted to state laboratories for parasitologic examination. During a 3-year period, approximately 4% of 1 million stool specimens submitted to state laboratories tested positive for *Giardia* (CDC, 1979). Other surveys have demonstrated *Giardia* prevalence rates ranging from 1 to 20%, depending on the location and ages of persons studied. Giardiasis ranks among the top 20 infectious diseases causing the greatest morbidity in Africa, Asia, and Latin America; it has been estimated that about 2 million infections occur per year in these regions (Walsh, 1981). People who are at highest risk for acquiring *Giardia* infection in the United States may be placed into five major categories:

1. People in cities whose drinking water originates from streams or rivers and whose water treatment process does not include filtration, or where filtration is ineffective because of malfunctioning equipment
2. Hikers, campers, and those who enjoy the outdoors
3. International travelers
4. Children who attend daycare centers, daycare center staff, and parents and siblings of children infected in daycare centers
5. Homosexual men

People in categories 1, 2, and 3 have in common the same general source of infection; that is, they acquire *Giardia* from fecally contaminated drinking water. The city resident usually becomes infected because the municipal water treatment process does not include the filter necessary to physically remove the parasite from the water. The number of people in the United States at risk (i.e., the number who receive municipal drinking water from unfiltered surface water) is estimated to be 20 million. International travelers may also acquire the parasite from improperly treated municipal waters in cities or villages in other parts of the world, particularly in developing countries. In Eurasia, only travelers to Leningrad appear to be at increased risk. In prospective studies, 88% of U.S. and 35% of Finnish travelers to Leningrad who had negative stool tests for *Giardia* on departure to the Soviet Union developed symptoms of giardiasis and had positive test for *Giardia* after they returned home (Brodsky et al., 1974). With the exception of visitors to Leningrad, however, *Giardia* has not been implicated as a major cause of traveler's diarrhea, as it has been detected in fewer than 2% of travelers developing diarrhea. Hikers and campers, however, risk infection every time they drink untreated raw water from a stream or river. Persons in categories 4 and 5 become exposed through more direct contact with feces or an infected person by exposure to the soiled diapers of an infected child in cases associated with daycare centers or through direct or indirect anal–oral sexual practices in the case of homosexual men.

Although community waterborne outbreaks of giardiasis have received the greatest publicity in the United States, about half of the *Giardia* cases discussed with the staff of the Centers for Disease Control and Prevention over a 3-year period had a daycare exposure as the most likely source of infection. Numerous outbreaks of *Giardia* in daycare centers have been reported. Infection rates for children in daycare center outbreaks range from 21 to 44% in the United States and from 8 to 27% in Canada (Black et al., 1981). The highest infection rates are usually observed in children who wear diapers (1 to 3 years of age). In a study of 18 randomly selected daycare centers in Atlanta, 10% of diapered children were found to be infected. Transmission from this age group to older children, daycare staff, and household contacts is also common. About 20% of parents caring for an infected child become infected.

Local health officials and managers of water utility companies need to realize that sources of *Giardia* infection other than municipal drinking water exist. Armed with this knowledge, they are less likely to make a quick (and sometimes wrong) assumption that a cluster of recently diagnosed cases in a city is related to municipal

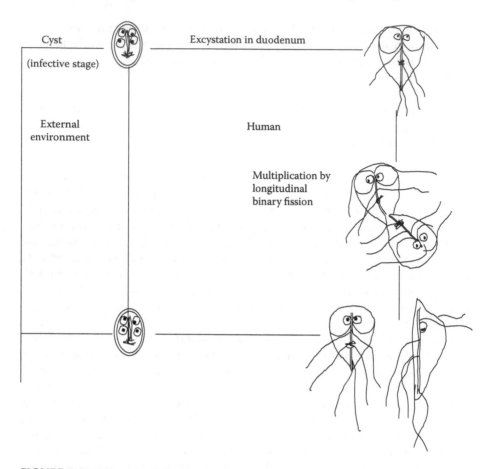

FIGURE 5.3   Life cycle of *Giardia lamblia*.

drinking water. Of course, drinking water must not be ruled out as a source of infection when a larger than expected number of cases is recognized in a community, but the possibility that the cases are associated with a daycare center outbreak, drinking untreated stream water, or international travel should also be entertained.

To understand the finer aspects of *Giardia* transmission and strategies for control, the drinking water practitioner must become familiar with several aspects of the biology of the parasite. Two forms of the parasite exist: a *trophozoite* and a *cyst*, both of which are much larger than bacteria (see Figure 5.3). Trophozoites live in the upper small intestine, where they attach to the intestinal wall by means of a disc-shaped suction pad on their ventral surface. Trophozoites actively feed and reproduce at this location. At some time during the trophozoite's life, it releases its hold on the bowel wall and floats in the fecal stream through the intestine. As it makes this journey, it undergoes a morphologic transformation into the egg-like cyst. The cyst, about 6 to 9 nm in diameter and 8 to 12 μm in length, has a thick exterior wall that protects the parasite against the harsh elements that it will encounter outside the body. This cyst form of parasite is infectious to other people or animals. Most people become infected either directly (by hand-to-mouth transfer of cysts from the feces of an infected individual) or indirectly (by drinking feces-contaminated water). Less common modes of transmission included ingestion of fecally contaminated food and hand-to-mouth transfer of cysts after touching a fecally contaminated surface. After the cyst is swallowed, the trophozoite is liberated through the action of stomach acid and digestive enzymes and becomes established in the small intestine.

Although infection after ingestion of only one *Giardia* cyst is theoretically possible, the minimum number of cysts shown to infect a human under experimental conditions is 10 (Rendtorff, 1954). Trophozoites divide by binary fission about every 12 hours. What this means in practical terms is that if a person swallowed only a single cyst then reproduction at this rate would result in more than 1 million parasites 10 days later—1 billion parasites by day 15.

The exact mechanism by which *Giardia* causes illness is not yet well understood, but it apparently is not necessarily related to the number of organisms present. Nearly all of the symptoms, however, are related to dysfunction of the gastrointestinal tract. The parasite rarely invades other parts of the body, such as the gall bladder or pancreatic ducts. Intestinal infection does not result in permanent damage.

Note: Giardia has an incubation period of 1 to 8 weeks.

Data reported by the CDC indicate that *Giardia* is the most frequently identified cause of diarrheal outbreaks associated with drinking water in the United States. The remainder of this section is devoted specifically to waterborne transmissions of *Giardia*. *Giardia* cysts have been detected in 16% of potable water supplies (lakes, reservoirs, rivers, springs, groundwater) in the United States at an average concentration of 3 cysts per 100 L (Rose et al., 1983). Waterborne epidemics of giardiasis are a relatively frequent occurrence. In 1983, for example, *Giardia* was identified as the cause of diarrhea in 68% of waterborne outbreaks in which the causal agent was identified. From 1965 to 1982, more than 50 waterborne outbreaks were reported (CDC, 1984). In 1984, about 250,000 people in Pennsylvania were advised to boil drinking water for 6 months because of *Giardia*-contaminated water.

Many of the municipal waterborne outbreaks of *Giardia* have been subjected to intense study to determine their cause. Several general conclusions can be made from data obtained in those studies. Waterborne transmission of *Giardia* in the United States usually occurs in mountainous regions where community drinking water obtained from clear running streams is chlorinated but not filtered before distribution. Although mountain streams appear to be clean, fecal contamination upstream by human residents or visitors, as well as by *Giardia*-infected animals such as beavers, has been well documented. Water obtained from deep wells is an unlikely source of *Giardia* because of the natural filtration of water as it percolates through the soil to reach underground cisterns. Wells that pose the greatest risk of fecal contamination are poorly constructed or improperly located ones. A few outbreaks have occurred in towns that included filtration in the water treatment process but the filtration was not effective in removing *Giardia* cysts because of defects in filter construction, poor maintenance of the filter media, or inadequate pretreatment of the water before filtration. Occasional outbreaks have also occurred because of accidental cross-connections between water and sewage systems.

Note: From these data, we can conclude that two major ingredients are necessary for a waterborne outbreak: *Giardia* cysts must be present in untreated source
water, and the water purification process must fail to either kill or remove *Giardia*
cysts from the water.

Although beavers are often blamed for contaminating water with *Giardia* cysts,
the suggestion that they are responsible for introducing the parasite into new areas
seems unlikely. Far more likely is that they are also victims: *Giardia* cysts may be
carried in untreated human sewage discharged into the water by small-town sewage disposal plants or originate from cabin toilets that drain directly into streams
and rivers. Backpackers, campers, and sports enthusiasts may also deposit *Giardia*-
contaminated feces in the environment which are subsequently washed into streams
by rain. In support of this concept is a growing amount of data indicating a higher
Giardia infection rate in beavers living downstream from U.S. national forest campgrounds when compared with beavers living in more remote areas that have a near
zero rate of infection.

Although beavers may be unwitting victims of the *Giardia* story, they still play
an important part in the contamination scheme because they can (and probably do)
serve as amplifying hosts. An *amplifying host* is one that is easy to infect, serves as a
good habitat for reproduction of the parasite, and, in the case of *Giardia*, returns millions of cysts to the water for every one ingested. Beavers are especially important
in this regard, because they tend to defecate in or very near the water, which ensures
that most of the *Giardia* cysts excreted are returned to the water.

The microbial quality of water resources and the management of the microbi-
ally laden wastes generated by the burgeoning animal agriculture industry are critical local, regional, and national problems. Animal wastes from cattle, hogs, sheep,
horses, poultry, and other livestock and commercial animals can contain high concentrations of microorganisms, such as *Giardia*, that are pathogenic to humans.

The contribution of other animals to waterborne outbreaks of *Giardia* is less
clear. Muskrats (another semiaquatic animal) have been found in several parts of
the United States to have high infection rates (30 to 40%) (Frost et al., 1984). Recent
studies have shown that muskrats can be infected with *Giardia* cysts from humans
and beavers. Occasional *Giardia* infections have been reported in coyotes, deer,
elk, cattle, dogs, and cats (but not in horses and sheep) encountered in mountainous
regions of the United States. Naturally occurring *Giardia* infections have not been
found in most other wild animals (bear, nutria, rabbit, squirrel, badger, marmot,
skunk, ferret, porcupine, mink, raccoon, river otter, bobcat, lynx, moose, and big-
horn sheep) (Frost et al., 1984).

Scientific knowledge about what is required to kill or remove *Giardia* cysts from
a contaminated water supply has increased considerably. We know, for example, that
cysts can survive in cold water (4°C) for at least 2 months and that they are killed
instantaneously by boiling water (100°C) (Frost et al., 1984). We do not know how
long the cysts will remain viable at other water temperatures (e.g., 0°C or in a can-
teen at 15 to 20°C), nor do we know how long the parasite will survive on various
environmental surfaces, such as under a pine tree, in the sun, on a diaper-changing
table, or in carpets in a daycare center.

The effect of chemical disinfection (chlorination, for example) on the viability of *Giardia* cysts is an even more complex issue. The number of waterborne outbreaks of *Giardia* that have occurred in communities where chlorination was employed as a disinfectant process demonstrates that the amount of chlorine used routinely for municipal water treatment is not effective against *Giardia* cysts. These observations have been confirmed in the laboratory under experimental conditions (Jarroll et al., 1979). This does not mean that chlorine does not work at all. It does work under certain favorable conditions. Without getting too technical, gaining some appreciation of the problem can be achieved by understanding a few of the variables that influence the efficacy of chlorine as a disinfectant:

- *Water pH*—At pH values above 7.5, the disinfectant capability of chlorine is greatly reduced.
- *Water temperature*—The warmer the water, the higher the efficacy. Chlorine does not work in ice-cold water from mountain streams.
- *Organic content of the water*—Mud, decayed vegetation, or other suspended organic debris in water chemically combines with chlorine, making it unavailable as a disinfectant.
- *Chlorine contact time*—The longer that *Giardia* cysts are exposed to chlorine, the more likely it is that the chemical will kill them.
- *Chlorine concentration*—The higher the chlorine concentration, the more likely it is that chlorine will kill *Giardia* cysts. Most water treatment facilities try to add enough chlorine to give a free (unbound) chlorine residual at the customer tap of 0.5 mg per liter of water.

These five variables are so closely interrelated that improving one can often compensate for another; for example, if chlorine efficacy is expected to be low because water is obtained from an icy stream, the chlorine contact time or chlorine concentration, or both, could be increased. In the case of *Giardia*-contaminated water, producing safe drinking water with a chlorine concentration of 1 mg per liter and contact time as short as 10 minutes might be possible if all the other variables are optimal—a pH of 7.0, water temperature of 25°C, and total organic content of the water close to zero. On the other hand, if all of these variables are unfavorable—pH of 7.9, water temperature of 5°C, and high organic content—chlorine concentrations in excess of 8 mg per liter with several hours of contact time may not be consistently effective. Because water conditions and water treatment plant operations (especially those related to water retention time and, therefore, to chlorine contact time) vary considerably in different parts of the United States, neither the USEPA nor the CDC has been able to identify a chlorine concentration that would be safe yet effective against *Giardia* cysts under all water conditions. For this reason, the use of chlorine as a preventive measure against waterborne giardiasis generally has been utilized under outbreak conditions when the amount of chlorine and contact time have been tailored to fit specific water conditions and the existing operational design of the water utility.

The bottom line is that filtration, reverse osmosis, and boiling are more effective at killing *Giardia*. In an outbreak, for example, the local health department and water utility may issue an advisory to boil water, and they may increase the

DID YOU KNOW?

Scientists estimate that between 1% and 5% of Americas are infected with *Cryptosporidium* at any one time.

chlorine residual at the consumer's tap from 0.5 mg/L to 1 or 2 mg/L. Also, if the physical layout and operation of the water treatment facility permit, they may increase the chlorine contact time, as well. These are emergency procedures intended to reduce the risk of transmission until a filtration device can be installed or repaired or until an alternative source of safe water (a well, for example) can be made operational.

The long-term solution to the problem of municipal waterborne outbreaks of giardiasis involves improvements in and more widespread use of filters in the municipal water treatment process. The sand filters most commonly used in municipal water treatment today cost millions of dollars to install, which makes them unattractive for many small communities. The pore sizes in these filters are not sufficiently small to remove *Giardia* (6 to 9 µm by 8 to 12 µm). For the sand filter to remove *Giardia* cysts from the water effectively, the water must receive some additional treatment before it reaches the filter. The flow of water through the filter bed must also be carefully regulated.

An ideal prefilter treatment for muddy water would include sedimentation (a holding pond where large suspended particles are allowed to settle out by the action of gravity) followed by flocculation or coagulation (the addition of chemicals such as alum or ammonium to cause microscopic particles to clump together). The sand filter easily removes the large particles resulting from the flocculation–coagulation process, including *Giardia* cysts bound to other microparticulates. Chlorine is then added to kill the bacteria and viruses that may escape the filtration process. If the water comes from a relatively clear source, chlorine may be added to the water before it reaches the filter.

The successful operation of a complete waterworks operation is a complex process that requires considerable training. Troubleshooting breakdowns or recognizing the potential problems in the system before they occur often requires the skills of an engineer. Unfortunately, most small water utilities with water treatment facilities that include filtration cannot afford the services of a full-time engineer. Filter operation or maintenance problems in such systems may not be detected until a *Giardia* outbreak is recognized in the community. The bottom line is that, although filtration is the best protection against waterborne giardiasis that water treatment technology has to offer for municipal water systems, it is not infallible. For municipal water filtration facilities to work properly, they must be properly constructed, operated, and maintained.

Whenever possible, persons outdoors should carry drinking water of known purity with them. When this is not practical and when water from streams, lakes, ponds, or other outdoor sources must be used, time should be taken to properly disinfect the water before drinking it.

Cryptosporidium

Ernest E. Tyzzer first described the protozoan parasite *Cryptosporidium* in 1907. Tyzzer frequently found a parasite in the gastric glands of laboratory mice. Tyzzer identified the parasite as a sporozoan but of uncertain taxonomic status, and he named it *Cryptosporidium muris*. Later, in 1910, after more detailed study, he proposed *Cryptosporidium* as a new genus and *muris* as the type of species. Amazingly, except for developmental stages, Tyzzer's original description of the life cycle (see Figure 5.4) was later confirmed by electron microscopy. In 1912, Tyzzer described another new species, *Cryptosporidium parvum* (Tyzzer, 1912).

For almost 50 years, Tyzzer's discovery of the genus *Cryptosporidium* remained (like himself) relatively obscure because it appeared to be of no medical or economic importance. Slight rumblings of the importance of the genus began to be felt in the medical community when Slavin (1955) wrote about a new species, *Cryptosporidium melagridis*, associated with illness and death in turkeys. Interest remained slight even when *Cryptosporidium* was found to be associated with bovine diarrhea (Panciera et al., 1971). Not until 1982 did worldwide interest focus on the study of organisms in the genus *Cryptosporidium*. At that time, the medical community and other interested parties were beginning a full-scale, frantic effort to find out as much as possible about acquired immune deficiency syndrome (AIDS), and the CDC reported that 21 AIDS-infected males from six large cities in the United States had severe protracted diarrhea caused by *Cryptosporidium*. It was in 1993, though, that *Cryptosporidium*—"the pernicious parasite"—made itself and Milwaukee famous (Mayo Foundation, 1996).

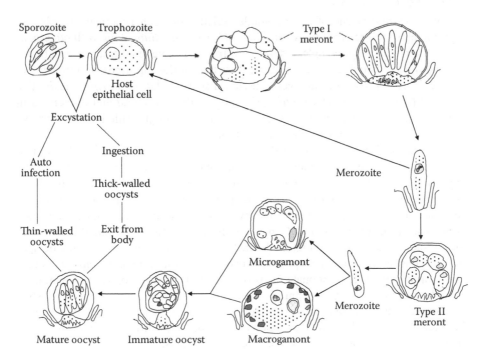

FIGURE 5.4 Life cycle of *Cryptosporidium parvum*.

Note: The *Cryptosporidium* outbreak in Milwaukee caused the deaths of 100 people—the largest episode of waterborne disease in the United States in the 70 years since health officials began tracking such outbreaks.

The massive waterborne outbreak in Milwaukee (more than 400,000 persons developed acute and often prolonged diarrhea or other gastrointestinal symptoms) increased interest in *Cryptosporidium* at an exponential level. The Milwaukee incident spurred both public interest and the interest of public health agencies, agricultural agencies and groups, environmental agencies and groups, and suppliers of drinking water. This increase in interest level and concern spurred new studies of *Cryptosporidium*, with an emphasis on developing methods for recovery, detection, prevention, and treatment (Fayer et al., 1997).

The USEPA is particularly interested in this pathogen. In its reexamination of regulations on water treatment and disinfection, the USEPA issued a maximum contaminant level goal (MCLG) and contaminant candidate list (CCL) for *Cryptosporidium*. Its similarity to *Giardia lamblia* and the need for an efficient conventional water treatment capable of eliminating viruses forced the USEPA to regulate surface water supplies in particular. The proposed Enhanced Surface Water Treatment Rule (ESWTR) included regulations from watershed protection to specialized operation of treatment plants (certification of operators and state overview) and effective chlorination. Protection against *Cryptosporidium* included control of waterborne pathogens such as *Giardia* and viruses (De Zuane, 1997).

Cryptosporidium Basics

Cryptosporidium is one of several single-celled protozoan genera in the phylum Apircomplexa (all referred to as coccidian). *Cryptosporidium* along with other genera in the phylum Apircomplexa develop in the gastrointestinal tract of vertebrates through all of their life cycle; in short, they live in the intestines of animals and people. This microscopic pathogen causes a disease called *cryptosporidiosis*. The dormant (inactive) form of *Cryptosporidium* is called an *oocyst* and is excreted in the feces (stool) of infected humans and animals. The tough-walled oocysts survive under a wide range of environmental conditions.

Several species of *Cryptosporidium* were incorrectly named after the host in which they were found, and subsequent studies have invalidated many species. Now, eight valid species of *Cryptosporidium* (see Table 5.6) have been named. Upton (1997) reported that *C. muris* infects the gastric glands of laboratory rodents and several other mammalian species but (even though several texts state otherwise) is not known to infect humans. *C. parvum*, however, infects the small intestine of an unusually wide range of mammals, including humans, and is the zoonotic species responsible for human cryptosporidiosis. In most mammals *C. parvum* is predominantly a parasite of neonate (newborn) animals. Upton pointed out that, even though exceptions occur, older animals generally develop poor infections, even when unexposed previously to the parasite. Humans are the one host that can be seriously infected at any time in their lives, and only previous exposure to the parasite results in either full or partial immunity to challenge infections.

TABLE 5.6

Valid Named Species of *Cryptosporidium*

Species	Host
Cryptosporidium baileyi	Chicken
Cryptosporidium felis	Domestic cat
Cryptosporidium meleagridis	Turkey
Cryptosporidium murishouse	House mouse
Cryptosporidium nasorium	Fish
Cryptosporidium parvum	House mouse
Cryptosporidium. serpentis	Corn snake
Cryptosporidium wrairi	Guinea pig

Source: Fayer, R., Ed., *Cryptosporidium and Cryptosporidiosis*, CRC Press, Boca Raton, FL, 1997. With permission.

Oocysts are present in most surface bodies of water across the United States, many of which supply public drinking water. Oocysts are more prevalent in surface waters when heavy rains increase runoff of wild and domestic animal wastes from the land or when sewage treatment plants are overloaded or break down. Only laboratories with specialized capabilities can detect the presence of *Cryptosporidium* oocysts in water. Unfortunately, current sampling and detection methods are unreliable. Recovering oocysts trapped on the material used to filter water samples is difficult. When a sample has been obtained, however, determining whether the oocyst is alive and if it is *C. parvum* and thus can infect humans is easily accomplished by looking at the sample under a microscope.

The number of oocysts detected in raw (untreated) water varies with location, sampling time, and laboratory methods. Water treatment plants remove most, but not always all, oocysts. Low numbers of oocysts are sufficient to cause cryptosporidiosis, but the low numbers of oocysts sometimes present in drinking water are not considered cause for alarm in the public.

Protecting water supplies from *Cryptosporidium* demands multiple barriers. Why? Because *Cryptosporidium* oocysts have tough walls that can withstand many environmental stresses and are resistant to chemical disinfectants such as chlorine that are traditionally used in municipal drinking water systems.

Physical removal of particles, including oocysts, from water by filtration is an important step in the water treatment process. Typically, water pumped from rivers or lakes into a treatment plant is mixed with coagulants, which help settle out particles suspended in the water. If sand filtration is used, even more particles are removed. Finally, the clarified water is disinfected and piped to customers. Filtration is the only conventional method now in use in the United States for controlling *Cryptosporidium*.

Ozone is a strong disinfectant that kills protozoa if sufficient doses and contact times are used, but ozone leaves no residual for killing microorganisms in the distribution system, as does chlorine. The high costs of new filtration or ozone treatment plants must be weighed against the benefits of additional treatment. Even well-operated water treatment plants cannot ensure that drinking water will be completely free of *Cryptosporidium* oocysts. Water treatment methods alone cannot solve the problem; watershed protection and monitoring of water quality are critical. As mentioned earlier, watershed protection is another barrier to *Cryptosporidium* in drinking water. Land use controls such as septic system regulations and best management practices to control runoff can help keep human and animal wastes out of water.

Under the Surface Water Treatment Rule of 1989, public water systems must filter surface water sources unless water quality and disinfection requirements are met and a watershed control program is maintained. This rule, however, did not address *Cryptosporidium*. The USEPA has now set standards for turbidity (cloudiness) and coliform bacteria (which indicate that pathogens are probably present) in drinking water. Frequent monitoring must occur to provide officials with early warning of potential problems to enable them to take steps to protect public health. Unfortunately, no water quality indicators can reliably predict the occurrence of cryptosporidiosis. More accurate and rapid assays of oocysts will make it possible to notify residents promptly if their water supply is contaminated with *Cryptosporidium* and thus avert outbreaks.

The bottom line is that the collaborative efforts of water utilities, government agencies, healthcare providers, and individuals are needed to prevent outbreaks of cryptosporidiosis.

Cryptosporidiosis

Cryptosporidium parvum is an important emerging pathogen in the U.S. and a cause of severe, life-threatening disease in patients with AIDS. No safe and effective form of specific treatment for cryptosporidiosis has been identified to date. The parasite is transmitted by ingestion of oocysts excreted in the feces of infected humans or animals. The infection can therefore be transmitted from person-to-person, through ingestion of contaminated water (drinking water and water used for recreational purposes) or food, from animal to person, or by contact with fecally contaminated environmental surfaces. Outbreaks associated with all of these modes of transmission have been documented. Patients with human immunodeficiency virus infection should be made more aware of the many ways that *Cryptosporidium* species are transmitted, and they should be given guidance on how to reduce their risk of exposure (Juranek, 1995).

Since the Milwaukee outbreak, concern about the safety of drinking water in the United States has increased, and new attention has been focused on determining and reducing the risk of acquiring cryptosporidiosis from community and municipal water supplies. Cryptosporidiosis is spread by putting something in the mouth that has been contaminated with the stool of an infected person or animal. In this way, people swallow the *Cryptosporidium* parasite. As mentioned earlier, a person can become infected by drinking contaminated water or eating raw or undercooked food contaminated with *Cryptosporidium* oocysts, by direct contact with the droppings of infected animals or stools of infected humans, or by hand-to-mouth transfer of oocysts from surfaces that may have become contaminated with microscopic amounts of stool from an infected person or animal.

The symptoms may appear 2 to 10 days after infection by the parasite. Although some persons may not have symptoms, others have watery diarrhea, headache, abdominal cramps, nausea, vomiting, and low-grade fever. These symptoms may lead to weight loss and dehydration. In otherwise healthy persons, these symptoms usually last 1 to 2 weeks, at which time the immune system is able to defeat the infection. In persons with suppressed immune systems, such as persons who have AIDS or who recently have had an organ or bone marrow transplant, the infection may continue and become life threatening.

Currently, no safe and effective cure for cryptosporidiosis exists. People with normal immune systems improve without taking antibiotic or antiparasitic medications. The treatment recommended for this diarrheal illness is to drink plenty of fluids and to get extra rest. Physicians may prescribe medication to slow the diarrhea during recovery.

The best way to prevent cryptosporidiosis is to

- Avoid water or food that may be contaminated.
- Wash hands after using the toilet and before handling food.
- Be sure, if you work in a daycare center, to wash your hands thoroughly with plenty of soap and warm water after every diaper change, even if you wear gloves when changing diapers.

During community-wide outbreaks caused by contaminated drinking water, drinking water practitioners should inform the public to boil drinking water for 1 minute to kill the *Cryptosporidium* parasite.

VIRUSES

Viruses are very different from the other microorganisms. Consider their size relationship, for example. Relative to size, if protozoans are the Goliaths of microorganisms, then viruses are the Davids. Stated more specifically and accurately, viruses are intercellular parasitic particles that are the smallest living infectious materials known—the midgets of the microbial world. Viruses are very simple life forms consisting of a central molecule of genetic material surrounded by a protein shell called a *capsid* and sometimes by a second layer called an *envelope*. Viruses occur in many shapes, including long slender rods, elaborate irregular shapes, and geometric polyhedrals (see Figure 5.5).

Viruses contain no mechanisms by which to obtain energy or reproduce on their own, thus viruses must have a host to survive. After they invade the cells of their specific host (animal, plant, insect, fish, or even bacteria), they take over the cellular machinery of the host and force it to make more viruses. In the process, the host cell is destroyed and hundreds of new viruses are released into the environment. The viruses of most concern to the waterworks operator are the pathogens that cause hepatitis, viral gastroenteritis, and poliomyelitis.

Smaller and different from bacteria, viruses are prevalent in water contaminated with sewage. Detecting viruses in water supplies is a major problem because of the complexity of the procedures involved, although experience has shown that the

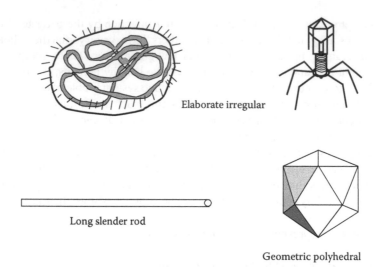

Elaborate irregular

Long slender rod

Geometric polyhedral

FIGURE 5.5   Virus shapes.

normal coliform index can be used as a rough guide for viruses as well as for bacteria. More attention must be paid to viruses, however, when surface water supplies have been used for sewage disposal. Viruses are difficult to destroy by normal disinfection practices, as they require increased disinfectant concentration and contact time for effective destruction. Reverse osmosis systems have a very high effectiveness in removing viruses (e.g., enteric viruses, hepatitis A, norovirus, rotavirus).

> *Note:* Viruses that infect bacterial cells cannot infect and replicate within cells of other organisms. It is possible to utilize this specificity to identify bacteria, a procedure called *phage typing*.

FUNGI

Fungi are of relatively minor importance in water/wastewater operations (except for biosolids composting, where they are critical). Fungi, like bacteria, are extremely diverse. They are multicellular, autotrophic, photosynthetic protists. They grow as filamentous, mold-like forms or as yeast-like (single-celled) organisms. They feed on organic material.

> *Note:* Aquatic fungi grow as *parasites* on living plants or animals and as *saprophytes* on those that are dead.

> *Note:* De Zuane (1997) reported that pathogenic parasites are not easily removed or eliminated by conventional treatment and disinfection unit processes. This is particularly true for *Giardia lamblia*, *Cryptosporidium*, and *Cyclospora*. Filtration facilities can be adjusted with regard to depth, prechlorination, filtration rate, and backwashing to become more effective in the removal of cysts. The pretreatment of protected watershed raw water is a major factor in the elimination of pathogenic protozoa.

REFERENCES AND RECOMMENDED READING

Abrahamson, D.E., Ed. (1988). *The Challenge of Global Warming*. Washington, DC: Island Press.

Asimov, I. (1989). *How Did We Find Out about Photosynthesis?* New York: Walker & Company.

Bingham, A.K., Jarroll, E.L, Meyer, E.A., and Radulescu, S. (1979). Introduction to *Giardia* excystation and the effect of temperature on cyst viability compared by eoxin exclusion and *in vitro* excystation. In: *Waterborne Transmission of Giardiasis*, EPA-600/9-79-001 (Jakubowski, J. and Hoff H.C., Eds.), pp. 217–229. Washington, DC: U.S. Environmental Protection Agency.

Black, R.E., Dykes, A.C., Anderson, K.E., Wells, J.G., Sinclair, S.P., Gary, G.W., Hatch, M.H., and Ginagaros, E.J. (1981). Handwashing to prevent diarrhea in day-care centers. *Am. J. Epidemiol.*, 113: 445–451.

Brodsky, R.E., Spencer, H.C., and Schultz, M.G. (1974). Giardiasis in American travelers to the Soviet Union. *J. Infect. Dis.*, 130: 319–323.

CDC. (1979). *Intestinal Parasite Surveillance, Annual Summary 1978*. Atlanta, GA: Centers for Disease Control and Prevention.

CDC. (1983). *Water-Related Disease Outbreaks Surveillance, Annual Summary 1983*. Atlanta, GA: Centers for Disease Control and Prevention.

CDC. (1995). *Giardiasis*. Atlanta, GA: Centers for Disease Control and Prevention.

De Zuane, J. (1997). *Handbook of Drinking Water Quality*. New York: John Wiley & Sons.

Fayer, R., Speer, C.A., and Dudley, J.P. (1997). The general biology of *Cryptosporidium*. In: *Cryptosporidium and Cryptosporidiosis* (Fayer, R., Ed.), pp. 1–42. Boca Raton, FL: CRC Press.

Frost, F., Plan, B., and Liechty, B. (1984). *Giardia* prevalence in commercially trapped mammals. *J. Environ. Health*, 42: 245–249.

Herwaldt, F.L. et al. (1997). An outbreak in 1996 of cyclosporasis associated with imported raspberries. *N. Engl. J. Med.*, 336: 1548–1556.

Huang, P., Weber, J.T., Sosin, D.M. et al. (1995). *Cyclospora. Ann. Intern. Med.*, 123: 401–414.

Jarroll, Jr., E.L., Gingham, A.K, and Meyer, E.A. (1979). *Giardia* cyst destruction: effectiveness of six small-quantity water disinfection methods. *Am. J. Trop. Med. Hyg.*, 29: 8–11.

Juranek, D.D. (1995). Cryptosporidiosis: sources of infection and guidelines for prevention. *Clin. Infect. Dis.*, 21: S37–S61.

Kemmer, F.N. (1979). *Water: The Universal Solvent*, 2nd ed. Oak Brook, IL: Nalco Chemical Co.

Kordon, C. (1992). *The Language of the Cell*. New York: McGraw-Hill.

Koren, H. (1991). *Handbook of Environmental Health and Safety: Principles and Practices*. Chelsea, MI: Lewis Publishers.

LeChevallier, M.W., Norton, W.D., and Less, R.G. (1991). Occurrences of *Giardia* and *Cryptosporidium* spp. in surface water supplies. *Appl. Environ. Microbiol.*, 57: 2610–2616.

Lozier, J.C., McKim T., and Rose, J. (1994). Meeting Stringent Surface Water Discharge Requirements for Reclaimed Water with Membrane Process, paper presented at ASCE/NCEE Conf., Boulder, CO, July 11–13.

Mayo Foundation. (1996). *The "Bug" That Made Milwaukee Famous*. Rochester, MN: Mayo Foundation.

Metcalf & Eddy. (2003). *Wastewater Engineering: Treatment, Disposal, and Reuse*. New York: McGraw-Hill.

Miller, G.T. (1988). *Environmental Science: An Introduction*. Belmont, CA: Wadsworth.

Odum, E.P. (1975). *Ecology: The Link Between the Natural and the Social Sciences*. New York: Holt, Rinehart and Winston.

Odum, E.P. (1983). *Basic Ecology*. Philadelphia, PA: Saunders.

Panciera, R.J., Thomassen, R.W., and Garner, R.M. (1971). Cryptosporidial infection in a calf. *Vet. Pathol.*, 8: 479.

Pennak, R.W. (1989). *Fresh-Water Invertebrates of the United States*, 3rd ed. New York: John Wiley & Sons.

Price, P.W. (1984). *Insect Ecology*. New York: John Wiley & Sons.

Rendtorff, R.C. (1954). The experimental transmission of human intestinal protozoan parasites. II. *Giardia lamblia* cysts given in capsules. *Am. J. Hyg.*, 59: 209–220.

Rose, J.B., Gerb, C.P., and Jakubowski, W. (1983). Survey of potable water supplies for *Cryptosporidium* and *Giardia. Environ. Sci. Technol.*, 25: 1393–1399.

Singleton, P. and Sainsbury, D. (1994). *Dictionary of Microbiology and Molecular Biology*, 2nd ed. New York: John Wiley & Sons.

Slavin, D. (1955). *Cryptosporidium melagridis. J. Comp. Pathol.*, 65: 262.

Smith, R.L. (1974). *Ecology and Field Biology*. New York: Harper & Row.

Spellman, F.R. (1996). *Stream Ecology and Self-Purification: An Introduction for Wastewater and Water Specialists*. Boca Raton, FL: CRC Press.

Spellman, F.R. (1997). *Microbiology for Water/Wastewater Operators*. Boca Raton, FL: CRC Press.

Tchobanoglous, G. and Schroeder, E.D. (1985). *Water Quality*. Reading, MA: Addison-Wesley.

Tyzzer, E.E. (1912). *Cryptosporidium parvum* sp.: a *Coccidium* found in the small intestine of the common mouse. *Arch. Protistenkd.*, 26: 394.

Upton, S.J. (1997). *Basic Biology of Cryptosporidium*. Manhattan: Kansas State University.

Walsh, J.D. and Warren, K.S. (1979). Selective primary health care: an interim strategy for disease control in developing countries, *N. Engl. J. Med.*, 301: 974–976.

Weller, P.F. (1985). Intestinal protozoa: giardiasis. *Sci. Am. Med.*, 12(4): 554–558.

WHO. (1990). *Guidelines for Drinking Water Quality*, Vol. 2, 2nd ed. Geneva: World Health Organization.

Wistriech, G.A. and Lechtman, M.D. (1980). *Microbiology*, 3rd ed. New York: Macmillan.

6 Water Quality Standards

I never drink water because of the disgusting things that fish do in it.

—**W.C. Fields**

INTRODUCTION

The effort to regulate drinking water and wastewater effluent has increased since the early 1900s. Beginning with an effort to control the discharge of wastewater into the environment, preliminary regulatory efforts focused on protecting public health. The goal of this early wastewater treatment program was to remove suspended and floatable material, treat biodegradable organics, and eliminate pathogenic organisms. Thus, regulatory efforts were directed toward constructing wastewater treatment plants in an effort to alleviate the problem. But a problem soon developed: *progress*. Progress in the sense that time marched on and with it so did proliferation of city growth in the United States, where it became increasingly difficult to find land required for wastewater treatment and disposal. Wastewater professionals soon recognized the need to develop methods of treatment that would accelerate nature's purification of water under controlled conditions in treatment facilities of comparatively smaller size. Regulatory influence on water quality improvements in both wastewater and drinking water took a giant step forward in the 1970s. The Water Pollution Control Act Amendments of 1972 (Clean Water Act) established national water pollution control goals. At about the same time, the Safe Drinking Water Act (SDWA) passed by Congress in 1974 was designed to maintain and protect the public drinking water supply.

CLEAN WATER ACT

In 1972, Congress adopted the Clean Water Act, which established a framework for achieving its national objective "to restore and maintain the chemical, physical, and biological integrity of the nation's waters." Congress decreed that, where attainable, water quality "provides for the protection and propagation of fish, shellfish, and wildlife and provides for recreation in and on the water." These goals are referred to as the "fishable and swimmable" goals of the Act.

Before the CWA, no specific national water pollution control goals or objectives existed. Current standards require that municipal wastewater be given secondary treatment (to be discussed in detail later) and that most effluents meet the conditions shown in Table 6.1. The goal, via secondary treatment (i.e., the biological treatment component of a municipal treatment plant), was set so the principal components of municipal wastewater—suspended solids, biodegradable material, and pathogens—could be reduced to acceptable levels. Industrial dischargers are required to

TABLE 6.1

Minimum National Standards for Secondary Treatment

Characteristic of Discharge	Unit of Measure	Average 30-Day	Average 7-Day
BOD$_5$	mg/L	30	45
Suspended solids	mg/L	30	45
Concentration	pH units	6.0–9.0	—

Source: 40 CFR Part 133, Secondary Treatment Regulations, U.S. Environmental Protection Agency, Washington, DC, 1988.

treat their wastewater to the level obtainable by the *best available technology* (BAT) for wastewater treatment in that particular type of industry. Moreover, a National Pollutant Discharge Elimination System (NPDES) program was established based on uniform technological minimums with which each point source discharger must comply. Under NPDES, each municipality and industry discharging effluent into streams is assigned discharge permits. These permits reflect the secondary treatment and BAT standards. Water quality standards are the benchmark against which monitoring data are compared to assess the health of waters to develop *total maximum daily loads* in impaired waters. They are also used to calculate water-quality-based discharge limits in permits issued under NPDES.

SAFE DRINKING WATER ACT

The Safe Drinking Water Act (SDWA) of 1974 mandated the U.S. Environmental Protection Agency (USEPA) to establish drinking water standards for all public water systems serving 25 or more people or having 15 or more connections. Pursuant to this mandate, the USEPA established maximum contaminant levels for drinking water delivered through public water distribution systems. The *maximum contaminant levels* (MCLs) of inorganics, organic chemicals, turbidity, and microbiological contaminants are shown in Table 6.2. The USEPA's primary regulations are mandatory and must be complied with by all public water systems to which they apply. If analysis of the water produced by a water system indicates that an MCL for a contaminant is being exceeded, the system must take steps to stop providing the water to the public or initiate treatment to reduce the contaminant concentration to below the MCL.

The USEPA has also issued guidelines to states with regard to secondary drinking water standards (Table 6.3). These guidelines apply to drinking water contaminants that may adversely affect the aesthetic qualities of the water (i.e., those qualities that make water appealing and useful), such as odor and appearance. These qualities have no known adverse health effects, so secondary regulations are not mandatory; however, most drinking water systems comply with the limits. They have learned through experience that the odor and appearance of drinking water are not problems until customers complain, and one thing is certain—they will complain.

TABLE 6.2
USEPA Primary Drinking Water Standards

1. Inorganic Contaminant Levels

Contaminants	Level (mg/L)
Arsenic	0.05
Barium	1.00
Cadmium	0.010
Chromium	0.05
Lead	0.05
Mercury	0.002
Nitrate	10.00
Selenium	0.01
Silver	0.05

2. Organic Contaminant Levels

Chemical		Maximum Contaminant Level (MCL) (mg/L)
Chlorinated hydrocarbons	Endrin	0.0002
	Lindane	0.004
	Mexthoxychlor	0.1
	Toxaphene	0.005
Chlorophenoxys	2,4-D	0.1
	2,4,5-TP Silvex®	0.01

3. Maximum Levels of Turbidity

Reading Basis	Maximum Contaminant Level Turbidity Units (TUs)
Turbidity reading (monthly average)	1 TU or up to 5 TUs if the water supplier can demonstrate to the state that the higher turbidity does not interfere with disinfection maintenance of an effective disinfection agent throughout the distribution system, or microbiological determinants
Turbidity reading (based on average of two consecutive days)	5 TUs

4. Microbiological Contaminants

		Individual Sample Basis	
Test Method Used	Monthly Basis	Fewer Than 20 Samples/Month	More Than 20 Samples/Month
Membrane filter technique	1/100 mL average daily	*Number of coliform bacteria not to exceed:* 4/100 mL in more than 1 sample	4/100 mL in more than 5% of samples
Fermentation		*Coliform bacteria shall not be present in:*	
10-mL standard portions	More than 10% of the portions	3 or more portions in more than 1 sample	3 or more portions in more than 5% of samples
100-mL standard portions	More than 60% of the portions	5 portions in more than 1 sample	5 portions in more than 20% of the samples

Source: Adapted from USEPA, *National Interim Primary Drinking Water Regulations*, U.S. Environmental Protection Agency, Washington, DC, 1975.

TABLE 6.3
Secondary Maximum Contaminant Levels

Contaminant	Level	Adverse Effect
Chloride	250 mg/L	Taste
Color	15 cu	Appearance
Copper	1 mg/L	Taste and odor
Corrosivity	Noncorrosive	Taste and odor
Fluoride	2 mg/L	Dental fluorosis
Foaming agents	0.5 mg/L	Appearance
Iron	0.3 mg/L	Appearance
Manganese	0.05 mg/L	Laundry discoloration
Odor	3 TON	Unappealing to drink
pH	6.5–8.5	Corrosion or scaling
Sulfate	250 mg/L	Laxative effect
Total dissolved solids	500 mg/L	Taste, corrosion
Zinc	5 mg/L	Taste, appearance

Note: cu, color unit; TON, threshold odor number
Source: Adapted from McGhee, T.J., *Water Supply and Sewerage*,
    6th ed., McGraw-Hill, New York, 1991.

REFERENCES AND RECOMMENDED READING

AWWA. (1995). *Water Quality*, 2nd ed. Denver, CO: American Water Works Association.

Bangs, R. and Kallen, C. (1985). *Rivergods: Exploring the World's Great Wild Rivers*. San Francisco, CA: Sierra Club Books.

Boyce, A. (1997). *Introduction to Environmental Technology*. New York: Van Nostrand Reinhold.

Coakley, P. (1975). Developments in our knowledge of sludge dewatering behavior. In: *Proceedings of the 8th Public Health Engineering Conference*, Department of Civil Engineering, University of Technology, Loughborough.

Fortner, B. and Schechter, D. (1996). U.S. water quality shows little improvement over 1992 inventory. *Water Environ. Technol.*, 8: 15–16.

Gilcreas, F.W., Sanderson, W.W., and Elmer, R.P. (1953). Two methods for the determination of grease in sewage. *Sewage Indust. Wastes*, 25: 1379.

Gleick, P.H. (2001). Freshwater forum. *U.S. Water News*, 18(6).

Ingram, C. (1991). *The Drinking Water Book*. Berkeley, CA: Ten Speed Press.

Koren, H. (1991). *Handbook of Environmental Health and Safety: Principles and Practices*. Chelsea, MI: Lewis Publishers.

Lewis, S.A. (1996). *Safe Drinking Water*. San Francisco, CA: Sierra Club Books.

Masters, G.M. (1991). *Introduction to Environmental Engineering and Science*. Englewood Cliffs, NJ: Prentice Hall.

McGhee, T.J. (1991). *Water Supply and Sewerage*, 6th ed. New York: McGraw-Hill.

Metcalf & Eddy. (2003). *Wastewater Engineering: Treatment, Disposal, and Reuse*, 4th ed. New York: McGraw-Hill.

Morrison, A. (1983). In Third World villages, a simple handpump saves lives. *Civil Eng. ASCE*, 53(10): 68–72.

Nathanson, J.A. (1997). *Basic Environmental Technology: Water Supply, Waste Management, and Pollution Control*. Upper Saddle River, NJ: Prentice Hall.

Peavy, H.S., Rowe, D.R., and Tchobanoglous, G. (1987). *Environmental Engineering*. New York: McGraw-Hill.

Rowe, D.R. and Abdel-Magid, I.M. (1995). *Handbook of Wastewater Reclamation and Reuse*. Boca Raton, FL: Lewis Publishers.

Sawyer, C.N., McCarty, A.L., and Parking, G.F. (1994). *Chemistry for Environmental Engineering*, New York: McGraw-Hill.

Spellman, F.R. (1996). *Stream Ecology and Self-Purification: An Introduction for Wastewater and Water Specialists*. Lancaster, PA: Technomic.

Spellman, F.R. (2015). *The Science of Water*, 3rd ed. Boca Raton, FL: CRC Press.

Sterritt, R.M. and Lester, J.M. (1988). *Microbiology for Environmental and Public Health Engineers*. London: E. & F.N. Spoon.

Tchobanoglous, G. and Schroeder, E.D. (1985). *Water Quality*. Reading, MA: Addison-Wesley.

USEPA. (2007). *Protecting America's Public Health*. Washington, DC: U.S. Environmental Protection Agency (www.epa.gov/safewater/publicoutreach.html.)

7 Conventional Water and Wastewater Treatment

Hercules, that great mythical giant and arguably the globe's first environmental engineer, discovered that the solution to stream pollution is dilution—that is, dilution is the solution. In reality, today's humans depend on various human-made water treatment processes to restore water to potable and palatable condition. However, it should be pointed out that Nature, as Hercules observed, is not defenseless in its fight against water pollution. For example, when a river or stream is contaminated, natural processes (including dilution) immediately kick in to restore the water body and its contents back to its natural state. If the level of contamination is not excessive, the stream or river can restore itself to normal conditions in a relatively short period of time. In this section, Nature's ability to purify and restore typical river systems to normal conditions is discussed.

> If you visit American city,
> You will find it very pretty.
> Just two things of which you must beware:
> Don't drink the water and don't breathe the air!
>
> **—Tom Lehrer (*Pollution*, 1965)**

Municipal water treatment operations and associated treatment unit processes are designed to provide reliable, high-quality water service for customers and to preserve and protect the environment for future generations. Water management officials and treatment plant operators are tasked with exercising responsible financial management, ensuring fair rates and charges, providing responsive customer service, providing a consistent supply of safe potable water for consumption by the user, and promoting environmental responsibility. While studying this chapter on water and wastewater treatment plant operation, keep in mind the major point in the chapter: Water and wastewater treatment plant design can be taught in school. But in operating the plants, experience, attention to detail, and operator common sense are most important.

In the past, water quality was described as "wholesome and delightful" and "sparkling to the eye." Today, we describe water quality as "safe and healthy to drink."

SHUTTING DOWN THE PUMP[*]

He wandered the foggy, filthy, garbage-strewn, corpse-ridden streets of 1854 London searching, making notes, always looking … seeking a murderous villain. No, not Jack the Ripper, but a killer just as insidious and unfeeling. He finally found the miscreant and took action; that is, he removed the handle from a water pump. And, fortunately for untold thousands of lives, his was a lifesaving action. He was a detective—of sorts. Not the real Sherlock Holmes, but absolutely as clever, as skillful, as knowledgeable, as intuitive—and definitely as driven. His name was Dr. John Snow. His middle name? Common Sense. The master criminal Snow sought? A mindless, conscienceless, brutal killer: cholera. Let's take a closer look at this medical super sleuth and his quarry. More to the point, let's look at Dr. Snow's impact on water treatment (disinfection) of raw water used for potable and other purposes.

DR. JOHN SNOW

Dr. Snow (1813–1858) was an unassuming but creative London obstetrician who achieved prominence in the mid-19th century for proving his theory (published in his *On the Mode of Communication of Cholera*) that cholera is a contagious disease caused by a "poison" that reproduces in the human body and is found in the vomitus and stools of cholera patients. He theorized that the main (though not the only) means of transmission was water contaminated with this poison. Many opinions about cholera's cause had been offered, but Dr. Snow's theory was not held in high regard at first because a commonly held and popular theory was that diseases are transmitted by the inhalation of vapors. In the beginning, Snow's argument did not cause a great stir; it was only one of the many hopeful theories proposed during a time when cholera was causing great distress. Eventually, Snow was able to prove his theory. We describe how Snow accomplished this later, but for now let's take a look again at Snow's target.

CHOLERA

According to the U.S. Centers for Disease Control and Prevention (CDC), cholera is an acute, diarrheal illness caused by infection of the intestine with the bacterium *Vibrio cholerae*. The infection is often mild or without symptoms, but it can sometimes be quite severe. Approximately 1 in 20 infected persons experience severe disease symptoms such as profuse watery diarrhea, vomiting, and leg cramps. In these persons, rapid loss of body fluids leads to dehydration and shock. Without treatment, death can occur within hours.

> **Note:** You don't need to be a rocket scientist to figure out just how deadly cholera was during the London cholera outbreak of 1854. Comparing the state of medicine at that time to ours is like comparing the speed potential of a horse and buggy to a state-of-the-art NASCAR race car today. Simply stated, cholera was

---

[*] Adapted from Spellman, F.R., *Choosing Disinfection Alternatives for Water/Wastewater Treatment*, CRC Press, Boca Raton, FL, 1999.

the classic epidemic disease of the 19th century, as the plague had been for the 14th century. Its defeat was a reflection of both common sense and of progress in medical knowledge—and of the enduring changes in European and American social thought.

How does a person contract cholera? Again, we refer to the CDC for our answer. A person may contract cholera (even today) by drinking water or eating food contaminated with the cholera bacterium. In an epidemic, the source of the contamination is usually feces of an infected person. The disease can spread rapidly in areas with inadequate treatment of sewage and drinking water. Disaster areas often pose special risks; for example, the aftermath of Hurricane Katrina in New Orleans raised concerns about a potential cholera outbreak.

The cholera bacterium also lives in brackish river and coastal waters. Raw shellfish have been a source of cholera, and a few people in the United States have contracted it from eating shellfish from the Gulf of Mexico. The disease is not likely to spread directly from one person to another; therefore, casual contact with an infected person is not a risk for transmission of the disease.

Flashback to 1854 London

Cholera is a waterborne disease. Today, we know quite a lot about cholera and its transmission, as well as how to prevent infection and how to treat it, but what did they know about cholera in the 1850s, other than that cholera was a deadly killer? Dr. Snow's theory about cholera being transmitted in contaminated water proved to be correct, of course, as we know today. The question is, how did he prove his theory 20 years before the development of germ theory? The answer to that question provides us with an account of one of the all-time legendary quests for answers in epidemiological research—and an interesting story!

Dr. Snow did his research during yet another severe cholera epidemic in London. Though ignorant of the concept of bacteria carried in water (germ theory), he traced an outbreak of cholera to a water pump located at the intersection of Cambridge and Broad Street. How did he do that? He began his investigation by determining where in London persons with cholera lived and worked. He then used this information to map the distribution of cases on what epidemiologists call a "spot map." His map indicated that the majority of the deaths occurred within 250 yards of that communal water pump. The water pump was used regularly by most of the area residents. Those who did not use the pump remained healthy. Suspecting that the Broad Street pump was the source of the plague, Snow had the water pump handle removed, which ended the cholera epidemic.

Sounds like a rather simple solution, doesn't it? Remember, though, that in that era aspirin had not yet been formulated, to say nothing of the development of other medical miracles we now take for granted (such as antibiotics). Dr. Snow, through the methodical process of elimination and linkage (Sherlock Holmes would have been impressed—and was), proved his point and his theory. Specifically, he painstakingly documented the cholera cases and correlated the comparative incidence of cholera among subscribers to the city's two water companies. He learned that one company drew water from the lower Thames River, and the other company obtained water from the upper Thames. Dr. Snow discovered that cholera was much more prevalent among customers of the water company that drew its water from the lower Thames, where the river had become contaminated with London sewage. Dr. Snow tracked the source of the Broad Street pump, and it was the contaminated lower Thames, of course.

Dr. Snow, an obstetrician, became the first effective practitioner of scientific epidemiology. His creative use of logic, common sense (removing the handle from the pump), and scientific observation enabled him to solve a major medical mystery—discerning the means by which cholera was transmitted.

From Pump Handle Removal to Water Treatment (Disinfection)

Dr. John Snow's major contribution to the medical profession, to society, and to humanity in general can be summarized rather succinctly: He determined and proved that the deadly disease cholera is a waterborne disease. (Incidentally, Dr. Snow's second medical accomplishment was that he was the first doctor to administer anesthesia during childbirth.)

What does all of this have to do with water treatment (disinfection)? Actually, Dr. Snow's discovery has quite a lot to do with water treatment. Combating any disease is difficult without a determination of how the disease is transmitted—how it travels from vector or carrier to receiver. Dr. Snow established this connection, and from his work, and the work of others, progress was made in understanding and combating many different waterborne diseases.

Today, sanitation problems in developed countries (those with the luxury of adequate financial and technical resources) deal more with the consequences that arise from inadequate commercial food preparation, and the results of bacteria becoming resistant to disinfection techniques and antibiotics. We simply flush our toilets to rid ourselves of unwanted wastes, and we turn on our taps to access a high-quality drinking water supply, from which we have all but eliminated cholera and epidemic diarrheal diseases. This is generally the case in most developed countries today—but it certainly wasn't true in Dr. Snow's time.

The progress made in water treatment from that notable day in 1854 when Snow made the connection between deadly cholera and its means of transmission to the present reads like a chronology of discoveries. This makes sense, of course, because with the passage of time, pivotal events and discoveries occur that have a profound effect on how we live today. Let's take a look at a few elements of the important chronological progression that evolved from the simple removal of a pump handle to the advanced water treatment (disinfection) methods we employ today to treat our water supplies.

After Dr. Snow's discovery, subsequent events began to drive the water/wastewater treatment process. In 1859, the British Parliament was suspended during the summer because the stench coming from the Thames was unbearable. According to one account, the river began to "seethe and ferment under a burning sun." As was the case in many cities at this time, storm sewers carried a combination of storm water, sewage, street debris, and other wastes to the nearest body of water.

In the 1890s, Hamburg, Germany, suffered a cholera epidemic. Detailed studies by Dr. Robert Koch tied the outbreak to a contaminated water supply. In response to the epidemic, Hamburg was among the first cities to use chlorine as part of a wastewater treatment regimen. About the same time, the town of Brewster, New York, became the first U.S. city to disinfect its treated wastewater. Chlorination of drinking water was used on a temporary basis in 1896, and its first known continuous use for water supply disinfection occurred in Lincoln, England, and in Chicago in 1905. Jersey City became one of the first routine users of chlorine in 1908.

Time marched on, and with it came increased realization of the need to treat and disinfect both water supplies and wastewater. Between 1910 and 1915, technological improvements in gaseous and then solution feed of elemental chlorine (Cl_2) made the process of disinfection more practical and efficient. Disinfection of water supplies and chlorination of treated wastewater for odor control increased over the next several decades. In the United States, disinfection, in one form or another, is now being used by more than 15,000 out of approximately 16,000 publicly owned treatment works (POTWs). The significance of this number becomes apparent when you consider that fewer than 25 of the 600 plus POTWs in the United States in 1910 were using disinfectants.

CONVENTIONAL WATER TREATMENT

This section focuses on water treatment operations and the various unit processes currently used to treat raw source water before it is distributed to the user. In addition, the reasons for water treatment and the basic theories associated with individual treatment unit processes are discussed. Moreover, this section is presented because it sets the stage to gain a better understanding of why reverse osmosis (RO) systems are needed and used in addition to conventional water purifying operations and treatment.

Note: Keep in mind that conventional water treatment, along with proper disinfection of water sources used for potable water, has served the consumer and public health quite well; it has saved countless numbers of lives. However, as explained later in Chapter 8, at the present time there exist contaminants of concern, emerging contaminants, that conventional water treatment processes do not adequately remove or neutralize.

Water treatment systems are installed to remove those materials that cause disease or create nuisances. At its simplest level, the basic goal of water treatment operations is to protect public health, with a broader goal to provide potable and palatable water. The bottom line is that the water treatment process functions to provide water that is safe to drink and is pleasant in appearance, taste, and odor.

In this text, *water treatment* is defined as any unit process that changes or alters the chemical, physical, or bacteriological quality of water with the purpose of making it safe for human consumption or appealing to the customer. Treatment also is used to protect the water distribution system components from corrosion.

Many water treatment unit processes are commonly used today. The treatment processes used depend on the evaluation of the nature and quality of the particular water to be treated and the desired quality of the finished water. For water treatment unit processes employed to treat raw water, one thing is certain: As new U.S. Environmental Protection Agency (USEPA) regulations take effect, many more processes will come into use in an attempt to produce water that complies with all current regulations, despite source water conditions.

Small water systems tend to use a smaller number of the wide array of unit treatment processes available, in part because they usually rely on groundwater as the source and because their small size makes many sophisticated processes impractical (e.g., too expensive to install, too expensive to operate, too sophisticated for limited operating staff). This section concentrates on those individual treatment unit processes usually found in conventional water treatment systems, corrosion control methods, and fluoridation.

Purpose of Water Treatment

The purpose of water treatment is to condition, modify, or remove undesirable impurities and to provide water that is safe, palatable, and acceptable to users. This may seem an obvious, expected purpose of treating water, but various regulations also require water treatment. Some regulations state that if the contaminants listed under the various regulations are found in excess of maximum contaminant levels (MCLs) then the water must be treated to reduce the levels. If a well or spring source is surface influenced, treatment is required, regardless of the actual presence of contamination. Some impurities affect the aesthetic qualities (taste, odor, color, and hardness) of the water; if they exceed secondary MCLs established by the USEPA and the state, the water may have to be treated.

If we assume that the water source used to feed a typical water supply system is groundwater (usually the case in the United States), a number of common groundwater problems may require water treatment. Keep in mind that water that must be treated for any one of these problems may also exhibit several other problems:

- Bacteriological contamination
- Hydrogen sulfide odors
- Hardness
- Corrosive water
- Iron and manganese

Stages of Water Treatment

A summary of basic water treatment processes (many of which are discussed in this chapter) are presented Table 7.1. Figure 7.1 presents the "conventional" model of water treatment discussed in this text and shows that water treatment is made up of various stages or unit processes combined to form one treatment system. Note that a given waterworks may contain all of the unit processes discussed in the following or any combination of them. One or more of these stages may be used to treat any one or more of the source water problems listed above. Also note that the model shown in Figure 7.1 does not necessarily apply to very small water systems. In some small systems, water treatment may consist of nothing more than removal of water

TABLE 7.1
Basic Water Treatment Processes

Process	Purpose
Screening	Removes large debris (leaves, sticks, fish) that can foul or damage plant equipment
Chemical pretreatment	Conditions the water for removal of algae and other aquatic nuisances
Presedimentation	Removes gravel, sand, silt, and other gritty materials
Microstraining	Removes algae, aquatic plants, and small debris
Chemical feed and rapid mix	Adds chemicals (e.g., coagulants, pH, adjusters) to water
Coagulation–flocculation	Converts nonsettleable or settable particles
Sedimentation	Removes settleable particles
Softening	Removes hardness-causing chemicals from water
Filtration	Removes particles of solid matter which can include biological contamination and turbidity
Disinfection	Kills disease-causing organisms
Adsorption using granular activated carbon	Removes radon and many organic chemicals such as pesticides, solvents, and trihalomethanes
Aeration	Removes volatile organic compounds (VOCs), radon H_2S, and other dissolved gases; oxidizes iron and manganese
Corrosion control	Prevents scaling and corrosion
Reverse osmosis, electrodialysis	Removes nearly all inorganic contaminants
Ion exchange	Removes some inorganic contaminants including hardness-causing chemicals
Activated alumina	Removes some inorganic contamination
Oxidation filtration	Removes some inorganic contaminants (e.g., iron, manganese, radium)

Source: Adapted from AWWA, *Introduction to Water Treatment*, Vol. 2, American Water Works Association, Denver, CO, 1984.

via pumping from a groundwater source to storage to distribution. In some small water supply operations, disinfection may be added because it is required. Although the water treatment model shown in Figure 7.1 may not exactly mimic the type of

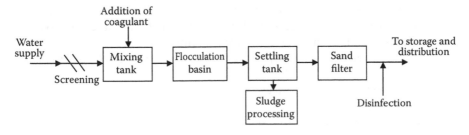

FIGURE 7.1 Conventional water treatment model.

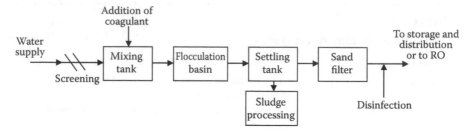

FIGURE 7.2   Conventional water treatment model used as pretreatment for an RO system.

treatment process used in most small systems, it is used in this text for illustrative and instructive purposes. Note that conventional water treatment is often used as pretreatment prior to an attached RO system (see Figure 7.2).

> *Note:* The drawing shown in Figure 7.2 is for instructional purposes only and should not be considered as or construed to be an exact representation of what every pretreatment system from conventional water treatment systems to RO is like.

Pretreatment for Conventional Water Treatment

Simply stated, water pretreatment (also called *preliminary treatment*) is any physical, chemical, or mechanical process used before main water treatment processes. It can include screening, presedimentation, and chemical addition (see Figure 7.1). Pretreatment in water treatment operations usually consists of oxidation or other treatment for the removal of tastes and odors, iron and manganese, trihalomethane precursors, or entrapped gases (such as hydrogen sulfide). Unit processes may include chlorine, potassium permanganate or ozone oxidation, activated carbon addition, aeration, and presedimentation. Pretreatment of surface water supplies accomplishes the removal of certain constituents and materials that interfere with or place an unnecessary burden on conventional water treatment facilities.

Typical pretreatment processes include the following (TWUA, 1988):

- Removal of debris from water from rivers and reservoirs that would clog pumping equipment
- Destratification of reservoirs to prevent anaerobic decomposition that could result in reducing iron and manganese in the soil to a state that would be soluble in water which can cause subsequent removal problems in the treatment plant; the production of hydrogen sulfide and other taste- and odor-producing compounds also results from stratification
- Chemical treatment of reservoirs to control the growth of algae and other aquatic growths that could result in taste and odor problems
- Presedimentation to remove excessively heavy silt loads prior to the treatment processes
- Aeration to remove dissolved odor-causing gases such as hydrogen sulfide and other dissolved gases or volatile constituents and to aid in the oxidation

of iron and manganese, although manganese or high concentrations of iron are not removed during detention provided in conventional aeration units

- Chemical oxidation of iron and manganese, sulfides, taste- and odor-producing compounds, and organic precursors that may produce trihalomethanes upon the addition of chlorine
- Adsorption for removal of tastes and odors

Note: An important point to keep in mind is that, in small systems using groundwater as a source, pretreatment may be the only treatment process used and/or required.

Note: Pretreatment generally involves aeration or the addition of chemicals to oxidize contaminants that exist in the raw water. It may be incorporated as part of the total treatment process or may be located adjacent to the source before the water is sent to the treatment facility.

Pretreatment for RO Systems

To prevent or minimize membrane fouling, scaling, and degradation of membrane materials and performance, pretreatment is one of the fundamental requirements for ensuring successful and cost-effective operation of an RO system (Kucera, 2010). Pretreatment technologies most commonly used and listed in this text are categorized as mechanical or chemical.

RO Mechanical Pretreatment

Mechanical pretreatment involves physical techniques to reduce turbidity, suspended solids, bacterial, hardness, and heavy metals present in the RO influent stream. Some of the mechanical treatments include the following:

- Clarification
- Multimedia filtration
- Carbon filters
- Iron filters
- High-efficiency filtration
- Sodium softeners
- Ultraviolet radiation
- Membrane

RO Chemical Pretreatment

Reverse osmosis chemical pretreatment focuses on bacteria, hardness, scale, and oxidizing agents (Kucera, 2010). Some of the chemical pretreatment techniques include

- Chlorine
- Ozone
- Antiscalants
- Sodium metabisulfite
- Non-oxidizing biocides

Aeration

Aeration is commonly used to treat water that contains trapped gases (such as hydrogen sulfide) that can impart an unpleasant taste and odor to the water. Just allowing the water to rest in a vented tank will (sometimes) drive off much of the gas, but usually some form of forced aeration is needed. Aeration works well (about 85% of the sulfides may be removed) whenever the pH of the water is less than 6.5. Aeration may also be useful in oxidizing iron and manganese, oxidizing humic substances that might form trihalomethanes when chlorinated, eliminating other sources of taste and odor, or imparting oxygen to oxygen-deficient water.

> *Note:* Iron is a naturally occurring mineral found in many water supplies. When the concentration of iron exceeds 0.3 mg/L, red stains will occur on fixtures and clothing. The customer then incurs costs for cleaning and replacement of damaged fixtures and clothing.

Manganese, like iron, is a naturally occurring mineral found in many water supplies. When the concentration of manganese exceeds 0.05 mg/L, black stains occur on fixtures and clothing. As with iron, this increases customer costs for cleaning and replacement of damaged fixtures and clothing. Iron and manganese are commonly found together in the same water supply. We discuss iron and manganese later.

Screening

Screening is usually the first major step in the water pretreatment process (see Figure 7.1). It is defined as the process whereby relatively large and suspended debris is removed from the water before it enters the plant. River water, for example, typically contains suspended and floating debris varying in size from small rocks to logs. Removing these solids is important, not only because these items have no place in potable water but also because this river trash may cause damage to downstream equipment (e.g., clogging and damaging pumps), increase chemical requirements, impede hydraulic flow in open channels or pipes, or hinder the treatment process. The most important criteria used in the selection of a particular screening system for water treatment technology are the screen opening size and flow rate. Other important criteria include costs related to operation and equipment, plant hydraulics, debris handling requirements, and operator qualifications and availability. Large surface water treatment plants may employ a variety of screening devices including rash screens (or trash rakes), traveling water screens, drum screens, bar screens, or passive screens.

Chemical Addition

Two of the major chemical pretreatment processes used in treating water for potable use are iron and manganese removal and hardness removal. Another chemical treatment process that is not necessarily part of the pretreatment process, but is also discussed in this section, is corrosion control. Corrosion prevention is effected by chemical treatment—not only in the treatment process but also in the distribution process. Before discussing each of these treatment methods in detail, however, it is important to describe chemical addition, chemical feeders, and chemical feeder calibration. When chemicals are used in the pretreatment process, they must be the proper ones, fed in the correct concentration and introduced to the water at the proper locations. Determining the proper amount of chemical to use is accomplished by testing. The operator must test the raw water periodically to determine if the chemical dosage should be adjusted. For surface supplies, checking must be done more frequently than for groundwater (remember, surface water supplies are subject to change on short notice, while groundwaters generally remain stable). The operator must be aware of the potential for interactions between various chemicals and how to determine the optimum dosage (e.g., adding both chlorine and activated carbon at the same point will minimize the effectiveness of both processes, as the adsorptive power of the carbon will be used to remove the chlorine from the water).

Note: Sometimes using too many chemicals can be worse than not using enough.

Prechlorination (distinguished from chlorination used in disinfection at the end of treatment) is often used as an oxidant to help with the removal of iron and manganese; however, a concern for systems that prechlorinate is the potential for the formation of total trihalomethanes (TTHMs), which form as a byproduct of the reaction between chlorine and naturally occurring compounds in raw water. The USEPA TTHM standard does not apply to water systems that serve fewer than 10,000 people, but operators should be aware of the impact and causes of TTHMs. Chlorine dosage or the application point may be changed to reduce problems with TTHMs.

Note: TTHMs such as chloroform are known or suspected to be carcinogenic and are limited by water and state regulations.

Note: To be effective, pretreatment chemicals must be thoroughly mixed with the water. Short circuiting or plug flows of chemicals that do not come in contact with most of the water will not result in proper treatment.

All chemicals intended for use in drinking water must meet certain standards. Thus, when ordering water treatment chemicals, the operator must be confident that they meet all appropriate standards for drinking water use. Chemicals are normally fed with dry chemical feeders or solution (metering) pumps. Operators must be familiar with all of the adjustments required to control the rate at which the chemical is fed to the water (wastewater). Some feeders are manually controlled and must be adjusted by the operator when the raw water quality or the flow rate changes; other feeders are paced by a flow meter to adjust the chemical feed so it matches the water flow rate. Operators must also be familiar with chemical solution and feeder calibration.

A significant part of the waterworks operator's daily functions is measuring quantities of chemicals and applying them to water at preset rates. This is normally accomplished semiautomatically by the use of electromechanical–chemical feed devices, but waterworks operators must still know what chemicals to add, how much to add to the water (wastewater), and the purpose of the chemical addition.

Chemical Solutions

A *water solution* is a homogeneous liquid made of the *solvent* (the substance that dissolves another substance) and the *solute* (the substance that dissolves in the solvent). Water is the solvent. The solute (whatever it may be) may dissolve up to a certain limit. This level is its *solubility*—that is, the solubility of the solute in the particular solvent (water) at a particular temperature and pressure.

Remember, in chemical solutions, the substance being dissolved is the solute, and the liquid present in the greatest amount in a solution (that does the dissolving) is the solvent. We should also be familiar with another term, *concentration*—the amount of solute dissolved in a given amount of solvent.

When working with chemical solutions, it is also necessary to be familiar with two important chemical properties: *density* and *specific gravity*. Density is defined as the weight of a substance per a unit of its volume—for example, pounds per cubic foot or pounds per gallon. Specific gravity is defined as the ratio of the density of a substance to a standard density.

Chemical Feeders

Simply put, a chemical feeder is a mechanical device for measuring a quantity of chemical and applying it to water at a preset rate. Two types of chemical feeders are commonly used: solution (or liquid) feeders and dry feeders. Liquid feeders apply chemicals in solutions or suspensions, and dry feeders apply chemicals in granular or powdered forms. In a solution feeder, the chemical enters and leaves the feeder in a liquid state; in a dry feeder, the chemical enters and leaves the feeder in a dry state.

Solution Feeders

Solution feeders are small, positive-displacement metering pumps of three types: (1) reciprocating (piston-plunger or diaphragm type), (2) vacuum type (e.g., gas chlorinator), or (3) gravity feed rotameter (e.g., drip feeder). Positive-displacement pumps are used in high-pressure, low-flow applications; they deliver a specific volume of liquid for each stroke of a piston or rotation of an impeller.

Dry Feeders

Two types of dry feeders are *volumetric* and *gravimetric*, depending on whether the chemical is measured by volume (volumetric) or weight (gravimetric). Simpler and less expensive than gravimetric pumps, volumetric dry feeders are also less accurate. Gravimetric dry feeders are extremely accurate, deliver high feed rates, and are more expensive than volumetric feeders.

Iron and Manganese Removal

Iron and manganese are frequently found in groundwater and in some surface waters. They do not cause health-related problems but are objectionable because they may cause aesthetic problems. Severe aesthetic problems may cause consumers to avoid an otherwise safe water supply in favor of one of unknown or questionable quality, or they may cause customers to incur unnecessary expenses for bottled water. Aesthetic problems associated with iron and manganese include the discoloration of water (iron, reddish water; manganese, brown or black water), staining of plumbing fixtures, a bitter taste, and the growth of microorganisms.

Although no health concerns are directly associated with iron and manganese, the growth of iron bacteria slimes may cause indirect health problems. Economic problems include damage to textiles, dye, paper, and food. Iron residue (or tuberculation) in pipes increases pumping head, decreases carrying capacity, may clog pipes, and may corrode through pipes.

Note: Iron and manganese are secondary contaminants. The secondary maximum contaminant levels (SMCLs) are 0.3 mg/L for iron and 0.05 mg/L for manganese.

Iron and manganese are most likely found in groundwater supplies, industrial waste, and acid mine drainage and are byproducts of pipeline corrosion. They may accumulate in lake and reservoir sediments, causing possible problems during lake/reservoir turnover. They are not usually found in running waters (e.g., streams, rivers).

The chemical precipitation treatments for iron and manganese removal are *deferrization* and *demanganization*, respectively. The usual process is *aeration*, where dissolved oxygen in the chemical causes precipitation; chlorine or potassium permanganate may also be required.

Precipitation

Precipitation (or pH adjustment) of iron or manganese from water in their solid forms can be performed in treatment plants by adjusting the pH of the water through the addition of lime or other chemicals. Some of the precipitate will settle out with time, while the rest is easily removed by sand filters. This process requires the pH of the water to be in the range of 10 to 11.

Note: Although the precipitation or pH adjustment technique for treating water containing iron and manganese is effective, note that the pH level must be adjusted higher (10 to 11) to cause the precipitation, which means that the pH level must also then be lowered (to 8.5 or a bit lower) to use the water for consumption.

Oxidation

One of the most common methods for removing iron and manganese is the process of oxidation (another chemical process), usually followed by settling and filtration. Air, chlorine, or potassium permanganate can oxidize these minerals. Each oxidant has advantages and disadvantages, as each operates slightly differently:

- *Air*—To be effective as an oxidant, the air must come in contact with as much of the water as possible. Aeration is often accomplished by bubbling diffused air through the water, by spraying the water up into the air, or by trickling the water over rocks, boards, or plastic packing materials in an aeration tower. The more finely divided the drops of water, the more oxygen comes in contact with the water and the dissolved iron and manganese.
- *Chlorine*—This is one of the most popular oxidants for iron and manganese control because it is also widely used as a disinfectant; controlling iron and manganese by prechlorination can be as simple as adding a new chlorine feed point in a facility already feeding chlorine. It also provides a predisinfecting step that can help control bacterial growth throughout the rest of the treatment system. The downside to using chlorine is that when chlorine reacts with the organic materials found in surface water and some groundwaters it forms TTHMs. This process also requires that the pH of the water be in the range of 6.5 to 7; because many groundwaters are more acidic than this, pH adjustment with lime, soda ash, or caustic soda may be necessary when oxidizing with chlorine.

- *Potassium permanganate*—This is the best oxidizing chemical to use for manganese control removal. An extremely strong oxidant, it has the additional benefit of producing manganese dioxide during the oxidation reaction. Manganese dioxide acts as an adsorbent for soluble manganese ions. This attraction for soluble manganese provides removal to extremely low levels.

The oxidized compounds form precipitates that are removed by a filter. Note that sufficient time should be allowed from the addition of the oxidant to the filtration step; otherwise, the oxidation process will be completed after filtration, creating insoluble iron and manganese precipitates in the distribution system.

Ion Exchange

The ion exchange process is used primarily to soften hard waters, but it will also remove soluble iron and manganese. The water passes through a bed of resin that adsorbs undesirable ions from the water, replacing them with less troublesome ions. When the resin has given up all of its donor ions, it is regenerated with strong salt brine (sodium chloride); the sodium ions from the brine replace the adsorbed ions and restore the ion exchange capabilities.

Sequestering

Sequestering or stabilization may be used when the water contains mainly low concentrations of iron and the volumes required are relatively small. This process does not actually remove the iron or manganese from the water but complexes (binds it chemically) it with other ions in a soluble form that is not likely to come out of solution (i.e., not likely oxidized).

Aeration

The primary physical process uses air to oxidize the iron and manganese. The water is either pumped up into the air or allowed to fall over an aeration device. The air oxidizes the iron and manganese, which are then removed by use of a filter. To raise the pH, lime is often added to the process. Although this is referred to as a physical process, removal is accomplished by chemical oxidation.

Potassium Permanganate Oxidation and Manganese Greensand

The continuous regeneration potassium greensand filter process is another commonly used filtration technique for iron and manganese control. Manganese greensand is a mineral (gluconite) that has been treated with alternating solutions of manganous chloride and potassium permanganate. The result is a sand-like (zeolite) material coated with a layer of manganese dioxide—an adsorbent for soluble iron and manganese. Manganese greensand has the ability to capture (adsorb) soluble iron and manganese that may have escaped oxidation, as well as the capability of physically filtering out the particles of oxidized iron and manganese. Manganese greensand filters are generally set up as pressure filters, totally enclosed tanks containing the greensand. The process of adsorbing soluble iron and manganese uses up the greensand by converting the manganese dioxide coating to manganic oxide, which does not have the adsorption property. The greensand can be regenerated in much the same way as ion exchange resins by washing the sand with potassium permanganate.

> **DID YOU KNOW?**
>
> Greensand can also be configured as pressure filters for the removal of arsenic from water; it requires a pH above 6.8.

Hardness Treatment

Hardness in water is caused by the presence of certain positively charged metallic ions in solution in the water, the most common of which are calcium and magnesium; others include iron, strontium, and barium. As a general rule, groundwaters are harder than surface waters, so hardness is frequently of concern to the small water system operator. This hardness is derived from contact with soil and rock formations such as limestone. Although rainwater itself will not dissolve many solids, the natural carbon dioxide in the soil enters the water and forms carbonic acid (HCO), which is capable of dissolving minerals. Where soil is thick (contributing more carbon dioxide to the water) and limestone is present, hardness is likely to be a problem. The total amount of hardness in water is expressed as the sum of its calcium carbonate ($CaCO_3$) and its magnesium hardness; however, for practical purposes, hardness is expressed as calcium carbonate. This means that, regardless of the amount of the various components that make up hardness, they can be related to a specific amount of calcium carbonate (e.g., hardness is expressed as "mg/L as $CaCO_3$," or milligrams per liter as calcium carbonate).

Note: The two types of water hardness are *temporary hardness* and *permanent hardness*. Temporary hardness is also known as *carbonate hardness* (hardness that can be removed by boiling), and permanent hardness is also known as *noncarbonate hardness* (hardness that cannot be removed by boiling).

Hardness is of concern in domestic water consumption because hard water increases soap consumption, leaves a soapy scum in the sink or tub, can cause water heater electrodes to burn out quickly, can cause discoloration of plumbing fixtures and utensils, and is perceived as being less desirable water. In industrial water use, hardness is a concern because it can cause boiler scale and damage to industrial equipment.

The objection of customers to hardness is often dependent on the amount of hardness they are used to. People familiar with water with a hardness of 20 mg/L might think that a hardness of 100 mg/L is too much. On the other hand, a person who has been using water with a hardness of 200 mg/L might think that 100 mg/L is very soft. Table 7.2 lists the classifications of hardness.

TABLE 7.2
Classification of Hardness

Classification	mg/L as $CaCO_3$
Soft	0–75
Moderately hard	75–150
Hard	150–300
Very hard	Over 300

Treatment Methods

Two common methods are used to reduce hardness:

- *Ion exchange*—The ion exchange process is the process most frequently used for softening water. As a result of charging a resin with sodium ions, the resin exchanges the sodium ions for calcium or magnesium ions. Naturally occurring and synthetic cation exchange resins are available. Natural exchange resins include such substances as aluminum silicate, zeolite clays (zeolites are hydrous silicates found naturally in the cavities of lavas [greensand], glauconite zeolites, or synthetic, porous zeolites), humus, and certain types of sediments. These resins are placed in a pressure vessel. Salt brine is flushed through the resins. The sodium ions in the salt brine attach to the resin. The resin is now said to be charged. Once charged, water is passed through the resin, and the resin exchanges the sodium ions attached to the resin for calcium and magnesium ions, thus removing them from the water. The zeolite clays are most common because they are quite durable, can tolerate extreme ranges in pH, and are chemically stable. They have relatively limited exchange capacities, however, so they should only be used for water with a moderate total hardness. One of the results is that the water may be more corrosive than before. Another concern is that addition of sodium ions to the water may increase the health risk of those with high blood pressure.

- *Cation exchange*—The cation exchange process takes place with little or no intervention from the treatment plant operator. Water containing hardness-causing cations (Ca^{2+}, Mg^{2+}, Fe^{3+}) is passed through a bed of cation exchange resin. The water coming through the bed contains hardness near zero, although it will have elevated sodium content. (The sodium content is not likely to be high enough to be noticeable, but it could be high enough to pose problems to people on highly restricted salt-free diets.) The total lack of hardness in the finished water is likely to make it very corrosive, so normal practice bypasses a portion of the water around the softening process. The treated and untreated waters are blended to produce an effluent with a total hardness around 50 to 75 mg/L as $CaCO_3$.

Corrosion Control

Water operators add chemicals (e.g., lime or sodium hydroxide) to water at the source or at the waterworks to control corrosion. Using chemicals to achieve a slightly alkaline chemical balance prevents the water from corroding distribution pipes and consumers' plumbing and keeps substances such as lead from leaching out of plumbing and into the drinking water. For our purposes, we define *corrosion* as the conversion of a metal to a salt or oxide with a loss of desirable properties such as mechanical strength. Corrosion may occur over an entire exposed surface or may be localized at micro- or macroscopic discontinuities in metal. In all types of corrosion, a gradual decomposition of the material occurs, often due to an electrochemical reaction. Corrosion may be caused by (1) stray current electrolysis, (2) dissimilar metals (i.e., galvanic corrosion), or (3) differential concentration cells. Corrosion begins at the surface of a material and moves inward.

The adverse effects of corrosion can be categorized according to health, aesthetics, and economic effects, among others. The corrosion of toxic metal pipe made from lead creates a serious health hazard. Lead tends to accumulate in the bones of humans and animals. Signs of lead intoxication include gastrointestinal disturbances, fatigue, anemia, and muscular paralysis. Lead is not a natural contaminant in either surface waters or groundwaters, and the MCL of 0.005 mg/L in source waters is rarely exceeded. It is a corrosion byproduct from high lead solder joints in copper and lead piping. Small dosages of lead can lead to developmental problems in children. The USEPA's Lead and Copper Rule addresses the matter of lead in drinking water exceeding specified action levels.

Note: The USEPA's Lead and Copper Rule requires that a treatment facility achieve optimum corrosion control.

Cadmium is the only other toxic metal found in samples from plumbing systems. Cadmium is a contaminant found in zinc. Its adverse health effects are best known for being associated with severe bone and kidney syndrome in Japan. The proposed maximum contaminant level (PMCL) for cadmium is 0.01 mg/L.

Aesthetic effects that are a result of corrosion of iron include pitting and are a consequence of the deposition of ferric hydroxide and other products and the solution of iron—*tuberculation*. Tuberculation reduces the hydraulic capacity of the pipe. Corrosion of iron can cause customer complaints of reddish or reddish-brown staining of plumbing fixtures and laundry. Corrosion of copper lines can cause customer complaints of bluish or blue–green stains on plumbing fixtures. Sulfide corrosion of copper and iron lines can cause a blackish color in the water. The byproducts of microbial activity (especially iron bacteria) can cause foul tastes or odors in the water.

The *economic effects* of corrosion may include water main replacement, especially when tuberculation reduces the flow capacity of the main. Tuberculation increases pipe roughness, causing an increase in pumping costs and a reduction in distribution system pressure. Tuberculation and corrosion can cause leaks in distribution mains and household plumbing. Corrosion of household plumping may require extensive treatment, public education, and other actions under the Lead and Copper Rule.

Other effects of corrosion include short service life of household plumbing caused by pitting. A build-up of mineral deposits in a hot water system may eventually restrict hot-water flow. Also, the structural integrity of steel water storage tanks may deteriorate, causing structural failures. Steel ladders in clearwells or water storage tanks may corrode, introducing iron into the finished water. Steel parts in flocculation tanks, sedimentation basins, clarifiers, and filters may also corrode.

Types of Corrosion

Three types of corrosion occur in water mains:

- *Galvanic* occurs when two dissimilar metals come into contact and are exposed to a conductive environment; a potential exists between them, and current flows. This type of corrosion is the result of an electrochemical reaction when the flow of electric current itself is an essential part of the reaction.

- *Tuberculation* refers to the formation of localized corrosion products scattered over the surface in the form of knob-like mounds. These mounds increase the roughness of the inside of the pipe, increasing resistance to water flow and decreasing the *C* factor of the pipe.
- *Pitting* is localized corrosion that is classified as pitting when the diameter of the cavity at the metal surface is the same or less than the depth of the cavity.

Factors Affecting Corrosion

The primary factors affecting corrosion are pH, alkalinity, hardness (calcium), dissolved oxygen, and total dissolved solids. Secondary factors include temperature, velocity of water in pipes, and carbon dioxide (CO_2).

Determination of Corrosion Problems

To determine if corrosion is taking place in water mains, materials removed from the distribution system should be examined for signs of corrosion damage. A primary indicator of corrosion damage is pitting. (Measure the depth of the pits to gauge the extent of damage.) Another common method used to determine if corrosion or scaling is taking place in distribution lines is to insert special steel specimens of known weight (called *coupons*) in the pipe and examine them for corrosion after a period of time. Detecting evidence of leaks, conducting flow tests and chemical tests for dissolved oxygen and toxic metals, and receiving customer complaints (e.g., red or black water, laundry and fixture stains) can also reveal corrosion problems.

Formulas can also be used to determine corrosion (to an extent). The *Langelier Saturation Index* (LSI) and the *Aggressive Index* (AI) are two of the most commonly used indices. The LSI determines whether water is corrosive. The AI is used for waters that have low natural pH, are high in dissolved oxygen, are low in total dissolved solids, and have low alkalinity and low hardness. These waters are very aggressive and can be corrosive. Both of these indices are typically used as starting points in determining the adjustments required to produce a protective film:

- LSI approximately 0.5
- AI value of 12 or higher

Note: The LSI and AI are based on the dissolving of and precipitation of calcium carbonate; therefore, the respective indices may not actually reflect the corrosive nature of the particular water for a specific pipe material. They can be useful tools, however, when selecting materials or treatment options for corrosion control.

Corrosion Control Method

One method used to reduce the corrosive nature of water is *chemical addition*. Selection of the chemicals depends on the characteristics of the water, where the chemicals can be applied, how they can be applied and mixed with water, and the cost of the chemicals. Another corrosion control method is *aeration*. Aeration works to remove carbon dioxide (CO_2), which can be reduced to about 5 mg/L. *Cathodic protection*, often employed to control corrosion, involves applying an outside electric current to the metal to reverse the electromechanical corrosion process. The application of DC current prevents normal electron flow. Cathodic protection uses a sacrificial metal electrode (a magnesium anode) that corrodes instead of the pipe or tank. *Linings*, *coatings*, and *paints* can also be used in corrosion control. Slip-line with a plastic liner, cement mortar, zinc or magnesium, polyethylene, epoxy, and coal tar enamels are some of the materials that can be used.

Note: Before using any protective coatings, consult the district engineer first!

Several *corrosive-resistant pipe materials* are used to prevent corrosion:

- PVC plastic pipe
- Aluminum
- Nickel
- Silicon
- Brass
- Bronze
- Stainless steel
- Reinforced concrete

In addition to internal corrosion problems, waterworks operators must also be concerned with external corrosion problems. The primary culprit involved with external corrosion of distribution system pipe is soil. The measure of corrosivity of the soil is the *soil resistivity*. If the soil resistivity is greater than 5000 ohm/cm, serious corrosion is unlikely. Steel pipe may be used under these conditions. If soil resistivity is less than 500 ohm/cm, plastic PVC pipe should be used. For intermediate ranges of soil resistivity (500 to 5000 ohm/cm), ductile iron pipe, linings, and coatings should be used.

Several operating problems are commonly associated with corrosion control:

- $CaCO_3$ not depositing a film is usually a result of poor pH control (out of the normal range of 6.5 to 8.5). This may also cause excessive film deposition.
- Persistence of red water problems are most probably a result of poor flow patterns, insufficient velocity, tuberculation of pipe surface, and the presence of iron bacteria:
  1. *Velocity*—Chemicals must make contact with the pipe surface. Dead ends and low-flow areas should have a flushing program; dead ends should be looped.
  2. *Tuberculation*—The best approach is to clean with *pig*. In extreme cases, clean pipe with metal scrapers and install cement-mortar lining.
  3. *Iron bacteria*—Slime prevents film contact with the pipe surface. Slime will grow and the coating will be lost. Pipe cleaning and disinfection programs are necessary.

Coagulation

The primary purpose in surface-water treatment is chemical clarification by coagulation and mixing, flocculation, sedimentation, and filtration. These unit processes, along with disinfection, work to remove particles; naturally occurring organic matter (NOM), such as bacteria, algae, zooplankton, and organic compounds; and microbes from water to produce water that is noncorrosive. Specifically, coagulation and flocculation work to destabilize particles and to agglomerate dissolved and particulate matter. Sedimentation removes solids and provides 1/2-log *Giardia* and 1-log virus removal. Filtration removes solids and provides 2-log *Giardia* and 1-log virus removal. Finally, disinfection provides microbial inactivation and 1/2-log *Giardia* and 2-log virus removal.

From Figure 7.1, it can be seen that following screening and the other pretreatment processes the next unit process in a conventional water treatment system is a mixer where chemicals are added in what is known as coagulation. The exception to this unit process configuration occurs in small systems using groundwater, when chlorine or other taste and odor control measures are introduced at the intake and are the extent of treatment.

Materials present in raw water may vary in size, concentration, and type. Dispersed substances in the water may be classified as *suspended, colloidal,* or *solution*. Suspended particles may vary in mass and size and are dependent on the flow of water. High flows and velocities can carry larger material. As velocities decrease, the suspended particles settle according to size and mass. Other material may be in solution; for example, salt dissolves in water. Matter in the colloidal state does not dissolve, but the particles are so small they will not settle out of the water. Color (as in tea-colored swamp water) is mainly due to colloids or extremely fine particles of matter in suspension. Colloidal and solute particles in water are electrically charged. Because most of the charges are alike (negative) and repel each other, the particles stay dispersed and remain in the colloidal or soluble state.

Suspended matter will settle without treatment, if the water is still enough to allow it to settle. The rate of settling of particles can be determined, as this settling follows certain laws of physics; however, much of the suspended matter may be so slow in settling that the normal settling processes become impractical, and if colloidal particles are present then settling will not occur. Moreover, water drawn from a raw water source often contains many small unstable (unsticky) particles; therefore, sedimentation alone is usually an impractical way to obtain clear water in most locations, and another method of increasing the settling rate must be used: coagulation, which is designed to convert stable (unsticky) particles to unstable (sticky) particles.

Coagulation is a series of chemical and mechanical operations by which coagulants are applied and made effective. These operations are comprised of two distinct phases: (1) rapid mixing to disperse coagulant chemicals by violent agitation into the water being treated, and (2) flocculation to agglomerate small particles into well-defined floc by gentle agitation for a much longer time.

The coagulant must be added to the raw water and perfectly distributed into the liquid; such uniformity of chemical treatment is reached through rapid agitation or mixing. Common coagulants (salts) include the following:

- Alum (aluminum sulfate)
- Sodium aluminate
- Ferric sulfate
- Ferrous sulfate
- Ferric chloride
- Polymers

Coagulation is the reaction between one of these salts and water. The simplest coagulation process occurs between alum and water. Alum, or aluminum sulfate, is produced by a chemical reaction between bauxite ore and sulfuric acid. The normal strength of liquid alum is adjusted to 8.3%, while the strength of dry alum is 17%.

When alum is placed in water, a chemical reaction occurs that produces positively charged aluminum ions. The overall result is the reduction of electrical charges and the formation of a sticky substance—the formation of *floc*, which when properly formed, will settle. These two destabilizing factors are the major contributions of coagulation toward the removal of turbidity, color, and microorganisms.

Liquid alum is preferred in water treatment because it has several advantages over other coagulants:

- Ease of handling
- Lower costs
- Less labor required to unload, store, and convey
- Elimination of dissolving operations
- Less storage space required
- Greater accuracy in measurement and control
- Elimination of the nuisance and unpleasantness of handling dry alum
- Easier maintenance

The formation of floc is the first step of coagulation; for greatest efficiency, rapid, intimate mixing of the raw water and the coagulant must occur. After mixing, the water should be slowly stirred so the very small, newly formed particles can attract and enmesh colloidal particles, holding them together to form larger floc. This slow mixing is the second stage of the process (flocculation), covered later.

A number of factors influence the coagulation process—pH, turbidity, temperature, alkalinity, and the use of polymers. The degree to which these factors influence coagulation depends on the coagulant use. The raw water conditions, optimum pH for coagulation, and other factors must be considered before deciding which chemical is to be fed and at what levels.

To determine the correct chemical dosage, a *jar test* or *coagulation test* is performed. Jar tests (widely used for many years by the water treatment industry) simulate full-scale coagulation and flocculation processes to determine optimum chemical dosages. It is important to note that jar testing is only an attempt to achieve a ballpark approximation of correct chemical dosage for the treatment process. The test conditions are intended to reflect the normal operation of a chemical treatment facility. The test can be used to

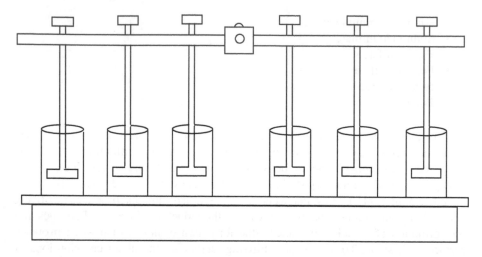

FIGURE 7.3   Variable-speed paddle mixer used in jar testing procedure.

- Select the most effective chemical.
- Select the optimum dosage.
- Determine the value of a flocculant aid and the proper dose.

The testing procedure requires a series of samples to be placed in testing jars (see Figure 7.3) and mixed at 100 ppm. Varying amounts of the process chemical or specified amounts of several flocculants are added (one volume/sample container). The mix is continued for 1 minute. Next, the mixing is slowed to 30 rpm to provide gentle agitation, and then the floc is allowed to settle. The flocculation period and settling process are observed carefully to determine the floc strength, settleability, and clarity of the *supernatant liquor* (the water that remains above the settled floc). Additionally, the supernatant can be tested to determine the efficiency of the chemical addition for removal of total suspended solids (TSS), biochemical oxygen demand (BOD_5), and phosphorus.

The equipment required for the jar test includes a six-position, variable-speed paddle mixer (see Figure 7.3); six 2-quart wide-mouthed jars; an interval timer; and assorted glassware, pipets, graduates, and so forth. The jar test procedure follows:

1. Place an appropriate volume of water sample in each of the jars (250- to 1000-mL samples may be used, depending on the size of the equipment being used). Start mixers and set for 100 rpm.
2. Add previously selected amounts of the chemical being evaluated. (Initial tests may use wide variations in chemical volumes to determine the approximate range; this is then narrowed in subsequent tests.)
3. Continue mixing for 1 minute.
4. Reduce the mixer speed to a gentle agitation (30 rpm), and continue mixing for 20 minutes. Again, time and mixer speed may be varied to reflect the particular facility.

Note: During this time, observe the floc formation—that is, how well the floc holds together during the agitation (floc strength).

5. Turn off the mixer and allow solids to settle for 20 to 30 minutes. Observe the settling characteristics, the clarity of the supernatant, the settleability of the solids, the flocculation of the solids, and the compactability of the solids.
6. Perform phosphate tests to determine removals.
7. Select the dose that provided the best treatment based on observations made during the analysis.

After initial ranges and chemical selections are determined, repeat the test using a smaller range of dosages to optimize performance.

Flocculation

As we see in Figure 7.1, flocculation follows coagulation in the conventional water treatment process. Flocculation is the physical process of slowly mixing the coagulated water to increase the probability of particle collision; unstable particles collide and stick together to form fewer larger flocs. Through experience, we have found that effective mixing reduces the required amount of chemicals and greatly improves the sedimentation process, resulting in longer filter runs and higher quality finished water.

The goal of flocculation is to form a uniform, feather like material similar to snowflakes—a dense, tenacious floc that entraps the fine, suspended, and colloidal particles and carries them down rapidly into the settling basin. Proper flocculation requires from 15 to 45 minutes. The time is based on water chemistry, water temperature, and mixing intensity. Temperature is the key component in determining the amount of time required for floc formation. To increase the speed of floc formation and the strength and weight of the floc, polymers are often added.

Settling (Sedimentation)

After raw water and chemicals have been mixed and the floc formed, the water containing the floc (because it has a higher specific gravity than water) flows to the sedimentation or settling basin (see Figure 7.1). Sedimentation is also called *clarification*. Sedimentation removes settleable solids by gravity. Water moves slowly though the sedimentation tank/basin with a minimum of turbulence at entry and exit points with minimum short-circuiting. Sludge accumulates at the bottom of the tank or basin. Typical tanks or basins used in sedimentation include conventional rectangular basins, conventional center-feed basins, peripheral-feed basins, and spiral-flow basins.

In conventional treatment plants, the amount of detention time required for settling can vary from 2 to 6 hours. Detention time should be based on the total filter capacity when the filters are passing 2 gpm per square foot of superficial sand area. For plants with higher filter rates, the detention time is based on a filter rate of 3 to 4 gpm per square foot of sand area. The time requirement is dependent on the weight of the floc, the temperature of the water, and how quiescent the basin is.

A number of conditions affect sedimentation: (1) uniformity of flow of water through the basin; (2) stratification of water due to a difference in temperature between water entering and water already in the basin; (3) release of gases that may collect in small bubbles on suspended solids, causing them to rise and float as scum rather than settle as sludge; (4) disintegration of previously formed floc; and (5) size and density of the floc.

Filtration

In the conventional water treatment process, *filtration* usually follows coagulation, flocculation, and sedimentation (see Figure 7.1). At present, filtration is not always used in small water systems; however, regulatory requirements under the USEPA Interim Enhanced Surface Water Treatment rules may make water filtering necessary at most water supply systems. Water filtration is a physical process of separating suspended and colloidal particles from water by passing water through a granular material. The process of filtration involves straining, settling, and adsorption. As floc passes into the filter, the spaces between the filter grains become clogged, reducing this opening and increasing removal. Some material is removed merely because it settles on a media grain. One of the most important processes is adsorption of the floc onto the surface of individual filter grains. This helps collect the floc and reduces the size of the openings between the filter media grains. In addition to removing silt and sediment, floc, algae, insect larvae, and any other large elements, filtration also contributes to the removal of bacteria and protozoa such as *Giardia lamblia* and *Cryptosporidium*. Some filtration processes are also used for iron and manganese removal.

The Surface Water Treatment Rule (SWTR) specifies four filtration technologies, although it also allows the use of alternative filtration technologies (e.g., cartridge filters). The specified technologies are (1) slow sand and rapid sand filtration, (2) pressure filtration, (3) diatomaceous earth filtration, and (4) direct filtration. Of these, all but rapid sand filtration are commonly employed in small water systems that use filtration. Each type of filtration system has advantages and disadvantages. Regardless of the type of filter, however, filtration involves the processes of *straining* (where particles are captured in the small spaces between filter media grains), *sedimentation* (where the particles land on top of the grains and stay there), and *adsorption* (where a chemical attraction occurs between the particles and the surface of the media grains).

Slow Sand Filters

The first slow sand filter was installed in London in 1829, and the technique was used widely throughout Europe, although not in the United States. By 1900, rapid sand filtration began taking over as the dominant filtration technology, although a few slow sand filters are still in operation today. With the advent of the Safe Drinking Water Act (SDWA) and its regulations (especially the Surface Water Treatment Rule) and recognition of the problems associated with *Giardia lamblia* and *Cryptosporidium* in surface water, the water industry is reexamining the use of slow sand filters. The low technology requirements may prevent many state water systems from using this type of equipment.

On the plus side, slow sand filtration is well suited for small water systems. It is a proven, effective filtration process with relatively low construction costs and low operating costs (it does not require constant operator attention). It is quite effective for water systems as large as 5000 people; beyond that, the surface area requirements and manual labor required to recondition the filters make rapid sand filters the more effective choice. The filtration rate is generally in the range of 45 to 150 gallons per day per square foot. Components of a slow sand filter include the following:

- A covered structure to hold the filter media
- An underdrain system
- Graded rock that is placed around and just above the underdrain
- The filter media, consisting of 30 to 55 inches of sand with a grain size of 0.25 to 0.35 mm
- Inlet and outlet piping to convey the water to and from the filter and the means to drain filtered water to waste

The area above the top of the sand layer is flooded with water to a depth of 3 to 5 feet, and the water is allowed to trickle down through the sand. An overflow device prevents excessive water depth. The filter must have provisions for filling it from the bottom up, and it must be equipped with a loss-of-head gauge, a rate-of-flow control device (such as an orifice or butterfly valve), a weir or effluent pipe that ensures that the water level cannot drop below the sand surface, and filtered waste sample taps.

When the filter is first placed in service, the head loss through the media caused by the resistance of the sand is about 0.2 feet (i.e., a layer of water 0.2 feet deep on top of the filter will provide enough pressure to push the water downward through the filter). As the filter operates, the media become clogged with the material being filtered out of the water, and the head loss increases. When it reaches about 4 to 5 feet, the filter must be cleaned. For efficient operation of a slow sand filter, the water being filtered should have a turbidity averaging less than 5 TU, with a maximum of 30 TU. Slow sand filters are not backwashed the way conventional filtration units are. One to 2 inches of material must be removed on a periodic basis to keep the filter operating.

Rapid Sand Filters

The rapid sand filter, which is similar in some ways to the slow sand filter, is one of the most widely used filtration units. The major difference is in the principle of operation—that is, in the speed or rate at which water passes through the media. In operation, water passes downward through a sand bed that removes the suspended particles. The suspended particles consist of the coagulated matter remaining in the water after sedimentation, as well as a small amount of uncoagulated suspended matter.

Some significant differences exist in construction, control, and operation between slow sand filters and rapid sand filters. Because of the design and construction of the rapid sand filtration, the land area required to filter the same quantity of water is reduced. Components of a rapid sand filter include the following:

- Structure to house media
- Filter media
- Gravel media support layer

- Underdrain system
- Valves and piping system
- Filter backwash system
- Waste disposal system

Usually 2 to 3 feet deep, the filter media are supported by approximately 1 foot of gravel. The media may be fine sand or a combination of sand, anthracite coal, and coal (dual- or multimedia filter). Water is applied to a rapid sand filter at a rate of 1.5 gallons per minute per square foot (gpm/ft^2) of filter media surface. When the rate is between 4 and 6 gpm/ft^2, the filter is referred to as a *high-rate filter*; at a rate over 6 gpm/ft^2, the filter is referred to as a *ultra-high-rate filter*. These rates compare to the slow sand filtration rate of 45 to 150 gallons per day per square foot. High-rate and ultra-high-rate filters must meet additional conditions to ensure proper operation.

Generally, raw water turbidity is not that high; however, even if raw water turbidity values exceed 1000 TU, properly operated rapid sand filters can produce filtered water with a turbidity of well under 0.5 TU. The time the filter is in operation between cleanings (filter runs) usually ranges from 12 to 72 hours, depending on the quality of the raw water; the end of the run is indicated by the head loss approaching 6 to 8 feet. Filter *breakthrough* (when filtered material is pulled through the filter into the effluent) can occur if the head loss becomes too great. Operation with head loss too high can also cause *air binding* (which blocks part of the filter with air bubbles), increasing the flow rate through the remaining filter area.

Rapid sand filters have the advantage of a lower land requirement, and they have other advantages, as well; for example, rapid sand filters cost less, are less labor intensive to clean, and offer higher efficiency with highly turbid waters. On the downside, the operation and maintenance costs of rapid sand filters are much higher in comparison because of the increased complexity of the filter controls and back-washing system.

When *backwashing* a rapid sand filter, the filter is cleaned by passing treated water backward (upward) through the filter media and agitating the top of the media. The need for backwashing is determined by a combination of filter run time (i.e., the length of time since the last backwashing), effluent turbidity, and head loss through the filter. Depending on the raw water quality, the run time varies from one filtration plant to another (and may even vary from one filter to another in the same plant).

Note: Backwashing usually requires 3 to 7% of the water produced by the plant.

Pressure Filter Systems

When raw water is pumped or piped from the source to a gravity filter, the head (pressure) is lost as the water enters the floc basin. When this occurs, pumping the water from the plant clearwell to the reservoir is usually necessary. One way to reduce pumping is to place the plant components into pressure vessels, thus maintaining the head. This type of arrangement is known as a pressure filter system. Pressure filters are also quite popular for iron and manganese removal and for filtration of water from wells. They may be placed directly in the pipeline from the well or pump with little head loss. Most pressure filters operate at a rate of about 3 gpm/ft^2.

Although pressure filtration is operationally the same as rapid sand filtration and consists of components similar to those of a rapid sand filter, the main difference between a rapid sand filtration system and a pressure filtration system is that the entire pressure filter is contained within a pressure vessel. These units are often highly automated and are usually purchased as self-contained units with all necessary piping, controls, and equipment contained in a single unit. They are backwashed in much the same manner as the rapid sand filter.

The major advantage of the pressure filter is its low initial cost. This type of filter is usually prefabricated, with a standardized design. A major disadvantage is that the operator is unable to observe the filter in the pressure filter and so is unable to determine the condition of the media. Unless the unit has an automatic shutdown feature for high effluent turbidity, driving filtered material through the filter is possible.

Diatomaceous Earth Filters

Diatomaceous earth is a white material made from the skeletal remains of diatoms. The skeletons are microscopic and in most cases porous. Diatomaceous earth is available in various grades, and the grade is selected based on filtration requirements. These diatoms are mixed in water slurry and fed onto a fine screen called a *septum*, usually made of stainless steel, nylon, or plastic. The slurry is fed at a rate of 0.2 lb/ft^2 of filter area. The diatoms collect in a precoat over the septum, forming an extremely fine screen. Diatoms are fed continuously with the raw water, causing the buildup of a filter cake approximately 1/8 to 1/5 inch thick. The openings are so small that the fine particles that cause turbidity are trapped on the screen. Coating the septum with diatoms gives it the ability to filter out very small microscopic material. The fine screen and the buildup of filtered particles cause a high head loss through the filter. When the head loss reaches a maximum level (30 psi on a pressure-type filter or 15 inches of mercury on a vacuum-type filter), the filter cake must be removed by backwashing.

The slurry of diatoms is fed with raw water during filtration in a process called *body feed*. The body feed prevents premature clogging of the septum cake. These diatoms are caught on the septum, increasing the head loss and preventing the cake from clogging too rapidly by the particles being filtered. Although the body feed increases head loss, head loss increases are more gradual than if body feed were not used.

Diatomaceous earth filters are relatively low in cost to construct, but they have high operating costs and can cause frequent operating problems if not properly operated and maintained. They can be used to filter raw surface waters or surface-influenced groundwaters with low turbidity (<5 NTU) and low coliform concentrations (no more than 50 coliforms per 100 mL) and may also be used for iron and manganese removal following oxidation. Filtration rates are between 1.0 and 1.5 gpm/ft^2.

Direct Filtration

Direct filtration is a treatment scheme that omits the flocculation and sedimentation steps prior to filtration. Coagulant chemicals are added, and the water is passed directly onto the filter. All solids removal takes place on the filter, which can lead to much shorter filter runs, more frequent backwashing, and a greater percentage of finished water used for backwashing. The lack of a flocculation process and sedimentation basin reduces construction costs but increases the requirement for skilled operators and high-quality instrumentation. Direct filtration must be used only where the water flow rate and raw water quality are fairly consistent and where the incoming turbidity is low.

Alternative Filters

A *cartridge filter system* can be employed as an alternative filtering system to reduce turbidity and remove *Giardia*. A cartridge filter is made of a synthetic media contained in a plastic or metal housing. These systems are normally installed in a series of three or four filters. Each filter contains media successively smaller than the previous filter. The media sizes typically range from 50 to 5 μm or less. The filter arrangement is dependent on the quality of the water, the capability of the filter, and the quantity of water needed. The USEPA and state agencies have established criteria for the selection and use of cartridge filters. Generally, cartridge filter systems are regulated in the same manner as other filtration systems.

Because of new regulatory requirements and the need to provide more efficient removal of pathogenic protozoans (e.g., *Giardia* and *Cryptosporidium*) from water supplies, *membrane filtration systems* are finding increased application in water treatment systems. A *membrane* is a thin film separating two different phases of a material acting as a selective barrier to the transport of matter operated by some driving force. Simply, a membrane can be regarded as a sieve with very small pores. Membrane filtration processes are typically pressure, electrically, vacuum, or thermally driven. The types of drinking water membrane filtration systems include microfiltration, ultrafiltration, nanofiltration, and reverse osmosis. A typical membrane filtration process has one input and two outputs. Membrane performance is largely a function of the properties of the materials to be separated and can vary throughout operation.

Common Filter Problems

Two common types of filter problems occur: those caused by filter runs that are too long (infrequent backwash) and those caused by inefficient backwash (cleaning). A filter run that is too long can cause *breakthrough* (the pushing of debris removed from the water through the media and into the effluent) and *air binding* (the trapping of air and other dissolved gases in the filter media). Air binding occurs when the rate at which water exits the bottom of the filter exceeds the rate at which the water penetrates the top of the filter. When this happens, a void and partial vacuum occur inside the filter media. The vacuum causes gases to escape from the water and fill the void. When the filter is backwashed, the release of these gases may cause a violent upheaval in the media and destroy the layering of the media bed, gravel, or underdrain. Two solutions to the problems are to (1) check the filtration rates to be sure they are within the design specifications, and (2) remove the top 1 inch of media and replace with new media. This keeps the top of the media from collecting the floc and sealing the entrance into the filter media.

Another common filtration problem that is associated with poor backwashing practices is the formation of *mudballs* that get trapped in the filter media. In severe cases, mudballs can completely clog a filter. Poor agitation of the surface of the filter can form a crust on top of the filter; the crust later cracks under the water pressure, causing uneven distribution of water through the filter media. Filter cracking can be corrected by removing the top 1-inch of the filter media, increasing the backwash rate, or checking the effectiveness of the surface wash (if installed). Backwashing at too high a rate can cause the filter media to wash out of the filter over the effluent troughs and may damage the filter underdrain system. Two possible solutions are to (1) check the backwash rate to be sure that it meets the design criteria, and (2) check the surface wash (if installed) for proper operation.

Disinfection*

The chemical or physical process used to control waterborne pathogenic organisms and prevent waterborne disease is called disinfection. The goal in proper disinfection in a water system is to destroy all disease-causing organisms. Disinfection should not be confused with sterilization. Sterilization is the complete killing of all living organisms. Waterworks operators disinfect by destroying organisms that might be dangerous; they do not attempt to sterilize water.

In water treatment, disinfection is almost always accomplished by adding chlorine or chlorine compounds after all other treatment steps (see Figure 7.1), although in the United States ultraviolet (UV) light, potassium permanganate, and ozone processes may be encountered.

The effectiveness of disinfection in a drinking water system is measured by testing for the presence or absence of coliform bacteria. Coliform bacteria found in water are generally not pathogenic, although they are good indicators of contamination. Their presence indicates the possibility of contamination, and their absence indicates the possibility that the water is potable—if the source is adequate, the waterworks history is good, and acceptable chlorine residual is present.

Desired characteristics of a disinfectant include the following:

- It must be able to deactivate or destroy any type or number of disease-causing microorganisms that may be in a water supply, in reasonable time, within expected temperature ranges, and despite changes in the character of the water (pH, for example).
- It must be nontoxic.
- It must not add unpleasant taste or odor to the water.
- It must be readily available at a reasonable cost and be safe and easy to handle, transport, store, and apply.
- It must be quick and easy to determine the concentration of the disinfectant in the treated water.
- It should persist within the disinfected water at a high enough concentration to provide residual protection through the distribution.

---

* Disinfection is a unit process used in both water and wastewater treatment. Many of the terms, practices, and applications discussed in this section apply to both water and wastewater treatment. There are also some differences, mainly in the types of disinfectants used and the applications of disinfection in water and wastewater treatment. In this section, we discuss disinfection only as it applies to water treatment.

Note: Disinfection is effective in reducing waterborne diseases because most pathogenic organisms are more sensitive to disinfection than are nonpathogens; however, disinfection is only as effective as the care used in controlling the process and ensuring that all of the water supply is continually treated with the amount of disinfectant required to produce safe water.

Methods of disinfection include the following:

- *Heat*—Possibly the first method of disinfection, which is accomplished by boiling water for 5 to 10 minutes; good, obviously, only for household quantities of water when bacteriological quality is questionable
- *Ultraviolet (UV) light*—A practical method of treating large quantities, but absorption of UV light is very rapid so this method is limited to nonturbid waters close to the light source
- *Metal ions*—Silver, copper, mercury
- *Alkalis and acids*
- *pH adjustment*—To under 3.0 or over 11.0
- *Oxidizing agents*—Bromine, ozone, potassium permanganate, and chlorine

The vast majority of drinking water systems in the United States use chlorine for disinfection. Along with meeting the desired characteristics listed above, chlorine has the added advantage of a long history of use and is fairly well understood. Although some small water systems may use other disinfectants, we concentrate on chlorine here.

Chlorination

Chlorination is the addition of chlorine or chlorine compounds to water. Chlorination is considered to be the single most important process for preventing the spread of waterborne disease. Chlorine has many attractive features that contribute to its wide use in industry. Five key attributes of chlorine are

1. It damages the cell wall.
2. It alters the permeability of the cell (the ability to pass water in and out through the cell wall).
3. It alters the cell protoplasm.
4. It inhibits the enzyme activity of the cell so it is unable to use its food to produce energy.
5. It inhibits cell reproduction.

Chlorine is available in a number of different forms:

- As pure elemental gaseous chlorine, a greenish-yellow gas possessing a pungent and irritating odor and is heavier than air, nonflammable, and nonexplosive; when released to the atmosphere, this form is toxic and corrosive
- As solid calcium hypochlorite (in tablets or granules)
- As a liquid sodium hypochlorite solution (in various strengths)

The selection of one form of chlorine over the others for a given water system depends on the amount of water to be treated, configuration of the water system, local availability of the chemicals, and skill of the operator.

One of the major advantages of using chlorine is the effective residual that it produces. A residual indicates that disinfection is completed and the system has an acceptable bacteriological quality. Maintaining a residual in the distribution system provides another line of defense against pathogenic organisms that could enter the distribution system and helps to prevent regrowth of those microorganisms that were injured but not killed during the initial disinfection stage.

Common chlorination terms include the following:

- *Chlorine reaction*—Regardless of the form of chlorine used for disinfection, the reaction in water is basically the same. The same amount of disinfection can be expected, provided the same amount of available chlorine is added to the water. The standard units used to express the concentration of chlorine in water are milligrams per liter (mg/L) and parts per million (ppm); these terms indicate the same quantity.
- *Chlorine dose*—The amount of chlorine added to the system. It can be determined by adding the desired residual for the finished water to the chlorine demand of the untreated water. Dosage can be either milligrams per liter (mg/L) or pounds per day. The most common is mg/L.
- *Chlorine demand*—The amount of chlorine used by iron, manganese, turbidity, algae, and microorganisms in the water. Because the reaction between chlorine and microorganisms is not instantaneous, demand is relative to time. For instance, the demand 5 minutes after applying chlorine will be less than the demand after 20 minutes. Demand, like dosage, is expressed in mg/L. The chlorine demand is determined as follows:

$$Cl_2 \text{ demand} = Cl_2 \text{ dose} - Cl_2 \text{ residual} \qquad (7.1)$$

- *Chlorine residual*—The amount of chlorine (determined by testing) that remains after the demand is satisfied. Residual, like demand, is based on time. The longer the time after dosage, the lower the residual will be, until all of the demand has been satisfied. Residual, like dosage and demand, is expressed in mg/L. The presence of a *free residual* of at least 0.2 to 0.4 ppm usually provides a high degree of assurance that the disinfection of the water is complete. *Combined residual* is the result of combining free chlorine with nitrogen compounds. Combined residuals are also called *chloramines*. The *total chlorine residual* is the mathematical combination of free and combined residuals. Total residual can be determined directly with standard chlorine residual test kits.
- *Chlorine contact time*—A key item in predicting the effectiveness of chlorine on microorganisms. It is the interval (usually only a few minutes) between the time when chlorine is added to the water and the time the water passes by the sampling point. Contact time is the "T" in CT. CT is

calculated based on the free chlorine residual prior to the first consumer multiplied by the contact time in minutes:

$$CT = \text{Concentration} \times \text{Contact time} = \text{mg/L} \times \text{minutes} \tag{7.2}$$

A certain minimum time period is required for the disinfecting action to be completed. The contact time is usually a fixed condition determined by the rate of flow of the water and the distance from the chlorination point to the first consumer connection. Ideally, the contact time should not be less than 30 minutes, but even more time is needed at lower chlorine doses, in cold weather, or under other conditions.

Pilot studies have shown that specific CT values are necessary for the inactivation of viruses and *Giardia*. The required CT value will vary depending on pH, temperature, and the organisms to be killed. Charts and formulas are available to make this determination. The USEPA has set a CT value of 3-log ($CT_{99.9}$) inactivation to ensure that the water is free of *Giardia*. State drinking water regulations provide charts containing CT values for various pH and temperature combinations. Filtration, in combination with disinfection, must provide 3-log removal or inactivation of *Giardia*. Charts in the USEPA Surface Water Treatment Rule guidance manual list the required CT values for various filter systems.

Under the 1996 Interim Enhanced Surface Water Treatment Rule, the USEPA requires systems that filter to remove 99% (2 log) of *Cryptosporidium* oocysts. To be sure that the water is free of viruses, a combination of filtration and disinfection that provides 4-log (99.99%) removal of viruses has been judged the best for drinking water safety. Viruses are inactivated more easily than cysts or oocysts.

Gas Chlorination

Gas chlorine is provided in 100-lb to 1-ton containers. Chlorine is placed in the container as a liquid. The liquid boils at room temperature and is reduced to a gas that builds pressure in the cylinder. At room temperature (70°F), a chlorine cylinder will have a pressure of 85 psi; 100- to 150-lb cylinders should be maintained in an upright position and chained to the wall. To prevent a chlorine cylinder from rupturing in a fire, the cylinder valves are equipped with special fusible plugs that melt between 158 and 164°F.

Chlorine gas is 99.9% chlorine. A gas chlorinator meters the gas flow and mixes it with water, which is then injected as a water solution of pure chlorine. As the compressed liquid chlorine is withdrawn from the cylinder, it expands as a gas, withdrawing heat from the cylinder. Care must be taken not to withdraw the chlorine at too fast a rate; if the operator attempts to withdraw more than about 40 lb of chlorine per day from a 150-lb cylinder, it will freeze up.

Note: All chlorine gas feed equipment sold today is vacuum operated. This safety feature ensures that, if a component of the chlorinator fails, the vacuum will be lost, and the chlorinator will shut down without allowing gas to escape.

Chlorine gas is a highly toxic lung irritant, and special facilities are required for storing and housing it. Chlorine gas will expand to 500 times its original compressed liquid volume at room temperature (1 gallon of liquid chlorine will expand to about

67 ft³). Its advantage as a drinking water disinfectant is the convenience afforded by having a relatively large quantity of chlorine available for continuous operation for several days or weeks without the need for mixing chemicals. Where water flow rates are highly variable, the chlorination rate can be synchronized with the flow.

Chlorine gas has a very strong, characteristic odor that can be detected by most people at concentrations as low as 3.5 ppm. Highly corrosive in moist air, it is extremely toxic and irritating in concentrated form. Its toxicity ranges from being a throat irritant at 15 ppm to causing rapid death at 1000 ppm. Although chlorine does not burn, it supports combustion, so open flames should never be used around chlorination equipment.

When changing chlorine cylinders, an accidental release of chlorine may occasionally occur. To handle this type of release, a National Institute for Occupational Safety and Health (NIOSH)-approved, self-contained breathing apparatus (SCBA) must be worn. Special emergency repair kits are available from the Chlorine Institute for use by emergency response teams to deal with chlorine leaks. Because chlorine gas is 2.5 times heavier than air, exhaust and inlet air ducts should be installed at floor level. A leak of chlorine gas can be found with a strong ammonia mist solution, as a white cloud develops when ammonia mist and chlorine combine.

Hypochlorination

Combining chlorine with calcium or sodium produces hypochlorites. Calcium hypochlorites are sold in powder or tablet forms and can contain chlorine concentrations up to 67%. Sodium hypochlorite is a liquid (bleach, for example) and is found in concentrations up to 16%. Chlorine concentrations of household bleach range from 4.75 to 5.25%. Most small system operators find using these liquid or dry chlorine compounds more convenient and safer than chlorine gas.

The compounds are mixed with water and fed into the water with inexpensive solution feed pumps. These pumps are designed to operate against high system pressures but can also be used to inject chlorine solutions into tanks, although injecting chlorine into the suction side of a pump is not recommended as the chlorine may corrode the pump impeller.

Calcium hypochlorite can be purchased as tablets or granules, with approximately 65% available chlorine (10 lb of calcium hypochlorite granules contain only 6.5 lb of chlorine). Normally, 6.5 lb of calcium hypochlorite will produce a concentration of 50 mg/L chlorine in 10,000 gal of water. Calcium hypochlorite can burn (at 350°F) if combined with oil or grease. When mixing calcium hypochlorite, operators must wear chemical safety goggles, a cartridge breathing apparatus, and rubberized gloves. Always place the powder in the water. Placing the water into the dry powder could cause an explosion.

Sodium hypochlorite is supplied as a clear, greenish-yellow liquid in strengths from 5.25 to 16% available chlorine. Often referred to as "bleach," it is, in fact, used for bleaching. Common household bleach is a solution of sodium hypochlorite containing 4.75 to 5.25% available chlorine. The amount of sodium hypochlorite required to produce a 50-mg/L chlorine concentration in 10,000 gal of water can be calculated using the solutions equation, as shown below:

$$C_1 \times V_1 = C_2 \times V_2 \qquad\qquad (7.3)$$

where

    C = Solution concentration (mg/L or %).
    V = Solution volume (e.g., liters, gallons, quarts).
    $1.0\% = 10,000$ mg/L.

In this example, C_1 and V_1 are associated with the sodium hypochlorite, and C_2 and V_2 are associated with the 10,000 gallons of water with a 50-mg/L chlorine concentration. Therefore:

$$C_1 = 5.25\% = \frac{5.25\% \times 10,000 \text{ mg/L}}{1.0\%} = 52,500 \text{ mg/L}$$

V_1 = Unknown volume of sodium hypochlorite

$C_2 = 50$ mg/L

$V_2 = 10,000$ gal

$C_1 \times V_1 = C_2 \times V_2$

$52,500 \text{ mg/L} \times V_1 = 50 \text{ mg/L} \times 10,000 \text{ gal}$

$$V_1 = \frac{50 \text{ mg/L} \times 10,000 \text{ gal}}{52,500 \text{ mg/L}} = 9.52 \text{ gal}$$

Sodium hypochlorite solutions are introduced to the water in the same manner as calcium hypochlorite solutions. The purchased stock bleach is usually diluted with water to produce a feed solution that is pumped into the water system.

Hypochlorites must be stored properly to maintain their strengths. Calcium hypochlorite must be stored in airtight containers in cool, dry, dark locations. Sodium hypochlorite degrades relatively quickly even when properly stored; it can lose more than half of its strength in 3 to 6 months. Operators should purchase hypochlorites in small quantities to be sure they are used while still strong. Old chemicals should be discarded safely.

The pumping rate of a chemical metering pump is usually manually adjusted by varying the length of the piston or the diaphragm stroke. Once the stroke has been set, the hypochlorinator feeds accurately at that rate; however, chlorine measurements must still be made occasionally at the beginning and end of the well pump cycle to ensure that the dosage is correct. A metering device may be used to vary the hypochlorinator feed rate, synchronized with the water flow rate. Where a well pump is used, the hypochlorinator is connected electrically with the on/off controls of the pump so the chlorine solution is not fed into the pipe when the well is not pumping.

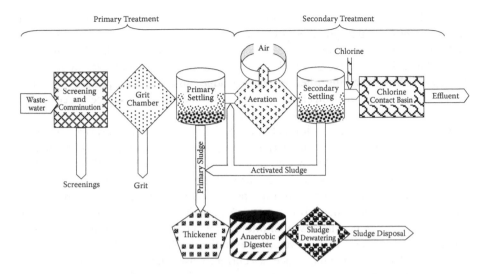

FIGURE 7.4 Schematic of an example wastewater treatment process providing primary and secondary treatment using the activated sludge process.

WASTEWATER TREATMENT

According to the Code of Federal Regulations (CFR) 40 CFR Part 403, regulations were established in the late 1970s and early 1980s to help publicly owned treatment works (POTWs) control industrial discharges to sewers. These regulations were designed to prevent pass-through and interference at the treatment plants and interference in the collection and transmission systems. Pass-through occurs when pollutants move through a POTW without being properly treated, which results in the POTW receiving an effluent violation or increases the magnitude or duration of such a violation. Interference occurs when a pollutant discharge causes a POTW to violate its permit by inhibiting or disrupting treatment processes, treatment operations, or processes related to sludge use or disposal.

WASTEWATER TREATMENT PROCESS MODEL

Unit operations (or processes) are the components that are linked together to form a process train (Figure 7.4). Keep in mind that the caboose attached to this train is treated and cleaned wastewater, which when outfalled is usually cleaner than the water in the receiving body. Unit operations are commonly divided on the basis of the fundamental mechanisms underlying them (i.e., physical, chemical, or biochemical). Physical operations, such as sedimentation are governed by the laws of physics (gravity). Chemical operations are those in which strictly chemical reactions occur, such as precipitation. Biochemical operations use living microorganisms to destroy or transform pollutants through enzymatically catalyzed chemical reactions (Grady et al., 2011).

The basic schematic shown in Figure 7.4 shows an example of a wastewater treatment process that provides primary and secondary treatment using the *activated sludge process*. Although secondary treatment (which provides BOD removal beyond what is achievable by simple sedimentation) commonly utilizes three different approaches—trickling filter, activated sludge, and oxidation ponds—we focus here, for instructive and illustrative purposes, on the activated sludge process. The purpose of Figure 7.4 is to allow the reader to follow the treatment process step by step as it is presented (and as it is actually configured in the real world) and to assist in demonstrating how all of the various unit processes sequentially follow and tie into each other. We will begin various sections with reference to Figure 7.4; it is important to follow through these sections in this manner because wastewater treatment is a series of individual steps (unit processes) that treat the waste stream as it makes its way through the entire process. A pictorial presentation of the treatment process, along with pertinent written information, enhances the learning process. It should also be pointed out, however, that, even though the model shown in Figure 7.4 does not include all of the unit processes currently used in wastewater treatment, we do not ignore the other major processes of trickling filters, rotating biological contactors (RBCs), oxidation ponds, and tertiary treatment components.

WASTEWATER TERMINOLOGY AND DEFINITIONS

Wastewater treatment technology, like many other technical fields, has its own unique terminology. Although some of the terms are unique to this field, many are common to other professions, as well. Remember that the science of wastewater treatment is a combination of engineering, biology, mathematics, hydrology, chemistry, physics, and other disciplines; therefore, many of the terms used in engineering, biology, mathematics, hydrology, chemistry, physics, and elsewhere are also used in wastewater treatment. Those terms not listed or defined below will be defined as they appear in the text.

> *Activated sludge*—The solids formed when microorganisms are used to treat wastewater using the activated sludge treatment process. It includes organisms, accumulated food materials, and waste products from the aerobic decomposition process.
>
> *Advanced wastewater treatment*—Treatment technology to produce an extremely high-quality discharge.
>
> *Aerobic*—Conditions in which free, elemental oxygen is present; also used to describe organisms, biological activity, or treatment processes that require free oxygen.
>
> *Anaerobic*—Conditions in which no oxygen (free or combined) is available; also used to describe organisms, biological activity, or treatment processes that function in the absence of oxygen.
>
> *Anoxic*—Conditions in which no free, elemental oxygen is present, and the only source of oxygen is combined oxygen, such as that found in nitrate compounds; also used to describe biological activity or treatment processes that function only in the presence of combined oxygen.

Average monthly discharge limitation—The highest allowable discharge over a calendar month.

Average weekly discharge limitation—The highest allowable discharge over a calendar week.

Biochemical oxygen demand (BOD_5)—Amount of organic matter that can be biologically oxidized under controlled conditions (5 days, 20°C, in the dark).

Biosolids—Solid organic matter that is recovered from a sewage treatment process and used in some way, especially as fertilizer.

Note: In this text, *biosolids* is used in many places (activated sludge being the exception) to replace the standard term *sludge*. The author and others in the field view the term sludge as an ugly four-letter word that is inappropriate to use when describing biosolids. Biosolids are a product that can be reused and has some value. Because biosolids have some value, they should not be classified as a waste product, and in this text they are not when their beneficial reuse is being addressed.

Buffer—Substance or solution that resists changes in pH.

Carbonaceous biochemical oxygen demand ($CBOD_5$)—The amount of biochemical oxygen demand that can be attributed to carbonaceous material.

Chemical oxygen demand (COD)—The amount of chemically oxidizable materials present in the wastewater.

Clarifier—A device designed to permit solids to settle or rise and be separated from the flow; also known as a settling tank or sedimentation basin.

Coliform—A type of bacteria used to indicate possible human or animal contamination of water.

Combined sewer—A collection system that carries both wastewater and stormwater flows.

Comminution—A process to shred solids into smaller, less harmful particles.

Composite sample—A combination of individual samples taken in proportion to flow.

Daily discharge—The discharge of a pollutant measured during a calendar day or any 24-hour period that reasonably represents a calendar day for the purposes of sampling. Limitations expressed as weight are the total mass (weight) discharged over the day. Limitations expressed in other units are average measurements of the day.

Daily maximum discharge—The highest allowable values for a daily discharge.

Detention time—The theoretical time water remains in a tank at a given flow rate.

Dewatering—The removal or separation of a portion of water present in a sludge or slurry.

Discharge Monitoring Report (DMR)—The monthly report required by the treatment plant's NPDES discharge permit.

Dissolved oxygen (DO)—Free or elemental oxygen dissolved in water.

Effluent—The flow leaving a tank, channel, or treatment process.

Effluent limitation—Any restriction imposed by the regulatory agency on quantities, discharge rates, or concentrations of pollutants that are discharged from point sources into state waters.

Facultative—Organisms that can survive and function in the presence or absence of free, elemental oxygen.

Fecal coliform—A type of bacteria found in the bodily discharges of warm-blooded animals; used as an indicator organism.

Floc—Solids that join together to form larger particles that will settle better.

Flume—A flow rate measurement device.

Food-to-microorganism ratio (F/M)—An activated sludge process control calculation based on the amount of food (BOD$_5$ or COD) available per pound of mixed liquor volatile suspended solids.

Grab sample—An individual sample collected at a randomly selected time.

Grit—Heavy inorganic solids such as sand, gravel, egg shells, or metal filings.

Industrial wastewater—Wastes associated with industrial manufacturing processes.

Infiltration/inflow—Extraneous flows in sewers; defined by Metcalf & Eddy (1991, pp. 29–31) as follows:

- *Infiltration*—Water entering the collection system through cracks, joints, or breaks.
- *Steady inflow*—Water discharged from cellar and foundation drains, cooling water discharges, and drains from springs and swampy areas. This type of inflow is steady and is identified and measured along with infiltration.
- *Direct flow*—Those types of inflow that have a direct stormwater runoff connection to the sanitary sewer and that cause an almost immediate increase in wastewater flows. Possible sources are roof leaders, yard and areaway drains, manhole covers, cross connections from storm drains and catch basins, and combined sewers.
- *Total inflow*—The sum of the direct inflow at any point in the system plus any flow discharged from the system upstream through overflows, pumping station bypasses, and the like.
- *Delayed inflow*—Stormwater that may require several days or more to drain through the sewer system. This category can include the discharge of sump pumps from cellar drainage as well as the slowed entry of surface water through manholes in ponded areas.

Influent—The wastewater entering a tank, channel, or treatment process.

Inorganic—Mineral materials such as salt, ferric chloride, iron, sand, or gravel.

License—A certificate issued by the State Board of Waterworks/Wastewater Works Operators authorizing the holder to perform the duties of a wastewater treatment plant operator.

Mean cell residence time (MCRT)—The average length of time a mixed liquor suspended solids particle remains in the activated sludge process; may also be known as sludge retention time.

Milligrams/liter (mg/L)—A measure of concentration equivalent to parts per million (ppm).

Mixed liquor—The combination of return activated sludge and wastewater in the aeration tank.

Mixed liquor suspended solids (MLSS)—The suspended solids concentration of the mixed liquor.

Mixed liquor volatile suspended solids (MLVSS)—The concentration of organic matter in the mixed liquor suspended solids.

Nitrogenous oxygen demand (NOD)—A measure of the amount of oxygen required to biologically oxidize nitrogen compounds under specified conditions of time and temperature.

NPDES permit—National Pollutant Discharge Elimination System permit that authorizes the discharge of treated wastes and specifies the conditions that must be met for discharge.

Nutrients—Substances required to support living organisms; usually refers to nitrogen, phosphorus, iron, and other trace metals.

Organic—Materials that consist of carbon, hydrogen, oxygen, sulfur, and nitrogen. Many organics are biologically degradable. All organic compounds can be converted to carbon dioxide and water when subjected to high temperatures.

Parts per million (ppm)—An alternative (but numerically equivalent) unit used in chemistry that is equal to milligrams per liter (mg/L). As an analogy, think of 1 ppm as being equivalent to a full shot glass in a swimming pool.

Pathogenic—Disease causing; a pathogenic organism is capable of causing illness.

Point source—Any discernible, defined, and discrete conveyance from which pollutants are or may be discharged.

Return activated sludge solids (RASS)—The concentration of suspended solids in the sludge flow being returned from the settling tank to the head of the aeration tank.

Sanitary wastewater—Wastes discharged from residences and from commercial, institutional, and similar facilities that include both sewage and industrial wastes.

Scum—The mixture of floatable solids and water that is removed from the surface of the settling tank.

Septic—A wastewater that has no dissolved oxygen present; generally characterized by a black color and rotten egg (hydrogen sulfide) odor.

Settleability—A process control test used to evaluate the settling characteristics of the activated sludge. Readings taken at 30 to 60 minutes are used to calculate the settled sludge volume (SSV) and the sludge volume index (SVI).

Settled sludge volume—The volume in percent occupied by an activated sludge sample after 30 to 60 minutes of settling; normally written as SSV with a subscript to indicate the time of the reading used for calculation (e.g., SSV_{60} or SSV_{30}).

Sewage—Wastewater containing human wastes.

Sludge—The mixture of settleable solids and water that is removed from the bottom of the settling tank.

Sludge retention time (SRT)—See *mean cell residence time*.

Sludge volume index (SVI)—A process control calculation used to evaluate the settling quality of the activated sludge; calculating the SVI requires the SSV_{30} and mixed liquor suspended solids test results.

Storm sewer—A collection system designed to carry only stormwater runoff.

Stormwater—Runoff resulting from rainfall and snowmelt.

Supernatant—In a digester, it is the amber-colored liquid above the sludge.

Waste activated sludge solids (WASS)—The concentration of suspended solids in the sludge being removed from the activated sludge process.

Wastewater—Water supply of a community after it has been soiled by use.

Weir—A device used to measure wastewater flow.

Zoogleal slime—The biological slime that forms on fixed-film treatment devices; it contains a wide variety of organisms essential to the treatment process.

MEASURING WASTEWATER PLANT PERFORMANCE

To evaluate how well a plant or treatment unit process is operating, performance efficiency or percent (%) removal is used. The results can be compared with those listed in the plant's operation and maintenance (O&M) manual to determine if the facility is performing as expected. In this chapter, sample calculations that are often used to measure plant performance/efficiency are presented. The calculation used to determine the performance (percent removal) of a digester is different from that used for performance (percent removal) for other processes. Care must be taken to select the correct formula.

$$\% \text{ Removal} = \frac{(\text{Influent concentration} - \text{Effluent concentration}) \times 100}{\text{Influent concentration}} \quad (7.4)$$

HYDRAULIC DETENTION TIME

The term *detention time* or *hydraulic detention time* (HDT) refers to the average length of time (theoretical time) a drop of water, wastewater, or suspended particles remains in a tank or channel. It is calculated by dividing the water/wastewater in the tank by the flow rate through the tank. The units of flow rate used in the calculation are dependent on whether the detention time is to be calculated in seconds, minutes, hours, or days. Detention time is used in conjunction with various treatment processes, including sedimentation and coagulation–flocculation. Generally, in practice, detention time is associated with the amount of time required for a tank to empty. The range of detention times varies with the process. For example, in a tank used for sedimentation, detention time is commonly measured in minutes.

WASTEWATER SOURCES AND CHARACTERISTICS

Wastewater treatment is designed to use the natural purification processes (self-purification processes of streams and rivers) to the maximum level possible. It is also designed to complete these processes in a controlled environment rather than over many miles of stream or river. Moreover, the treatment plant is also designed to remove other contaminants that are not normally subjected to natural processes, in addition to treating the solids that are generated through the treatment unit steps. The typical wastewater treatment plant is designed to achieve many different purposes:

- Protect public health.
- Protect public water supplies.
- Protect aquatic life.
- Preserve the best uses of the waters.
- Protect adjacent lands.

Wastewater treatment is a series of steps. Each of the steps can be accomplished using one or more treatment processes or types of equipment. The major categories of treatment steps are as follows:

1. *Preliminary treatment*—Removes materials that could damage plant equipment or would occupy treatment capacity without being treated
2. *Primary treatment*—Removes settleable and floatable solids (may not be present in all treatment plants)
3. *Secondary treatment*—Removes BOD_5 and dissolved and colloidal suspended organic matter by biological action; organics are converted to stable solids, carbon dioxide, and more organisms
4. *Advanced wastewater treatment*—Uses physical, chemical, and biological processes to remove additional BOD_5, solids, and nutrients (not present in all treatment plants)
5. *Disinfection*—Removes microorganisms to eliminate or reduce the possibility of disease when the flow is discharged
6. *Sludge treatment*—Stabilizes the solids removed from wastewater during treatment, inactivates pathogenic organisms, and reduces the volume of the sludge by removing water

The various treatment processes described above are discussed in detail later.

Wastewater Sources

The principal sources of domestic wastewater in a community are the residential areas and commercial districts. Other important sources include institutional and recreational facilities, stormwater (runoff), and groundwater (infiltration). Each source produces wastewater with specific characteristics. In this section, wastewater sources and the specific characteristics of wastewater are described.

Generation of Wastewater

Wastewater is generated by human and animal wastes, household wastes, industrial wastes, stormwater runoff, and groundwater infiltration:

- *Human and animal wastes*—Wastes that contain the solid and liquid discharges of humans and animals and are considered by many to be the most dangerous from a human health viewpoint. The primary health hazard is presented by the millions of bacteria, viruses, and other microorganisms (some of which may be pathogenic) present in the waste stream.
- *Household wastes*—Wastes, other than human and animal wastes, discharged from the home. Household wastes usually contain paper, household cleaners, detergents, trash, garbage, and other substances homeowners discharge into the sewer system.
- *Industrial wastes*—Materials discharged from industrial processes into the collection system. Industrial wastes typically contain chemicals, dyes, acids, alkalis, grit, detergents, and highly toxic materials.
- *Stormwater runoff*—Many collection systems are designed to carry both the wastes of the community and stormwater runoff. In this type of system, when a storm event occurs the waste stream can contain large amounts of sand, gravel, and other grit as well as excessive amounts of water.
- *Groundwater infiltration*—Groundwater will enter older, improperly sealed collection systems through cracks or unsealed pipe joints. This can add not only large amounts of water to wastewater flows but also additional grit.

Classification of Wastewater

Wastewater can be classified according to the sources of flows:

- *Domestic (sewage) wastewater*—Mainly contains human and animal wastes, household wastes, small amounts of groundwater infiltration, and small amounts of industrial wastes.
- *Sanitary wastewater*—Consists of domestic wastes and significant amounts of industrial wastes. In many cases, the industrial wastes can be treated without special precautions; however, in some cases the industrial wastes will require special precautions or a pretreatment program to ensure that the wastes do not cause compliance problems for the wastewater treatment plant.
- *Industrial wastewater*—Industrial wastes only; often, the industry will determine that it is safer and more economical to treat its waste independent of domestic waste.
- *Combined wastewater*—The combination of sanitary wastewater and stormwater runoff. All of the wastewater and stormwater of the community is transported through one system to the treatment plant.
- *Stormwater*—A separate collection system (no sanitary waste) that carries stormwater runoff including street debris, road salt, and grit.

Wastewater Characteristics

Wastewater contains many different substances that can be used to characterize it. The specific substances and amounts or concentrations of each will vary, depending on the source; thus, it is difficult to precisely characterize wastewater. Instead, wastewater characterization is usually based on and applied to an average domestic wastewater. Wastewater is characterized in terms of its physical, chemical, and biological characteristics.

Note: Keep in mind that other sources and types of wastewater can dramatically change the characteristics.

Physical Characteristics

The physical characteristics of wastewater are based on color, odor, temperature, and flow:

- *Color*—Fresh wastewater is usually a light brownish-gray color; however, typical wastewater is gray and has a cloudy appearance. The color of the wastewater will change significantly if allowed to go septic (if travel time in the collection system increases). Typical septic wastewater will have a black color.
- *Odor*—Odors in domestic wastewater are usually caused by gases produced by the decomposition of organic matter or by other substances added to the wastewater. Fresh domestic wastewater has a musty odor. If the wastewater is allowed to go septic, this odor will change significantly to a rotten-egg odor associated with the production of hydrogen sulfide (H_2S).
- *Temperature*—The temperature of wastewater is commonly higher than that of the water supply because of the addition of warm water from households and industrial plants; however, significant amounts of infiltration or stormwater flow can cause major temperature fluctuations.
- *Flow*—The actual volume of wastewater is commonly used as a physical characterization of wastewater and is normally expressed in terms of gallons per person per day. Most treatment plants are designed using an expected flow of 100 to 200 gallons per person per day. This figure may have to be revised to reflect the degree of infiltration or storm flow the plant receives. Flow rates will vary throughout the day. This variation, which can be as much as 50 to 200% of the average daily flow, is known as the *diurnal flow variation*.

Note: Diurnal means "occurs in a day or each day; daily."

Chemical Characteristics

When describing the chemical characteristics of wastewater, the discussion generally includes topics such as organic matter, the measurement of organic matter, inorganic matter, and gases. For the sake of simplicity, in this text we specifically describe chemical characteristics in terms of alkalinity, biochemical oxygen demand (BOD), chemical oxygen demand (COD), dissolved gases, nitrogen compounds, pH, phosphorus, solids (organic, inorganic, suspended, and dissolved solids), and water:

- *Alkalinity* is a measure of the capability of the wastewater to neutralize acids. It is measured in terms of bicarbonate, carbonate, and hydroxide alkalinity. Alkalinity is essential to buffer (hold the neutral pH) of the wastewater during the biological treatment processes.
- *Biochemical oxygen demand (BOD)* is a measure of the amount of biodegradable matter in the wastewater. It is normally measured by a 5-day test conducted at 20°C. The BOD_5 domestic waste is normally in the range of 100 to 300 mg/L.
- *Chemical oxygen demand (COD)* is a measure of the amount of oxidizable matter present in the sample. The COD is normally in the range of 200 to 500 mg/L. The presence of industrial wastes can increase this significantly.
- *Dissolved gases* are gases that are dissolved in wastewater. The specific gases and normal concentrations are based on the composition of the wastewater. Typical domestic wastewater contains oxygen in relatively low concentrations, carbon dioxide, and hydrogen sulfide (if septic conditions exist).
- The type and amount of *nitrogen compounds* present will vary from the raw wastewater to the treated effluent. Nitrogen follows a cycle of oxidation and reduction. Most of the nitrogen in untreated wastewater will be in the form of organic nitrogen or ammonia nitrogen. Laboratory tests exist for determination of both of these forms. The sum of these two forms of nitrogen is also measured and is known as *total Kjeldahl nitrogen* (TKN). Wastewater will normally contain between 20 to 85 mg/L of nitrogen. Organic nitrogen will normally be in the range of 8 to 35 mg/L, and ammonia nitrogen will be in the range of 12 to 50 mg/L.
- *pH* is used to express the acid condition of the wastewater. pH is expressed on a scale of 1 to 14. For proper treatment, wastewater pH should normally be in the range of 6.5 to 9.0 (ideal is 6.5 to 8.0).
- *Phosphorus* is essential to biological activity and must be present in at least minimum quantities or secondary treatment processes will not perform. Excessive amounts can cause stream damage and excessive algal growth. Phosphorus will normally be in the range of 6 to 20 mg/L. The removal of phosphate compounds from detergents has had a significant impact on the amounts of phosphorus in wastewater.
- Most pollutants found in wastewater can be classified as *solids*. Wastewater treatment is generally designed to remove solids or to convert solids to a form that is more stable or can be removed. Solids can be classified by their chemical composition (organic or inorganic) or by their physical characteristics (settleable, floatable, and colloidal). Concentrations of total solids in wastewater are normally in the range of 350 to 1200 mg/L.
  - *Organic solids* consist of carbon, hydrogen, oxygen, and nitrogen and can be converted to carbon dioxide and water by ignition at 550°C; they are also known as *fixed solids* or *loss on ignition*.
  - *Inorganic solids* are mineral solids that are unaffected by ignition; they are also known as *fixed solids* or *ash*.
  - *Suspended solids* will not pass through a glass-fiber filter pad; they can be further classified as total suspended solids (TSS), volatile suspended

solids, and fixed suspended solids. They can also be separated into three components based on settling characteristics: settleable solids, floatable solids, and colloidal solids. Total suspended solids in wastewater are normally in the range of 100 to 350 mg/L.

* *Dissolved solids* will pass through a glass-fiber filter pad. They can also be classified as total dissolved solids (TDS), volatile dissolved solids, and fixed dissolved solids. Total dissolved solids are normally in the range of 250 to 850 mg/L.

* *Water* is always the major constituent of wastewater. In most cases, water makes up 99.5 to 99.9% of the wastewater. Even in the strongest wastewater, the total amount of contamination present is less than 0.5% of the total, and in average-strength wastes it is usually less than 0.1%.

Biological Characteristics and Processes

(Note that the biological characteristics of water were discussed in detail earlier in this text.) After undergoing the physical aspects of treatment (i.e., screening, grit removal, and sedimentation) in preliminary and primary treatment, wastewater still contains some suspended solids and other solids that are dissolved in the water. In a natural stream, such substances are a source of food for protozoa, fungi, algae, and several varieties of bacteria. In secondary wastewater treatment, these same microscopic organisms (which are one of the main reasons for treating wastewater) are allowed to work as fast as they can to biologically convert the dissolved solids to suspended solids that will physically settle out at the end of secondary treatment.

Raw wastewater influent typically contains millions of organisms. The majority of these organisms are not pathogenic; however, several pathogenic organisms may also be present (these may include the organisms responsible for diseases such as typhoid, tetanus, hepatitis, dysentery, gastroenteritis, and others). Many of the organisms found in wastewater are microscopic (microorganisms); they include algae, bacteria, protozoa (e.g., amoebae, flagellates, free-swimming ciliates, stalked ciliates), rotifers, and viruses. Table 7.3 provides a summary of typical domestic wastewater characteristics.

TABLE 7.3
Typical Domestic Wastewater Characteristics

Characteristic	Typical Characteristic
Color	Gray
Odor	Musty
Dissolved oxygen	>1.0 mg/L
pH	6.5–9.0
TSS	100–350 mg/L
BOD$_5$	100–300 mg/L
COD	200–500 mg/L
Flow	100–200 gallons per person per day
Total nitrogen	20–85 mg/L
Total phosphorus	6–20 mg/L
Fecal coliform	500,000–3,000,000 MPN/100 mL

WASTEWATER COLLECTION SYSTEMS

Wastewater collection systems collect and convey wastewater to the treatment plant. The complexity of the system depends on the size of the community and the type of system selected. Methods of collection and conveyance of wastewater include gravity systems, force main systems, vacuum systems, and combinations of all three types of systems.

GRAVITY COLLECTION SYSTEM

In a gravity collection system, the collection lines are sloped to permit the flow to move through the system with as little pumping as possible. The slope of the lines must keep the wastewater moving at a velocity (speed) of 2 to 4 feet per second (fps); otherwise, at lower velocities, solids will settle out causing clogged lines, overflows, and offensive odors. To keep collection system lines at a reasonable depth, wastewater must be lifted (pumped) periodically so it can continue flowing downhill to the treatment plant. Pump stations are installed at selected points within the system for this purpose.

FORCE MAIN COLLECTION SYSTEM

In a typical force main collection system, wastewater is collected to central points and pumped under pressure to the treatment plant. The system is normally used for conveying wastewater long distances. The use of the force main system allows the wastewater to flow to the treatment plant at the desired velocity without using sloped lines. It should be noted that the pump station discharge lines in a gravity system are considered to be force mains, as the contents of the lines are under pressure.

> **Note:** Extra care must be taken when performing maintenance on force main systems because the contents of the collection system are under pressure.

VACUUM SYSTEM

In a vacuum collection system, wastewaters are collected to central points and then drawn toward the treatment plant under vacuum. The system consists of a large amount of mechanical equipment and requires a large amount of maintenance to perform properly. Generally, the vacuum type of collection systems are not economically feasible.

PUMPING STATIONS

Pumping stations provide the motive force (energy) to keep the wastewater moving at the desired velocity. They are used in both the force main and gravity systems. They are designed in several different configurations and may use different sources of energy to move the wastewater (i.e., pumps, air pressure, or vacuum). One of the more commonly used types of pumping station designs is the wet well/dry well design.

Wet Well/Dry Well Pumping Stations

The wet well/dry well pumping station consists of two separate spaces or sections separated by a common wall. Wastewater is collected in one section (wet well section) and the pumping equipment (and, in many cases, the motors and controllers) are located in a second section known as the dry well. Many different designs for this type of system are available, but in most cases the pumps selected for this system are of a centrifugal design. Among the major considerations when selecting the centrifugal design are that (1) it allows for the separation of mechanical equipment (pumps, motors, controllers, wiring, etc.) from the potentially corrosive atmosphere (sulfides) of the wastewater, and (2) it is usually safer for workers because they can monitor, maintain, operate, and repair equipment without entering the pumping station wet well.

> *Note:* Most pumping station wet wells are confined spaces. To ensure safe entry into such spaces, compliance with OSHA's 29 CFR 1910.146 (Confined Space Entry Standard) is required.

Wet Well Pumping Stations

Another type of pumping station design is the wet well type. The wet well consists of a single compartment that collects the wastewater flow. The pump is submerged in the wastewater with motor controls located in the space or has a weatherproof motor housing located above the wet well. In this type of station, a submersible centrifugal pump is normally used.

Pneumatic Pumping Stations

The pneumatic pumping station consists of a wet well and a control system that controls the inlet and outlet valve operations and provides pressurized air to force or push the wastewater through the system. The exact method of operation depends on the system design. When wastewater in the wet well reaches a predetermined level, an automatic valve is activated which closes the influent line. The tank (wet well) is then pressurized to a predetermined level. When the pressure reaches the predetermined level, the effluent line valve is opened and the pressure pushes the waste stream out the discharge line.

Pumping Station Wet Well Calculations

Calculations normally associated with pumping station wet well design (such as determining design lift or pumping capacity) are usually left up to design and mechanical engineers; however, on occasion, wastewater operators or interceptor technicians may be called upon to make certain basic calculations. Usually these calculations deal with determining either pump capacity without influent (to check the pumping rate of the constant speed pump) or pump capacity with influent (to check how many gallons per minute the pump is discharging).

PRELIMINARY WASTEWATER TREATMENT

The initial stage in the wastewater treatment process (following collection and influent pumping) is *preliminary treatment*. Raw influent entering the treatment plant may contain many kinds of materials (trash). The purpose of preliminary treatment is to protect plant equipment by removing these materials, which can cause clogs, jams, or excessive wear to plant machinery. In addition, the removal of various materials at the beginning of the treatment process saves valuable space within the treatment plant.

Preliminary treatment may include many different processes; each is designed to remove a specific type of material that poses a potential problem for the treatment process. Processes include wastewater collection—influent pumping, screening, shredding, grit removal, flow measurement, preaeration, chemical addition, and flow equalization; the major processes are shown in Figure 7.4. In this section, we describe and discuss each of these processes and their importance in the treatment process.

Note: Not all treatment plants will include all of the processes shown in Figure 7.4. Specific processes have been included to facilitate discussion of major potential problems with each process and its operation; this is information that may be important to the wastewater operator.

SCREENING

The purpose of screening is to remove large solids such as rags, cans, rocks, branches, leaves, or roots from the flow before the flow moves on to downstream processes.

Note: Typically, a treatment plant will remove anywhere from 0.5 to 12 ft³ of screenings for each million gallons of influent received.

A *bar screen* traps debris as wastewater influent passes through. Typically, a bar screen consists of a series of parallel, evenly spaced bars or a perforated screen placed in a channel. The waste stream passes through the screen and the large solids (*screenings*) are trapped on the bars for removal.

Note: The screenings must be removed frequently enough to prevent accumulation that will block the screen and cause the water level in front of the screen to build up.

The bar screen may be coarse (2- to 4-in. openings) or fine (0.75- to 2.0-in. openings). The bar screen may be manually cleaned (bars or screens are placed at an angle of 30° for easier solids removal; see Figure 7.4) or mechanically cleaned (bars are placed at an angle of 45° to 60° to improve mechanical cleaner operation).

The screening method employed depends on the design of the plant, the amount of solids expected, and whether the screen is for constant or emergency use only.

Manually Cleaned Screens

Manually cleaned screens are cleaned at least once per shift (or often enough to prevent buildup that may cause reduced flow into the plant) using a long tooth rake. Solids are manually pulled to the drain platform and allowed to drain before being stored in a covered container. The area around the screen should be cleaned frequently to prevent a buildup of grease or other materials that can cause odors, slippery conditions, and insect and rodent problems. Because screenings may contain organic matter as well as large amounts of grease, they should be stored in a covered container. Screenings can be disposed of by burial in approved landfills or by incineration. Some treatment facilities grind the screenings into small particles, which are then returned to the wastewater flow for further processing and removal later in the process.

Mechanically Cleaned Screens

Mechanically cleaned screens use a mechanized rake assembly to collect the solids and move them out of the wastewater flow for discharge to a storage hopper. The screen may be continuously cleaned or cleaned on a time- or flow-controlled cycle. As with the manually cleaned screen, the area surrounding the mechanically operated screen must be cleaned frequently to prevent the buildup of materials that can cause unsafe conditions. As with all mechanical equipment, operator vigilance is required to ensure proper operation and that proper maintenance is performed. Maintenance includes lubricating equipment and maintaining it in accordance with the manufacturer's recommendations or the plant's operation and maintenance (O&M) manual. Screenings from mechanically operated bar screens are disposed of in the same manner as screenings from manually operated screens: landfill disposal, incineration, or being ground into smaller particles for return to the wastewater flow.

SHREDDING

As an alternative to screening, shredding can be used to reduce solids to a size that can enter the plant without causing mechanical problems or clogging. Shredding processes include comminution (*comminute* means "to cut up") and barminution devices.

Comminution

The comminutor is the most common shredding device used in wastewater treatment. In this device, all of the wastewater flow passes through the grinder assembly. The grinder consists of a screen or slotted basket, a rotating or oscillating cutter, and a stationary cutter. Solids pass through the screen and are chopped or shredded between the two cutters. The comminutor will not remove solids that are too large to fit through the slots, and it will not remove floating objects. These materials must be removed manually. Maintenance requirements for comminutors include aligning, sharpening, and replacing cutters in addition to corrective and preventive maintenance performed in accordance with the plant O&M manual.

Barminution

In barminution, the barminutor uses a bar screen to collect solids, which are then shredded and passed through the bar screen for removal at a later process. The cutter alignment and sharpness of each device are critical factors in effective operation. Cutters must be sharpened or replaced and alignment must be checked in accordance with the manufacturer's recommendations. Solids that are not shredded must be removed daily, stored in closed containers, and disposed of by burial or incineration. Barminutor operational problems are similar to those listed above for comminutors. Preventive and corrective maintenance as well as lubrication must be performed by qualified personnel and in accordance with the plant O&M manual. Because of higher maintenance requirements, the barminutor is less frequently used.

GRIT REMOVAL

The purpose of grit removal is to remove the heavy inorganic solids that could cause excessive mechanical wear. Grit is heavier than inorganic solids and includes sand, gravel, clay, egg shells, coffee grounds, metal filings, seeds, and other similar materials. Several processes or devices are used for grit removal. All of the processes are based on the fact that grit is heavier than the organic solids that should be kept in suspension for treatment in following processes. Grit removal may be accomplished in grit chambers or by the centrifugal separation of sludge. Processes use gravity/velocity, aeration, or centrifugal force to separate the solids from the wastewater.

Gravity/Velocity-Controlled Grit Removal

Gravity/velocity-controlled grit removal is normally accomplished in a channel or tank where the speed or velocity of the wastewater is controlled to about 1 foot per second (ideal), so the grit will settle while organic matter remains suspended. As long as the velocity is controlled in the range of 0.7 to 1.4 fps, the grit removal will remain effective. Velocity is controlled by the amount of water flowing through the channel, the depth of the water in the channel, the width of the channel, or the cumulative width of channels in service.

Aeration

Aerated grit removal systems use aeration to keep the lighter organic solids in suspension while allowing the heavier grit particles to settle out. Aerated grit removal systems may be manually or mechanically cleaned; however, the majority of the systems are mechanically cleaned. In normal operation, the aeration rate is adjusted to produce the desired separation, which requires observation of mixing and aeration and sampling of fixed suspended solids. Actual grit removal is controlled by the rate of aeration. If the rate is too high, all of the solids remain in suspension. If the rate is too low, both grit and organics will settle out. The operator observes the same kinds of conditions as those listed for the gravity/velocity-controlled system, but must also pay close attention to the air distribution system to ensure proper operation.

Centrifugal Force

The cyclone degritter uses a rapid spinning motion (centrifugal force) to separate the heavy inorganic solids or grit from the light organic solids. This unit process is normally used on primary sludge rather than the entire wastewater flow. The critical control factor for the process is the inlet pressure. If the pressure exceeds the recommendations of the manufacturer, the unit will flood, and grit will be carried through with the flow. Grit is separated from flow, washed, and discharged directly to a storage container. Grit removal performance is determined by calculating the percent removal for inorganic (fixed) suspended solids. The operator observes the same kinds of conditions listed for the gravity/velocity-controlled and aerated grit removal systems, with the exception of the air distribution system. Typical problems associated with grit removal include mechanical malfunctions and rotten egg odor in the grit chamber (hydrogen sulfide formation), which can lead to metal and concrete corrosion problems. Low recovery rate of grit is another typical problem. Bottom scour, overaeration, or not enough detention time normally causes this. When these problems occur, the operator must make the required adjustments or repairs to correct the problem.

PREAERATION

In the preaeration process (diffused or mechanical), we aerate wastewater to achieve and maintain an aerobic state (to freshen septic wastes), strip off hydrogen sulfide (to reduce odors and corrosion), agitate solids (to release trapped gases and improve solids separation and settling), and reduce BOD_5. All of this can be accomplished by aerating the wastewater for 10 to 30 min. To reduce BOD_5, preaeration must be conducted from 45 to 60 min.

CHEMICAL ADDITION

Chemical addition is made (either via dry chemical metering or solution feed metering) to the waste stream to improve settling, reduce odors, neutralize acids or bases, reduce corrosion, reduce BOD_5, improve solids and grease removal, reduce loading on the plant, add or remove nutrients, add organisms, or aid subsequent downstream processes. The particular chemical and amount used depends on the desired result. Chemicals must be added at a point where sufficient mixing will occur to obtain maximum benefit. Chemicals typically used in wastewater treatment include chlorine, peroxide, acids and bases, mineral salts (e.g., ferric chloride, alum), and bioadditives and enzymes.

EQUALIZATION

The purpose of flow equalization (whether by surge, diurnal, or complete methods) is to reduce or remove the wide swings in flow rates normally associated with wastewater treatment plant loading; it minimizes the impact of storm flows. The process can be designed to prevent flows that are above the maximum plant design hydraulic capacity, to reduce the magnitude of diurnal flow variations, and to eliminate flow variations. Flow equalization is accomplished using mixing or aeration equipment, pumps, and flow measurement. Normal operation depends on the purpose and requirements of the flow equalization system. Equalized flows allow the plant to perform at optimum levels by providing stable hydraulic and organic loading. The downside to flow equalization is in additional costs associated with construction and operation of the flow equalization facilities.

AERATED SYSTEMS

Aerated grit removal systems use aeration to keep the lighter organic solids in suspension while allowing the heavier grit particles to settle out. Aerated grit removal systems may be manually or mechanically cleaned; however, the majority of the systems are mechanically cleaned. In normal operation, the aeration rate is adjusted to produce the desired separation, which requires observation of mixing and aeration and sampling of fixed suspended solids. Actual grit removal is controlled by the rate of aeration. If the rate is too high, all of the solids remain in suspension. If the rate is too low, both grit and organics will settle out.

CYCLONE DEGRITTER

The cyclone degritter uses a rapid spinning motion (centrifugal force) to separate the heavy inorganic solids or grit from the light organic solids. This unit process is normally used on primary sludge rather than the entire wastewater flow. The critical control factor for the process is the inlet pressure. If the pressure exceeds the recommendations of the manufacturer, the unit will flood, and grit will be carried through with the flow. Grit is separated from flow, washed, and discharged directly to a storage container. Grit removal performance is determined by calculating the percent removal for inorganic (fixed) suspended solids.

PRIMARY WASTEWATER TREATMENT (SEDIMENTATION)

The purpose of primary treatment (primary sedimentation or primary clarification) is to remove settleable organic and floatable solids. Normally, each primary clarification unit can be expected to remove 90 to 95% settleable solids, 40 to 60% total suspended solids, and 25 to 35% BOD_5.

> *Note:* Performance expectations for settling devices used in other areas of plant operation are normally expressed as overall unit performance rather than settling unit performance.

Sedimentation may be used throughout the plant to remove settleable and floatable solids. It is used in primary treatment, secondary treatment, and advanced wastewater treatment processes. In this section, we focus on primary treatment or primary clarification, which uses large basins where primary settling is achieved under relatively quiescent conditions (see Figure 7.4). Within these basins, mechanical scrapers collect the primary settled solids into a hopper, from which they are pumped to a sludge processing area. Oil, grease, and other floating materials (scum) are skimmed from the surface. The effluent is discharged over weirs into a collection trough. In primary sedimentation, wastewater enters a settling tank or basin. Velocity is reduced to approximately 1 foot per minute.

> *Note:* Notice that the velocity is based on minutes instead of seconds, as was the case in the grit channels. A grit channel velocity of 1 ft/sec would be 60 ft/min.

Solids that are heavier than water settle to the bottom, while solids that are lighter than water float to the top. Settled solids are removed as sludge, and floating solids are removed as scum. Wastewater leaves the sedimentation tank over an effluent weir and moves on to the next step in treatment. Detention time, temperature, tank design, and condition of the equipment control the efficiency of the process.

OVERVIEW OF PRIMARY TREATMENT

- Primary treatment reduces the organic loading on downstream treatment processes by removing a large amount of settleable, suspended, and floatable materials.
- Primary treatment reduces the velocity of the wastewater through a clarifier to approximately 1 to 2 ft/min so settling and flotation can take place. Slowing the flow enhances removal of suspended solids in wastewater.
- Primary settling tanks remove floated grease and scum, as well as the settled sludge solids, and collect them for pumped transfer to disposal or further treatment.
- Clarifiers may be rectangular or circular. In rectangular clarifiers, wastewater flows from one end to the other, and the settled sludge is moved to a hopper at the one end, either by flights set on parallel chains or by a single bottom scraper set on a traveling bridge. Floating material (mostly grease and oil) is collected by a surface skimmer. In circular tanks, the wastewater usually enters at the middle and flows outward. Settled sludge is pushed to a hopper in the middle of the tank bottom, and a surface skimmer removes floating material.
- Factors affecting primary clarifier performance include
  - Rate of flow through the clarifier
  - Wastewater characteristics (strength, temperature, amount and type of industrial waste, and the density, size, and shapes of particles)
  - Performance of pretreatment processes
  - Nature and amount of any wastes recycled to the primary clarifier

Types of Sedimentation Tanks

Sedimentation equipment includes septic tanks, two-story tanks, and plain settling tanks or clarifiers. All three devices may be used for primary treatment, but plain settling tanks are normally used for secondary or advanced wastewater treatment processes.

Septic Tanks

Septic tanks are prefabricated tanks that serve as a combined settling and skimming tank and as an unheated, unmixed anaerobic digester. Septic tanks provide long settling times (6 to 8 hr or more) but do not separate decomposing solids from the wastewater flow. When the tank becomes full, solids will be discharged with the flow. The process is suitable for small facilities (e.g., schools, motels, homes), but, due to the long detention times and lack of control, it is not suitable for larger applications.

Two-Story (Imhoff) Tank

The two-story or Imhoff tank is similar to a septic tank with regard to the removal of settleable solids and the anaerobic digestion of solids. The difference is that the two-story tank consists of a settling compartment where sedimentation is accomplished, a lower compartment where settled solids and digestion takes place, and gas vents. Solids removed from the wastewater by settling pass from the settling compartment into the digestion compartment through a slot in the bottom of the settling compartment. The design of the slot prevents solids from returning to the settling compartment. Solids decompose anaerobically in the digestion section. Gases produced as a result of the solids decomposition are released through the gas vents running along each side of the settling compartment.

Plain Settling Tanks (Clarifiers)

The plain settling tank or clarifier optimizes the settling process. Sludge is removed from the tank for processing in other downstream treatment units. Flow enters the tank, is slowed and distributed evenly across the width and depth of the unit, passes through the unit, and leaves over the effluent weir. Detention time within the primary settling tank can vary from 1 to 3 hr (2 hr on average). Sludge removal is accomplished frequently on either a continuous or an intermittent basis. Continuous removal requires additional sludge treatment processes to remove the excess water resulting from the removal of sludge containing less than 2 to 3% solids. Intermittent sludge removal requires the sludge be pumped from the tank on a schedule frequent enough to prevent large clumps of solids rising to the surface but infrequent enough to obtain 4 to 8% solids in the sludge withdrawn.

Scum must be removed from the surface of the settling tank frequently. This is normally a mechanical process but may require manual start-up. The system should be operated frequently enough to prevent excessive buildup and scum carryover but not so frequent as to cause hydraulic overloading of the scum removal system. Settling tanks require housekeeping and maintenance. Baffles (which prevent floatable solids and scum from leaving the tank), scum troughs, scum collectors, effluent troughs, and effluent weirs require frequent cleaning to prevent heavy biological growths and solids accumulations. Mechanical equipment must be lubricated and maintained as specified in the manufacturer's recommendations or in accordance with procedures listed in the plant O&M manual. Process control sampling and testing are used to evaluate the performance of the settling process. Settleable solids, dissolved oxygen, pH, temperature, total suspended solids, and BOD_5, as well as sludge solids and volatile matter, testing is routinely carried out.

Effluent from Settling Tanks

Upon completion of screening, degritting, and settling in sedimentation basins, large debris, grit, and many settleable materials have been removed from the waste stream. What is left is referred to as *primary effluent*. Usually cloudy and frequently gray in color, primary effluent still contains large amounts of dissolved food and other chemicals (nutrients). These nutrients are treated in the next step in the treatment process, secondary treatment, which is discussed in the next section.

Note: Two of the most important nutrients left to remove are phosphorus and ammonia. Although we want to remove these two nutrients from the waste stream, we do not want to remove too much. Carbonaceous microorganisms in secondary treatment (biological treatment) require both phosphorus and ammonia.

SECONDARY WASTEWATER TREATMENT

The main purpose of secondary treatment (sometimes referred to as *biological treatment*) is to provide biochemical oxygen demand (BOD) removal beyond what is achievable by primary treatment (see Figure 7.4). Three commonly used approaches all take advantage of the ability of microorganisms to convert organic wastes (via biological treatment) into stabilized, low-energy compounds. Two of these approaches, the *trickling filter* or its variation, the *rotating biological contactor* (RBC), and the *activated sludge process*, sequentially follow normal primary treatment. The third approach, *ponds* (oxidation ponds or lagoons), however, can provide equivalent results without preliminary treatment. In this section, we present a brief overview of the secondary treatment process followed by a detailed discussion of wastewater treatment ponds (used primarily in smaller treatment plants), trickling filters, and RBCs. We then shift focus to the activated sludge process—the secondary treatment process used primarily in large installations.

Secondary treatment refers to those treatment processes that use biological processes to convert dissolved, suspended, and colloidal organic wastes to more stable solids that can be either removed by settling or discharged to the environment without causing harm. Exactly what is secondary treatment? As defined by the Clean Water Act (CWA), secondary treatment produces an effluent with no more than 30 mg/L BOD_5 and 30 mg/L total suspended solids.

Note: The Clean Water Act also states that ponds and trickling filters will be included in the definition of secondary treatment even if they do not meet the effluent quality requirements continuously.

Most secondary treatment processes decompose solids aerobically, producing carbon dioxide, stable solids, and more organisms. Because solids are produced, all of the biological processes must include some form of solids removal (e.g., settling tank, filter). Secondary treatment processes can be separated into two large categories: fixed-film systems and suspended-growth systems.

Fixed-film systems are processes that use a biological growth (biomass or slime) attached to some form of media. Wastewater passes over or around the media and the slime. When the wastewater and slime are in contact, the organisms remove and oxidize the organic solids. The media may be stone, redwood, synthetic materials, or any other substance that is durable (capable of withstanding weather conditions for many years), provides a large area for slime growth while providing open space for ventilation, and is not toxic to the organisms in the biomass. Fixed-film devices include trickling filters and rotating biological contactors. *Suspended-growth systems* are processes that use a biological growth mixed with the wastewater. Typical suspended-growth systems consist of various modifications of the activated sludge process.

TREATMENT PONDS

Wastewater treatment can be accomplished using *ponds*. Ponds are relatively easy to build and manage, they accommodate large fluctuations in flow, and they can also provide treatment that approaches the effectiveness of conventional systems (producing a highly purified effluent) at a much lower cost. It is the cost factor that drives many managers to decide on the pond option. The actual degree of treatment provided depends on the type and number of ponds used. Ponds can be used as the sole type of treatment or they can be used in conjunction with other forms of wastewater treatment; that is, other treatment processes can be followed by a pond or a pond can be followed by other treatment processes.

Stabilization (treatment) ponds have been used for the treatment of wastewater for over 3000 years. The first recorded construction of a pond system in the United States was in San Antonio, Texas, in 1901. Today, over 8000 wastewater treatment ponds are in place, involving more than 50% of the wastewater treatment facilities in the United States. Facultative ponds account for 62%, aerated ponds 25%, anaerobic ponds 0.04%, and total containment 12% of pond treatment systems. They treat a variety of wastewaters from domestic wastewater to complex industrial wastes, and they function under a wide range of weather conditions, from tropical to arctic. Ponds can be used alone or in combination with other wastewater treatment processes. As our understanding of pond operating mechanisms has increased, different types of ponds have been developed for application in specific types of wastewater under local environmental conditions.

Although the tendency in the United States has been for smaller communities to build ponds, in other parts of the world, including Australia, New Zealand, Mexico and Latin America, Asia, and Africa, treatment ponds have been built to serve large cities. As a result, our understanding of the biological, biochemical, physical, and climatic factors that interact to transform the organic compounds, nutrients, and pathogenic organisms found in sewage into less harmful chemicals and unviable organisms (i.e., dead or sterile) has grown since 1983. A wealth of experience has been developed as civil, sanitary, and environmental engineers, operators, public works managers, and public health and environmental agencies have worked with these systems. Although some of this information makes its way into technical journals and textbooks, there is a need for a less formal presentation of the subject for those working in the field every day (USEPA, 2011).

DID YOU KNOW?

A pond can be judged as an "attractive nuisance." The term "attractive nuisance" is a legal expression that implies that the pond could be attractive to potential users, such as duck hunters, fisherman, or playing children. Because ponds have fairly steep slopes, the potential for someone falling in and drowning is a significant legal problem that must be addressed. It is important that adequate fencing and signing be provided.

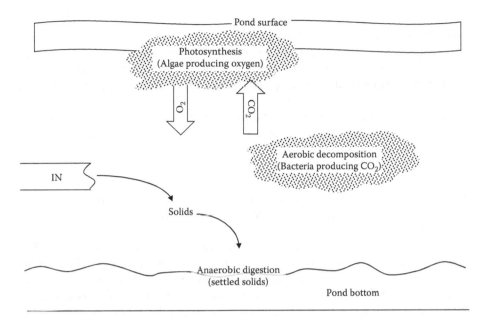

FIGURE 7.5 Stabilization pond processes.

Ponds are designed to enhance the growth of natural ecosystems that are anaerobic (providing conditions for bacteria that grow in the absence of oxygen environments), aerobic (promoting the growth of oxygen-producing and/or oxygen-requiring organisms, such as algae and bacteria), or facultative (a combination of the two). Ponds are managed to reduce concentrations of biochemical oxygen demand, total suspended solids, and coliform numbers (fecal or total) to meet water quality requirements.

Ponds can be classified based on their location in the system, by the type of wastes they receive, and by the main biological process occurring in the pond. First, we will take a look at the types of ponds according to their location and the type of wastes they receive: *raw sewage stabilization ponds* (see Figure 7.5), *oxidation ponds*, and *polishing ponds*. Then, we will look at ponds classified by the type of processes occurring within the pond: *aerobic ponds, anaerobic ponds, facultative ponds*, and *aerated ponds*.

Ponds Based on Location and Types of Wastes They Receive

Raw Sewage Stabilization Pond

The raw sewage stabilization pond is the most common type of pond used. With the exception of screening and shredding, this type of pond receives no prior treatment. Generally, raw sewage stabilization ponds are designed to provide a minimum of 45 days of detention time and to receive no more than 50 pounds of BOD_5 per day per acre. The quality of the discharge is dependent on the time of the year. Summer months produce high BOD_5 removal but low suspended solids removal. The pond consists of an influent structure, pond berm or walls, and an effluent structure designed to permit selection of the best quality effluent. The normal operating depth of the pond is 3 to 5 feet. The processes occurring in the pond involve bacteria decomposing the organics in the wastewater (aerobically and anaerobically) and algae using the products of the bacterial action to produce oxygen (photosynthesis). Because this type of pond is the most commonly used in wastewater treatment, the processes that occur within the pond are described in greater detail in the following text.

When wastewater enters the stabilization pond, several processes begin to occur. These include settling, aerobic decomposition, anaerobic decomposition, and photosynthesis (see Figure 7.5). Solids in the wastewater will settle to the bottom of the pond. In addition to the solids present in the wastewater entering the pond, solids that are produced by the biological activity will also settle to the bottom. Eventually this will reduce the detention time and the performance of the pond. When this occurs (normally 20 to 30 years), the pond will have to be replaced or cleaned.

Bacteria and other microorganisms use the organic matter as a food source. They use oxygen (aerobic decomposition), organic matter, and nutrients to produce carbon dioxide, water, and stable solids, which may settle out, as well as more organisms. The carbon dioxide is an essential component of the photosynthesis process occurring near the surface of the pond. Organisms also use the solids that settle out as food material; however, the oxygen levels at the bottom of the pond are extremely low so the process used is anaerobic decomposition. The organisms use the organic matter to produce gases (hydrogen sulfide, methane, etc.) dissolved in the water, stable solids, and more organisms. Near the surface of the pond a population of green algae develops that can use the carbon dioxide produced by the bacterial population, nutrients, and sunlight to produce more algae and oxygen, which is dissolved into the water. The dissolved oxygen (DO) is then used by organisms in the aerobic decomposition process.

When compared with other wastewater treatment systems involving biological treatment, a stabilization pond treatment system is the simplest to operate and maintain. Operation and maintenance activities include collecting and testing samples for dissolved oxygen and pH, removing weeds and other debris (scum) from the pond, mowing the berms, repairing erosion, and removing burrowing animals.

Note: Dissolved oxygen and pH levels in the pond will vary throughout the day. Normal operation will result in very high DO and pH levels due to the natural processes occurring.

Note: When operating properly, the stabilization pond will exhibit a wide variation in both dissolved oxygen and pH. This is due to photosynthesis occurring in the system.

Oxidation Pond

An oxidation pond, which is normally designed using the same criteria as the stabilization pond, receives flows that have passed through a stabilization pond or primary settling tank. This type of pond provides biological treatment, additional settling, and some reduction in the number of fecal coliform present.

Polishing Pond

A polishing pond, which uses the same equipment as a stabilization pond, receives flow from an oxidation pond or from other secondary treatment systems. Polishing ponds remove additional BOD_5, solids, fecal coliform, and some nutrients. They are designed to provide 1 to 3 days of detention time and normally operate at a depth of 5 to 10 ft. Excessive detention time or too shallow a depth will result in algae growth, which increases influent suspended solids concentrations.

Ponds Based on the Type of Processes Occurring Within

The type of processes occurring within the pond may also classify ponds. These include the aerobic, anaerobic, facultative, and aerated processes.

Aerobic Ponds

Not widely used, aerobic ponds (also known as *oxidation ponds* or *high-rate aerobic ponds*) have oxygen present throughout the pond. All biological activity is aerobic decomposition. They are usually 30 to 45 cm deep, which allows light to penetrate throughout the pond. Mixing is often provided, keeping algae at the surface to maintain maximum rates of photosynthesis and O_2 production and to prevent algae from settling and producing an anaerobic bottom layer. The rate of photosynthetic production of O_2 may be enhanced by surface reaeration; O_2 and aerobic bacteria biochemically stabilize the waste. Detention time is typically 2 to 6 days.

These ponds are appropriate for treatment in warm, sunny climates. They are used where a high degree of BOD_5 removal is desired but land area is limited. The chief advantage of these ponds is that they produce a stable effluent during short detention times with low land and energy requirements. However, their operation is somewhat more complex than that of facultative ponds and, unless the algae are removed, the effluent will contain high TSS. While the shallow depths allow penetration of ultraviolet (UV) light that may reduce pathogens, shorter detention times may work against effective coliform and parasite die-off. Because they are shallow, bottom paving or veering is usually necessary to prevent aquatic plants from colonizing the ponds. The Advanced Integrated Wastewater Pond System® (AIWPS®) uses the high-rate pond to maximize the growth of microalgae using a low-energy paddle-wheel (USEPA, 2011).

Anaerobic Ponds

Anaerobic ponds are normally used to treat high-strength industrial wastes; that is, they receive heavy organic loading, so much so that there is no aerobic zone—no oxygen is present and all biological activity is anaerobic decomposition. They are usually 2.5 to 4.5 m in depth and have detention times of 5 to 50 days. The predominant biological treatment reactions are bacterial acid formation and methane fermentation. Anaerobic ponds are usually used for treatment of strong industrial and agricultural (food processing) wastes, as a pretreatment step in municipal systems, or where an industry is a significant contributor to a municipal system. The biochemical reactions in an anaerobic pond produce hydrogen sulfide (H_2S) and other odorous compounds. To reduce odors, the common practice is to recirculate water from a downstream facultative or aerated pond. This provides a thin aerobic layer at the surface of the anaerobic pond which prevents odors from escaping into the air. A cover may also be used to contain odors. The effluent from anaerobic ponds usually requires further treatment prior to discharge (USEPA, 2011).

Facultative Pond

The facultative pond, which may also be called an *oxidation* or *photosynthetic pond*, is the most common type of pond (based on processes occurring). Oxygen is present in the upper portions of the pond and aerobic processes occur. No oxygen is present in the lower levels of the pond where processes occurring are anoxic and anaerobic. Facultative ponds are usually 0.9 to 2.4 m deep or deeper, with an aerobic layer overlying an anaerobic layer. Recommended detention times vary from 5 to 50 days in warm climates and 90 to 180 days in colder climates (NEIWPCC, 1998). Aerobic treatment processes in the upper layer provide odor control as well as nutrient and BOD removal. Anaerobic fermentation processes, such as sludge digestion, denitrification, and some BOD removal, occur in the lower layer. The key to successful operation of this type of pond is O_2 production by photosynthetic algae and/or reaeration at the surface. Facultative ponds are used to treat raw municipal wastewater in small communities and for primary or secondary effluent treatment for small or large cities. They are also used in industrial applications, usually in the process line after aerated or anaerobic ponds, to provide additional treatment prior to discharge. Commonly achieved effluent BOD values, as measured in the BOD_5 test, range from 20 to 60 mg/L, and TSS levels may range from 30 to 150 mg/L. The size of the pond needed to treat BOD loadings depends on specific conditions and regulatory requirements.

Aerated Pond

Facultative ponds overloaded due to unplanned additional sewage volume or higher strength influent from a new industrial connection may be modified by the addition of mechanical aeration. Ponds originally designed for mechanical aeration are generally 2 to 6 m deep with detention times of 3 to 10 days. For colder climates, a detention time of 20 to 40 days is recommended. Mechanically aerated ponds require less land area but have greater energy requirements. When aeration is used, the depth of the pond and/or the acceptable loading levels may increase. Mechanical or diffused aeration is often used to supplement natural oxygen production or to replace it.

Elements of Pond Processes

Although our understanding of wastewater pond ecology is far from complete, general observations about the interactions of macro- and microorganisms in these biologically driven systems support our ability to design, operate, and maintain them.

Bacteria

In this section, we discuss types of bacteria found in the pond; these organisms help to decompose complex, organic constituents in the influent to simple, nontoxic compounds. Certain pathogen bacteria and other microbial organisms (viruses, protozoa) associated with human waste enter into the system with the influent; the wastewater treatment process is designed so that their numbers will be reduced adequately to meet public health standards.

- *Aerobic bacteria* are found in the aerobic zone of a wastewater pond and are primarily the same type as those found in an activated sludge process or in the zoogleal mass of a trickling filter. The most frequently isolated bacteria include *Beggiatoa alba*, *Sphaerotilus natans*, *Achromobacter*, *Alcaligenes*, *Flavobacterium*, *Pseudomonas*, and *Zoogoea* spp. (Lynch and Poole, 1979; Pearson, 2005; Spellman, 2000). These organisms decompose the organic materials present in the aerobic zone into oxidized end products.
- *Anaerobic bacteria* are hydrolytic bacteria that convert complex organic material into simple alcohols and acids, primarily amino acids, glucose, fatty acid, and glycerols (Brockett, 1976; Paterson and Curtis, 2005; Pearson, 2005; Spellman, 2000). Acidogenic bacteria convert the sugars and amino acids into acetate, ammonia (NH_3), hydrogen, and carbon dioxide (CO_2). Methanogenic bacteria break down these products further to methane (CH_4) and CO_2 (Gallert and Winter, 2005).
- *Cyanobacteria*, formerly classified as blue–green algae, are autotrophic organisms that are able to synthesize organic compounds using CO_2 as the major carbon source. Cyanobacteria produce O_2 as a byproduct of photosynthesis, providing an O_2 source for other organisms in the ponds. They are found in very large numbers as blooms when environmental conditions are suitable (Gaudy and Gaudy, 1980). Commonly encountered cyanobacteria include *Oscillatoria*, *Arthrospira*, *Spirulina*, and *Microcystis* (Spellman, 2000; Vasconcelos and Pereira, 2001).
- *Purple sulfur bacteria* (Chromatiaceae) may grow in any aquatic environment to which light of the required wavelength penetrates, provided that CO_2, nitrogen, and a reduced form of sulfur or hydrogen are available. Purple sulfur bacteria occupy the anaerobic layer below the algae, cyanobacteria, and other aerobic bacteria in a pond. They are commonly found at a specific depth, in a thin layer where light and nutrient conditions are at an optimum (Gaudy and Gaudy, 1980; Pearson, 2005). Their biochemical conversion of odorous sulfide compounds to elemental sulfur or sulfate (SO_4) helps to control odor in facultative and anaerobic ponds.

Algae

Algae constitute a group of aquatic organisms that may be unicellular or multicellular, motile or immotile, and, depending on the phylogenetic family, have different combinations of photosynthetic pigments. As autotrophs, algae need only inorganic nutrients, such as nitrogen, phosphorus, and a suite of microelements, to fix CO_2 and grow in the presence of sunlight. Algae do not fix atmospheric nitrogen; they require an external source of inorganic nitrogen in the form of nitrate (NO_3) or NH_3. Some algal species are able to use amino acids and other organic nitrogen compounds. Oxygen is a byproduct of these reactions.

DID YOU KNOW?

A pond should be drawn down in fall after the first frost and when the algae concentration drops off, the BOD is still low, and the receiving stream temperature is low with accompanying high dissolved oxygen. A pond should be drawn down in spring before the algae concentration increases, when the BOD level is acceptable, and when the receiving stream flows are high (low temperature with high dissolved oxygen helps). During the actual discharge, the effluent must be sampled for BOD, suspended solids, and pH at a frequency specified in the discharge permit.

Algae can be divided into three groups, based on the color reflected from the cells by the chlorophyll and other pigments involved in photosynthesis. Green and brown algae are common to wastewater ponds; red algae occur infrequently. The algal species dominant at any particular time is thought to be primarily a function of temperature, although the effects of predation, nutrient availability, and toxins are also important. Green algae (Chlorophyta) include unicellular, filamentous, and colonial forms. Some green algal genera commonly found in facultative and aerobic ponds are *Euglena*, *Phacus*, *Chlamydomonas*, *Ankistrodesmus*, *Chlorella*, *Micractinium*, *Scenedesmus*, *Selenastrum*, *Dictyosphaerium*, and *Volvox*. Chrysophytes, or brown algae, are unicellular and may be flagellated; they include the diatoms. Certain brown algae are responsible for toxic red blooms. Brown algae found in wastewater ponds include the diatoms *Navicula* and *Cyclotella*. Red algae (Rhodophyta) include a few unicellular forms but are primarily filamentous (Gaudy and Gaudy, 1980; Pearson, 2005).

Importance of Interactions between Bacteria and Algae

It is generally accepted that the presence of both algae and bacteria is essential for the proper functioning of a treatment pond. Bacteria break down the complex organic waste components found in anaerobic and aerobic pond environments into simple compounds, which are then available for uptake by the algae. Algae, in turn, produce the O_2 necessary for the survival of aerobic bacteria. In the process of pond reactions of biodegradation and mineralization of waste material by bacteria and the synthesis of new organic compounds in the form of algal cells, a pond effluent might contain a higher than acceptable TSS. Although this form of TSS does not contain the same constituents as the influent TSS, it does contribute to turbidity and must be removed before the effluent is discharged. Once concentrated and removed, depending on regulatory requirements, algal TSS may be used as a nutrient for use in agriculture or as a feed supplement (Grönlund, 2002).

Invertebrates

Although bacteria and algae are the primary organisms through which waste stabilization is accomplished, predator life forms do play a role in wastewater pond ecology. It has been suggested that the planktonic invertebrate *Cladocera* spp. and the benthic invertebrate family Chironomidae are the most significant fauna in the pond community in terms of stabilizing organic material. The cladocerans feed on the algae and promote flocculation and settling of particulate matter, resulting in better light penetration and algal growth at greater depths. Settled matter is further broken down and stabilized by the benthic feeding Chironomidae. Predators, such as rotifers, often control the population levels of certain of the smaller life forms in the pond, thereby influencing the succession of species throughout the seasons.

Mosquitoes can present a problem in some ponds. Aside from their nuisance characteristics, certain mosquitoes are also vectors for such diseases as encephalitis, malaria, and yellow fever and constitute a hazard to public health that must be controlled. *Gambusia*, commonly called mosquito fish, have been introduced to eliminate mosquito problems in some ponds in warm climates (Pearson, 2005; Pipes, 1961; Ullrich, 1967), but their introduction has been problematic as they can outcompete native fish that also feed on mosquito larvae. There are also biochemical controls, such as the larvicides *Bacillus thuringiensis israelensis* (Bti), and Abate®, which may be effective if the product is applied directly to the area containing mosquito larvae. The most effective means of control of mosquitoes in ponds is the control of emergent vegetation (USEPA, 2011).

Biochemistry in a Pond

Photosynthesis

Photosynthesis is the process whereby organisms use solar energy to fix CO_2 and obtain the reducing power to convert it to organic compounds. In wastewater ponds, the dominant photosynthetic organisms include algae, cyanobacteria, and purple sulfur bacteria (Pearson, 2005; Pipes, 1961). Photosynthesis may be classified as oxygenic or anoxygenic, depending on the source of reducing power used by a particular organism. In oxygenic photosynthesis, water serves as the source of reducing power, with O_2 as a byproduct. Oxygenic photosynthetic algae and cyanobacteria convert CO_2 to organic compounds, which serve as the major source of chemical energy for other aerobic organisms. Aerobic bacteria need the O_2 produced to function in their role as primary consumers in degrading complex organic waste material. Anoxygenic photosynthesis does not produce O_2 and, in fact, occurs in the complete absence of O_2. The bacteria involved in anoxygenic photosynthesis are largely strict anaerobes, unable to function in the presence of O_2. They obtain energy by reducing inorganic compounds. Many photosynthetic bacteria utilize reduced sulfur compounds or element sulfur in anoxygenic photosynthesis.

Respiration

Respiration is a physiological process by which organic compounds are oxidized into CO_2 and water. Respiration is also an indicator of cell material synthesis. It is a complex process that consists of many interrelated biochemical reactions (Pearson, 2005). Aerobic respiration is common to species of bacteria, algae, protozoa, invertebrates, and higher plants and animals. The bacteria involved in aerobic respiration are primarily responsible for degradation of waste products. In the presence of light, respiration and photosynthesis can occur simultaneously in algae. However, the respiration rate is low compared to the photosynthesis rate, which results in a net

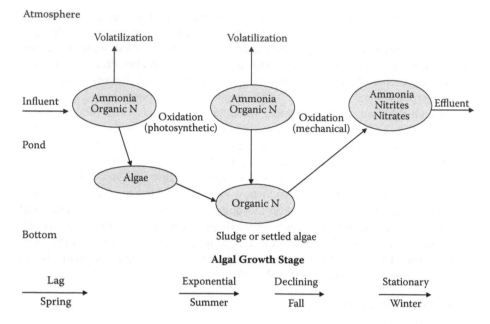

FIGURE 7.6 The nitrogen cycle in wastewater pond system.

consumption of CO_2 and production of O_2. In the absence of light, on the other hand, algal respiration continues while photosynthesis stops, resulting in a net consumption of O_2 and production of CO_2 (USEPA, 2011).

Nitrogen Cycle

The nitrogen cycle occurring in a wastewater treatment pond consists of a number of biochemical reactions mediated by bacteria. A schematic representation of the changes in nitrogen speciation in wastewater ponds over a year is shown in Figure 7.6. Organic nitrogen and NH_3 enter with the influent wastewater. Organic nitrogen in fecal matter and other organic materials undergoes conversion to NH_3 and ammonium ion NH_4^+ by microbial activity. The NH_3 may volatize into the atmosphere. The rate of gaseous NH_3 losses to the atmosphere is primarily a function of pH, surface-to-volume ratio, temperature, and the mixing conditions. An alkaline pH shifts the equilibrium of NH_3 gas and NH_4^+ toward gaseous NH_3 production, while the mixing conditions affect the magnitude of the mass transfer coefficient.

Ammonium is nitrified to nitrite (NO_2^-) by the bacterium *Nitrosomonas* and then to NO_3^- by *Nitrobacter*. The NO_3^- produced in the nitrification process, as well as a portion of the NH_4^- produced from ammonification, can be assimilated by organisms to produce cell protein and other nitrogen-containing compounds. The NO_3^- may also be denitrified to form NO_2^{2-} and then nitrogen gas. Several species of bacteria may be involved in the denitrification process, including *Pseudomonas, Micrococcus, Achromobacter,* and *Bacillus*. Nitrogen gas may be fixed by certain species of cyanobacteria when nitrogen is limited. This may occur in nitrogen-poor industrial ponds, but rarely in municipal or agricultural ponds (USEPA, 1975, 1993).

Nitrogen removal in facultative wastewater ponds can occur through any of the following processes: (1) gaseous NH_3 stripping to the atmosphere, (2) NH_4^- assimilation in algal biomass, (3) NO_3^- uptake by floating vascular plants and algae, and (4) biological nitrification–denitrification. Whether NH_4^- is assimilated into algal biomass depends on the biological activity in the system and is affected by several factors such as temperature, organic load, detention time, and wastewater characteristics.

Dissolved Oxygen

Oxygen is a partially soluble gas. Its solubility varies in direct proportion to the atmospheric pressure at any given temperature. Dissolved oxygen (DO) concentrations of approximately 8 mg/L are generally considered to be the maximum available under local ambient conditions. In mechanically aerated ponds, the limited solubility of O_2 determines its absorption rate (Sawyer et al., 1994). The natural sources of DO in ponds are photosynthetic oxygenation and surface reaeration. In areas of low wind activity, surface reaeration may be relatively unimportant, depending on the water depth. Where surface turbulence is created by excessive wind activity, surface reaeration can be significant. Experiments have shown that DO in wastewater ponds varies almost directly with the level of photosynthetic activity, which is low at night and early morning and rises during daylight hours to a peak in the early afternoon. At increased depth, the effects of photosynthetic oxygenation and surface reaeration decrease, as the distance from the water–atmosphere interface increases and light penetration decreases. This can result in the establishment of a vertical gradient. The microorganisms in the pond will segregate along the gradient.

pH and Alkalinity

In wastewater ponds, the hydrogen ion concentration, expressed as pH, is controlled through the carbonate buffering system. The equilibrium of this system is affected by the rate of algal photosynthesis. In photosynthetic metabolism, CO_2 is removed from the dissolved phase, forcing equilibrium. This tends to decrease the hydrogen ion (H^+) concentration and the bicarbonate (HCO_3^-) alkalinity. The effect of the decrease in HCO_3^- concentration is to force a decrease in total alkalinity. The decreased alkalinity associated with photosynthesis will simultaneously reduce the carbonate hardness present in the waste. Because of the close correlation between pH and photosynthetic activity, there is a diurnal fluctuation in pH when respiration is the dominant metabolic activity.

Physical Factors

Light

The intensity and spectral composition of light penetrating a pond surface significantly affect all resident microbial activity. In general, activity increases with increasing light intensity until the photosynthetic system becomes light saturated. The rate at which photosynthesis increases in proportion to an increase in light intensity, as well as the level at which an organism's photosynthetic system becomes light saturated, depends on the particular biochemistry of the species (Lynch and Poole, 1979; Pearson, 2005). In ponds, photosynthetic O_2 production has been shown to be

DID YOU KNOW?

The variation in pH in a facultative pond normally occurs in the upper aerobic zone, while the anaerobic and facultative zones will be relatively constant. This variation happens due to the changes that occur in the concentration of dissolved carbon dioxide. When carbon dioxide is dissolved in water it forms a weak carbonic acid, which would tend to lower pH. The relationship between algae and bacteria affect the carbon dioxide levels. During intense photosynthesis, algae use carbon dioxide and produce oxygen to be used by bacteria to assimilate organic wastes. The algae use much of the carbon dioxide and the pH can rise significantly (pH in the 11 to 12 range is not uncommon). During the night or during cloudy weather, the algae respire and active photosynthesis does not occur. The bacteria continue to use up oxygen and produce carbon dioxide. This can cause a significant drop in the pond pH, especially if the influent wastewater has low alkalinity. This same pH swing can occur in natural ponds, lakes, and stream impoundments. During peak summer algae activity, the dissolved oxygen levels of stream impoundments can vary from dawn levels of less than 1 mg/L to later afternoon values of 13 to 15 mg/L (supersaturation).

relatively constant with the range of 5380 to 53,800 lumens/m² light intensity with a reduction occurring at higher and lower intensities (Paterson and Curtis, 2005; Pipes, 1961).

The spectral composition of available light is also crucial in determining photosynthetic activity. The ability of photosynthetic organisms to utilize available light energy depends primarily upon their ability to absorb the available wavelengths. This absorption ability is determined by the specific photosynthetic pigment of the organism. The main photosynthetic pigments are chlorophylls and phycobilins. Bacterial chlorophyll differs from algal chlorophyll in both chemical structure and absorption capacity. These differences allow the photosynthetic bacteria to live below dense algal layers where they can utilize light not absorbed by the algae (Lynch and Poole, 1979; Pearson, 2005).

The quality and quantity of light penetrating the pond surface to any depth depend on the presence of dissolved and particulate matter as well as the water absorption characteristics. The organisms themselves contribute to water turbidity,

DID YOU KNOW?

Duckweed must be physically removed with a rake, pushboard, or broom. With sufficient wind, the duckweed will be pushed to one side or corner of a pond. This is an ideal time to rake the duckweed out. It is important that duckweed not be allowed to become too abundant, as it reduces oxygen transfer at the water surface, reduces light penetration and photosynthesis, and, upon decomposing, can cause both odor and BOD problems.

further limiting the depth of light penetration. Given the light penetration interferences, photosynthesis is significant only in the upper pond layers. This region of net photosynthetic activity is called the *euphotic zone* (Lynch and Poole, 1979; Pearson, 2005). Light intensity from solar radiation varies with the time of day and difference in latitudes. In cold climates, light penetration can be reduced during the winter by ice and snow cover. Supplementing the treatment ponds with mechanical aeration may be necessary in these regions during that time of year.

Temperature

Temperature at or near the surface of the aerobic environment of a pond determines the succession of predominant species of algae, bacteria, and other aquatic organisms. Algae can survive at temperatures of 5 to 40°C. Green algae show most efficient growth and activity at temperatures of 30 to 35°C. Aerobic bacteria are viable within a temperature range of 10 to 40°C; 35 to 40°C is optimum for cyanobacteria (Anderson and Zwieg, 1962; Crites et al., 2006; Gloyna, 1976; Paterson and Curtis, 2005).

As the major source of heat for these systems is solar radiation, a temperature gradient can develop in a pond with depth. This will influence the rate of anaerobic decomposition of solids that have settled at the bottom of the pond. The bacteria responsible for anaerobic degradation are active in temperatures from 15 to 65°C. When they are exposed to lower temperatures, their activity is reduced.

The other major source of heat is the influent water. In sewerage systems with no major inflow or infiltration problems, the influent temperature is higher than that of the pond contents. Cooling influences are exerted by evaporation, contact with cooler groundwater, and wind action. The overall effect of temperature in combination with light intensity is reflected in the fact that nearly all investigators report improved performance during summer and autumn months when both temperature and light are at their maximum. The maximum practical temperature of wastewater ponds is likely less than 30°C, indicating that most ponds operate at less than optimum temperature for anaerobic activity (Crites et al., 2006; Oswald, 1996; Paterson and Curtis, 2005; USEPA, 2011).

DID YOU KNOW?

Common maintenance issues associated with pond systems include the following:

- Weed control (cattails and other rooted aquatic plants)
- Algae control (blue–green and associated floating algae mats)
- Burrowing animals (e.g., muskrats, turtles)
- Duckweed control and removal
- Floating sludge mats
- Dike vegetation requiring mowing and removing woody plants
- Dike erosion affecting riprap and proper vegetation
- Fence maintenance to restrict access
- Mechanical equipment maintenance (e.g., pumps, blowers)

During certain times of the year, cooler, denser water remains at depth, while the warmer water stays at the surface. Water temperature differences may cause ponds to stratify throughout their depth. As the temperature decreases during the fall and the surface water cools, stratification decreases and the deeper water mixes with the cooling surface water. This phenomenon is called *mixis*, or pond or lake overturn. As the density of water decreases and the temperature falls below 4°C, winter stratification can develop. When the ice cover breaks up and the water warms, a spring overturn can also occur (Spellman, 1996).

Pond overturn, which releases odorous compounds into the atmosphere, can generate complaints from property owners living downwind of the pond. The potential for pond overturn during certain times of the year is the reason why regulations may specify that ponds be located downwind, based on prevailing winds during overturn periods, and away from dwellings.

Wind

Prevailing and storm-generated wind should be factored into pond design and siting as they influence performance and maintenance in several significant ways:

- *Oxygen transfer and dispersal*—By producing circulatory flows, winds provide the mixing needed for O_2 transfer and diffusion below the surface of facultative ponds. This mixing action also helps disperse microorganisms and augments the movement of algae, particularly green algae.
- *Prevention of short-circuiting and reduction of odor events*—Care must be taken during design to position the pond inlet/outlet axis perpendicular to the direction of prevailing winds to reduce short-circuiting, which is the most common cause of poor performance. Consideration must also be made for the transport and fate of odors generated by treatment byproducts in anaerobic and facultative ponds.
- *Disturbance of pond integrity*—Waves generated by strong prevailing or storm winds are capable of eroding or overtopping embankments. Some protective material should extend one or more feet above and below the water level to stabilize earthen berms.
- *Hydraulic detention time*—Wind effects can reduce hydraulic detention time.

Pond Nutritional Requirements

In order to function as designed, the wastewater pond must provide sufficient macro- and micronutrients for the microorganisms to grow and populate the system adequately. It should be understood that a treatment pond system should be neither overloaded nor underloaded with wastewater nutrients.

Nitrogen

Nitrogen can be a limiting nutrient for primary productivity in a pond. The conversion of organic nitrogen to various other nitrogen forms results in a total net loss (Assenzo and Reid, 1966; Craggs, 2005; Middlebrooks and Pano, 1983; Middlebrooks et al., 1982; Pano and Middlebrooks, 1982). This nitrogen loss may be due to algal uptake or bacterial action. It is likely that both mechanisms contribute to the overall total nitrogen reduction. Another factor contributing to the reduction of total nitrogen is the removal of gaseous NH_3 under favorable environmental conditions. Regardless of the specific removal mechanism involved, NH_3 removal in facultative wastewater ponds has been observed at levels greater than 90%, with the major removal occurring in the primary cell of a multicell pond system (Crites et al., 2006; Middlebrooks et al., 1982; Shilton, 2005; USEPA, 2011).

Phosphorus

Phosphorus is most often the growth-limiting nutrient in aquatic environments. Municipal wastewater in the United States is normally enriched in phosphorus even though restrictions on phosphorus-containing compounds in laundry detergents in some states have resulted in reduced concentrations since the 1970s. As of 1999, 27 states and the District of Columbia had passed laws prohibiting the manufacture and use of laundry detergents containing phosphorus. However, phosphate (PO_4^{3-}) content limits in automatic dishwashing detergents and other household cleaning agents containing phosphorus remain unchanged in most states. With a contribution of approximately 15%, the concentration of phosphorus from wastewater treatment plants is still adequate to promote growth in aquatic organisms.

In aquatic environments, phosphorus occurs in three forms: (1) particulate phosphorus, (2) soluble organic phosphorus, and (3) inorganic phosphorus. Inorganic phosphorus, primarily in the form of orthophosphate ($OP(OR)_3$), is readily utilized by aquatic organisms. Some organisms may store excess phosphorus as polyphosphate. At the same time, some PO_4^{3-} is continuously lost to sediments, where it is locked up in insoluble precipitates (Craggs, 2005; Crites et al., 2006; Lynch and Poole, 1979). Phosphorus removal in ponds occurs via physical mechanisms such as adsorption, coagulation, and precipitation. The uptake of phosphorus by organisms in metabolic function as well as for storage can also contribute to its removal. Removal in wastewater ponds has been reported to range from 30 to 95% (Assenzo and Reid, 1966; Crites et al., 2006; Pearson, 2005). Algae discharged in the final effluent may introduce organic phosphorus to receiving waters. Excessive algal "afterblooms" observed in waters receiving effluents have, in some cases, been attributed to nitrogen and phosphorus compounds remaining in the treated wastewater.

Sulfur

Sulfur is a required nutrient for microorganisms, and it is usually present in sufficient concentration in natural waters. Because sulfur is rarely limiting, its removal from wastewater is usually not considered necessary. Ecologically, sulfur compounds such as hydrogen sulfide (H_2S) and sulfuric acid (H_2SO_4) are toxic, while the oxidation of certain sulfur compounds is an important energy source for some aquatic bacteria (Lynch and Poole, 1979; Pearson, 2005).

Carbon

The decomposable organic carbon content of a waste is traditionally measured in terms of its BOD_5, or the amount of O_2 required under standardized conditions for the aerobic biological stabilization of the organic matter over a certain period of time.

Because complete treatment by biological oxidation can take several weeks, depending on the organic material and the organism present, standard practice is to use the BOD_5 as an index of the organic carbon content or organic strength of a waste. The removal of BOD_5 is a primary criterion by which treatment efficiency is evaluated. BOD_5 reduction in wastewater ponds ranging from 50 to 95% has been reported in the literature. Various factors affect the rate of reduction of BOD_5. A very rapid reduction occurs in a wastewater pond during the first 5 to 7 days. Subsequent reductions take place at a sharply reduced rate. BOD_5 removals are generally much lower during winter and early spring than in summer and early fall. Many regulatory agencies recommend that pond operations do not include discharge during cold periods.

TRICKLING FILTERS

Trickling filters have been used to treat wastewater since the 1890s. It was found that if settled wastewater was passed over rock surfaces, slime grew on the rocks and the water became cleaner. Today we still use this principle, but, in many installations, instead of rocks we use plastic media. In most wastewater treatment systems, the trickling filter follows primary treatment and includes a secondary settling tank or clarifier as shown in Figure 7.7. Trickling filters are widely used for the treatment of domestic and industrial wastes. The process is a fixed-film biological treatment method designed to remove BOD_5 and suspended solids.

A trickling filter consists of a rotating distribution arm that sprays and evenly distributes liquid wastewater over a circular bed of fist-sized rocks, other coarse materials, or synthetic media (see Figure 7.8). The spaces between the media allow air to circulate easily so aerobic conditions can be maintained. The spaces also allow wastewater to trickle down through, around, and over the media. A layer of biological slime that absorbs and consumes the wastes trickling through the bed covers the media material. The organisms aerobically decompose the solids, producing more organisms and stable wastes, which either become part of the slime or are discharged back into the wastewater flowing over the media. This slime consists mainly of bacteria, but it may also include algae, protozoa, worms, snails, fungi, and insect larvae. The accumulating slime occasionally sloughs off (*sloughings*) individual media materials (see Figure 7.9); it is collected at the bottom of the filter, along

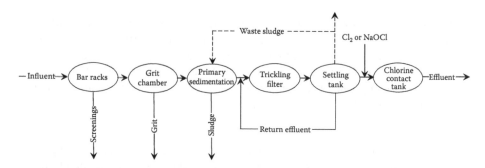

FIGURE 7.7　Simplified flow diagram of trickling filter used for wastewater treatment.

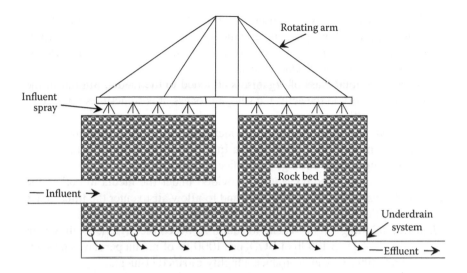

FIGURE 7.8 Schematic of cross-section of a trickling filter.

with the treated wastewater, and is passed on to the secondary settling tank where it is removed. The overall performance of the trickling filter is dependent on hydraulic and organic loading, temperature, and recirculation.

Trickling Filter Definitions

To clearly understand the correct operation of the trickling filter, the operator must be familiar with certain terms.

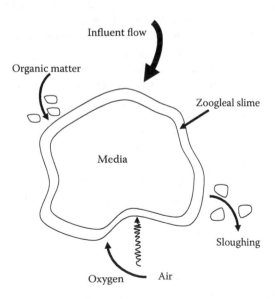

FIGURE 7.9 Filter media showing biological activities that take place on the surface area.

Biological towers—A type of trickling filter that is very deep (10 to 20 ft). Filled with a lightweight synthetic media, these towers are also known as *oxidation* or *roughing towers* or (because of their extremely high hydraulic loading) *super-rate trickling filters*.

Biomass—The total mass of organisms attached to the media. Similar to the solids inventory in the activated sludge process, it is sometimes referred to as *zoogleal slime*.

Distribution arm—The device most widely used to apply wastewater evenly over the entire surface of the media. In most cases, the force of the wastewater being sprayed through the orifices moves the arm.

Filter underdrain—The open space provided under the media to collect the liquid (wastewater and sloughings) and to allow air to enter the filter. It has a sloped floor to collect the flow to a central channel for removal.

High-rate trickling filters—A classification in which the organic loading is in the range of 25 to 100 lb of BOD_5 per 1000 ft^3 of media per day. The standard rate filter may also produce a highly nitrified effluent.

Hydraulic loading—The amount of wastewater flow applied to the surface of the trickling filter media. It can be expressed in several ways: flow per square foot of surface per day (gpd/ft^2), flow per acre per day (MGAD), or flow per acre foot per day (MGAFD). The hydraulic loading includes all flow entering the filter.

Media—An inert substance placed in the filter to provide a surface for the microorganism to grow on. The media can be field stone, crushed stone, slag, plastic, or redwood slats.

Organic loading—The amount of BOD_5 or chemical oxygen demand (COD) applied to a given volume of filter media. It does not include the BOD_5 or COD contributed to any recirculated flow and is commonly expressed as pounds of BOD_5 or COD per 1000 ft^3 of media.

Recirculation—The return of filter effluent back to the head of the trickling filter. It can level flow variations and assist in solving operational problems, such as ponding, filter flies, and odors.

Roughing filters—Type of trickling filter in which the organic loading exceeds 200 lb of BOD_5 per 1000 ft^3 of media per day. A roughing filter is used to reduce the loading on other biological treatment processes to produce an industrial discharge that can be safely treated in a municipal treatment facility.

Sloughing—The process in which the excess growths break away from the media and wash through the filter to the underdrains with the wastewater. These sloughings must be removed from the flow by settling.

Staging—The practice of operating two or more trickling filters in series. The effluent of one filter is used as the influent of the next. This practice can produce a higher quality effluent by removing additional BOD_5 or COD.

Trickling Filter Equipment

The trickling filter distribution system is designed to spread wastewater evenly over the surface of the entire media. The most common system is the *rotary distributor*, which moves above the surface of the media and sprays the wastewater on the surface. The force of the water leaving the orifices drives the rotary system. The distributor arms usually have small plates below each orifice to spread the wastewater into a fan-shaped distribution system. The second type of distributor is the *fixed-nozzle system*. In this system, the nozzles are fixed in place above the media and are designed to spray the wastewater over a fixed portion of the media. This system is used frequently with deep-bed synthetic media filters.

> *Note:* Trickling filters that use ordinary rock are normally only about 3 m in depth because of structural problems caused by the weight of rocks—which also requires the construction of beds that are quite wide—in many applications, up to 60 ft in diameter. When synthetic media are used, the bed can be much deeper.

No matter what type of media is selected, the primary consideration is that it must be capable of providing the desired film location for the development of the biomass. Depending on the type of media used and the filter classification, the media may be 3 to 20 ft or more in depth.

The underdrains are designed to support the media, collect the wastewater and sloughings, carry them out of the filter, and provide ventilation to the filter.

> *Note:* To ensure sufficient airflow to the filter, the underdrains should never be allowed to flow more than 50% full of wastewater.

The effluent channel is designed to carry the flow from the trickling filter to the secondary settling tank. The secondary settling tank provides 2 to 4 hr of detention time to separate the sloughing materials from the treated wastewater. The design, construction, and operation are similar to those for the primary settling tank. Longer detention times are provided because the sloughing materials are lighter and settle more slowly.

Recirculation pumps and piping are designed to recirculate (thus improving the performance of the trickling filter or settling tank) a portion of the effluent back to be mixed with the filter influent. When recirculation is used, pumps and metering devices must be provided.

Filter Classifications

Trickling filters are classified by hydraulic and organic loading. Moreover, the expected performance and the construction of the trickling filter are determined by the filter classification. Filter classifications include standard-rate, intermediate-rate, high-rate, super-high-rate (plastic media), and roughing. Standard-rate, high-rate, and roughing filters are the ones most commonly used. The *standard-rate filter* has a hydraulic loading that varies from 25 to 90 gpd/ft^3. It has a seasonal sloughing frequency and does not employ recirculation. It typically has an 80 to 85% BOD$_5$ removal rate and an 80 to 85% TSS removal rate. The *high-rate filter* has a hydraulic loading of 230 to 900 gpd/ft^3. It has a continuous sloughing frequency and always employs recirculation. It typically has a 65 to 80% BOD$_5$ removal rate and a 65 to 80% TSS removal rate. The *roughing filter* has a hydraulic loading of >900 gpd/ft^3. It has a continuous sloughing frequency and does not normally include recirculation. It typically has a 40 to 65% removal rate and a 40 to 65% TSS removal rate.

Overview and Brief Summary of Trickling Filter Process

Note: Trickling filters that use ordinary rock are normally only about 10 ft in depth because of structural problems caused by the weight of rocks, which also requires the construction of beds that are quite wide—in many applications, up to 60 ft in diameter. When synthetic media are used, the bed can be much deeper.

- The wastewater is applied to the media at a controlled rate, using a rotating distributor arm or fixed nozzles. Organic material is removed by contact with the microorganisms as the wastewater trickles down through the media openings. The treated wastewater is collected by an underdrain system.
- The trickling filter is usually built into a tank that contains the media. The filter may be square, rectangular, or circular.
- The trickling filter does not provide any actual filtration. The filter media provides a large amount of surface area that the microorganisms can cling to and grow in a slime that forms on the media as they feed on the organic material in the wastewater.
- The slime growth on the trickling filter media periodically sloughs off and is settled and removed in a secondary clarifier that follows the filter.

ROTATING BIOLOGICAL CONTACTORS

The rotating biological contactor (RBC) is a biological treatment system (see Figure 7.10) and is a variation of the attached-growth idea provided by the trickling filter. Still relying on microorganisms that grow on the surface of a medium, the RBC is instead a fixed-film biological treatment device, but the basic biological process is similar to that occurring in the trickling filter. An RBC consists of a series of closely spaced (mounted side by side), circular, plastic (synthetic) disks that are typically about 3.5 m in diameter and attached to a rotating horizontal shaft (see Figure 7.10). Approximately 40% of each disk is submersed in a tank containing the wastewater to

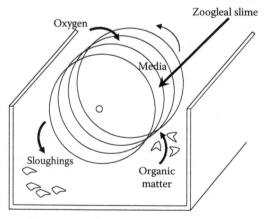

FIGURE 7.10   Rotating biological contactor (RBC) cross-section and treatment system.

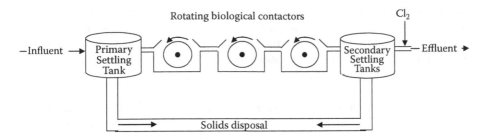

FIGURE 7.11   Rotating biological contactor (RBC) treatment system.

be treated. As the RBC rotates, the attached biomass film (zoogleal slime) that grows on the surface of the disk moves into and out of the wastewater. While submerged in the wastewater, the microorganisms absorb organics; when they are rotated out of the wastewater, they are supplied with the oxygen required for aerobic decomposition. As the zoogleal slime reenters the wastewater, excess solids and waste products are stripped off the media as sloughings. These sloughings are transported with the wastewater flow to a settling tank for removal.

Modular RBC units are placed in series (see Figure 7.11) simply because a single contactor is not sufficient to achieve the desired level of treatment; the resulting treatment achieved exceeds conventional secondary treatment. Each individual contactor is called a *stage* and the group is known as a *train*. Most RBC systems consist of two or more trains with three or more stages in each. The key advantage in using RBCs instead of trickling filters is that RBCs are easier to operate under varying load conditions, as it is easier to keep the solid medium wet at all times. Moreover, the level of nitrification that can be achieved by a RBC system is significant, especially when multiple stages are employed.

RBC Equipment

The equipment that makes up a RBC includes the rotating biological contactor (the media, either standard or high density), a center shaft, drive system, tank, baffles, housing or cover, and a settling tank. The rotating biological contactor consists of circular sheets of synthetic material (usually plastic) mounted side by side on a shaft. The *sheets* (media) contain large amounts of surface area for growth of the biomass. The *center shaft* provides the support for the disks of media and must be strong enough to support the weight of the media and the biomass; experience has indicated that a major problem is collapse of the support shaft. The *drive system* provides the motive force to rotate the disks and shaft. The drive system may be mechanical or air driven or a combination of each. When the drive system does not provide uniform movement of the RBC, major operational problems can arise.

The *tank* holds the wastewater in which the RBC rotates. It should be large enough to permit variation of the liquid depth and detention time. *Baffles* are required to permit proper adjustment of the loading applied to each stage of the RBC process. Adjustments can be made to increase or decrease the submergence of the RBC. RBC stages are normally enclosed in some type of protective structure (*cover*) to prevent loss of biomass due to severe weather changes (e.g., snow, rain, temperature, wind, sunlight). In many instances, this housing greatly restricts access to the RBC. The *settling tank* is provided to remove the sloughing material created by the biological activity and is similar in design to the primary settling tank. The settling tank provides 2- to 4-hr detention times to permit settling of lighter biological solids.

RBC Operation

During normal operation, operator vigilance is required to observe the RBC movement, slime color, and appearance; however, if the unit is covered, observations may be limited to that portion of the media that can be viewed through the access door. Slime color and appearance can indicate process condition; for example,

- Gray, shaggy slime growth indicates normal operation.
- Reddish brown, golden shaggy growth indicates nitrification.
- White chalky appearance indicates high sulfur concentrations.
- No slime indicates severe temperature or pH changes.

Sampling and testing should be conducted daily for dissolved oxygen content and pH. BOD_5 and suspended solids testing should also be performed to aid in assessing performance.

ACTIVATED SLUDGE

The biological treatment systems discussed to this point—ponds, trickling filters, and rotating biological contactors—have been around for years. The trickling filter, for example, has been around and successfully used since the late 1800s. The problem with ponds, trickling filters, and RBCs is that they are temperature sensitive, remove less BOD, and cost more to build (particularly trickling filters) than the activated sludge systems that were later developed.

> *Note:* Although trickling filters and other systems cost more to build than activated sludge systems, it is important to point out that activated sludge systems cost more to operate because of the need for energy to run pumps and blowers.

As shown in Figure 7.4, the activated sludge process follows primary settling. The basic components of an activated sludge sewage treatment system include an *aeration tank* and a *secondary basin, settling basin,* or *clarifier* (see Figure 7.12). Primary effluent is mixed with settled solids recycled from the secondary clarifier and is then introduced into the aeration tank. Compressed air is injected continuously into the mixture through porous diffusers located at the bottom of the tank, usually along one side.

Wastewater is fed continuously into an aerated tank, where the microorganisms metabolize and biologically flocculate the organics. Microorganisms (activated sludge) are settled from the aerated mixed liquor under quiescent conditions in the final clarifier and are returned to the aeration tank. Left uncontrolled, the number of organisms would eventually become too great; therefore, some must periodically be removed (wasted). A portion of the concentrated solids from the bottom of the settling tank must be removed from the process (waste activated sludge, or WAS). Clear supernatant from the final settling tank is the plant effluent.

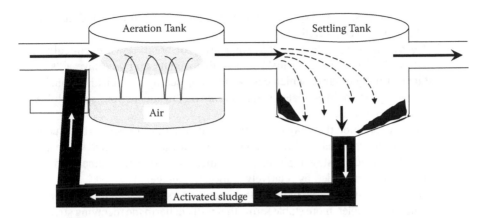

FIGURE 7.12 Activated sludge process.

ACTIVATED SLUDGE TERMINOLOGY

To better understand the discussion of the activated sludge process presented in the following sections, it is necessary to understand the terms associated with the process. Some of these terms have been used and defined earlier in the text, but we list them here again to refresh your memory. Review these terms and remember them, as they are used throughout the discussion:

Absorption—Taking in or reception of one substance into the body of another by molecular or chemical actions and distribution throughout the absorber.

Activated—To speed up reaction. When applied to sludge, it means that many aerobic bacteria and other microorganisms are in the sludge particles.

Activated sludge—A floc or solid formed by the microorganisms. It includes organisms, accumulated food materials, and waste products from the aerobic decomposition process.

Activated sludge process—Biological wastewater treatment process in which a mixture of influent and activated sludge is agitated and aerated. The activated sludge is then separated from the treated mixed liquor by sedimentation and is returned to the process as needed. The treated wastewater overflows the weir of the settling tank in which separation from the sludge takes place.

Adsorption—The adherence of dissolved, colloidal, or finely divided solids to the surface of solid bodies when they are brought into contact.

Aeration—Mixing air and a liquid by one of the following methods: spraying the liquid in the air, diffusing air into the liquid, or agitating the liquid to promote surface adsorption of air.

Aerobic—A condition in which free or dissolved oxygen is present in the aquatic environment. Aerobic organisms must be in the presence of dissolved oxygen to be active.

Bacteria—Single-cell plants with a vital role in stabilization of organic waste.

Biochemical oxygen demand (BOD)—A measure of the amount of food available to the microorganisms in a particular waste. It is measured by the amount of dissolved oxygen used up during a specific time period (usually 5 days, expressed as BOD$_5$).

Biodegradable—From *degrade* ("to wear away or break down chemically") and *bio* ("by living organisms"). Put it all together, and you have a "substance, usually organic, which can be decomposed by biological action."

Bulking—A problem in activated sludge plants that results in poor settleability of sludge particles.

Coning—A condition that may be established in a sludge hopper during sludge withdrawal when part of the sludge moves toward the outlet while the remainder tends to stay in place; development of a cone or channel of moving liquids surrounded by relatively stationary sludge.

Decomposition—Generally, in waste treatment, refers to the changing of waste matter into simpler, more stable forms that will not harm the receiving stream.

Diffuser—A porous plate or tube through which air is forced and divided into tiny bubbles for distribution in liquids; commonly made of carborundum, aluminum, or silica sand.

Diffused air aeration—A diffused-air-activated sludge plant takes air, compresses it, then discharges the air below the water surface to the aerator through some type of air diffusion device.

Dissolved oxygen—Atmospheric oxygen dissolved in water or wastewater; usually abbreviated as DO. The typical required DO for a well-operated activated sludge plant is between 2.0 and 2.5 mg/L.

Facultative bacteria—Bacteria that can use molecular (dissolved) oxygen or oxygen obtained from food materials. In other words, facultative bacteria can live under aerobic or anaerobic conditions.

Filamentous bacteria—Organisms that grow in thread or filamentous form.

Food-to-microorganism ratio—A process control calculation used to evaluate the amount of food (BOD or COD) available per pound of mixed liquor volatile suspended solids.

Fungi—Multicellular aerobic organisms.

Gould sludge age—A process control calculation used to evaluate the amount of influent suspended solids available per pound of mixed liquor suspended solids.

Mean cell residence time (MCRT)—The average length of time particles of mixed liquor suspended solids remain in the activated sludge process; may also be referred to as the *sludge retention time* (SRT).

Mixed liquor—The contribution of return activated sludge and wastewater (either influent or primary effluent) that flows into the aeration tank.

Mixed liquor suspended solids (MLSS)—The suspended solids concentration of the mixed liquor. Many references use this concentration to represent the amount of organisms in the activated sludge process.

Mixed liquor volatile suspended solids (MLVSS)—The organic matter in the mixed liquor suspended solids; can also be used to represent the amount of organisms in the process.

Nematodes—Microscopic worms that may appear in biological waste treatment systems.

Nutrients—Substances required to support plant organisms. Major nutrients are carbon, hydrogen, oxygen, sulfur, nitrogen, and phosphorus.

Protozoa—Single-cell animals that are easily observed under the microscope at a magnification of 100×. Bacteria and algae are prime sources of food for advanced forms of protozoa.

Return activated sludge (RAS)—The solids returned from the settling tank to the head of the aeration tank.

Rising sludge—Occurs in the secondary clarifiers or activated sludge plant when the sludge settles to the bottom of the clarifier, is compacted, and then rises to the surface in relatively short time.

Rotifers—Multicellular animals with flexible bodies and cilia near the mouth used to attract food. Bacteria and algae are their major sources of food.

Secondary treatment—A wastewater treatment process used to convert dissolved or suspended materials into a form that can be removed.

Settleability—A process control test that is used to evaluate the settling characteristics of the activated sludge. Readings taken at 30 to 60 min are used to calculate the settled sludge volume (SSV) and the sludge volume index (SVI).

Settled sludge volume (SSV)—The volume (mL/L or percent) occupied by an activated sludge sample after 30 or 60 min of settling. Normally written as SSV with a subscript to indicate the time of the reading used for calculation (e.g., SSV_{30} or SSV_{60}).

Shock load—The arrival at a plant of a waste toxic to organisms in sufficient quantity or strength to cause operating problems, such as odor or sloughing off of the growth of slime on the trickling filter media. Organic overloads also can cause a shock load.

Sludge volume index (SVI)—A process control calculation used to evaluate the settling quality of the activated sludge.

Solids—Material in the solid state.
- *Dissolved solids*—Solids present in solution; solids that will pass through a glass-fiber filter.
- *Fixed solids*—Also known as *inorganic solids*; the solids left after a sample is ignited at 550°C for 15 min.
- *Floatable solids (scum)*—Solids that will float to the surface of still water, sewage, or other liquid; usually composed of grease particles, oils, light plastic material, etc.
- *Nonsettleable solids*—Finely divided suspended solids that will not sink to the bottom in still water, sewage, or other liquid in a reasonable period, usually two hours; also known as *colloidal solids*.
- *Suspended solids*—Solids that will not pass through a glass-fiber filter.
- *Total solids*—Solids in water, sewage, or other liquids, including suspended solids and dissolved solids.
- *Volatile solids*—Organic solids; measured as the solids that are lost on ignition of the dry solids at 550°C.

Waste activated sludge (WAS)—The solids being removed from the activated sludge process.

ACTIVATED SLUDGE PROCESS EQUIPMENT

Equipment requirements for the activated sludge process are more complex than other processes discussed. The equipment includes an *aeration tank, aeration system, system settling tank, return sludge system,* and *waste sludge system.* These are discussed in the following.

Aeration Tank

The aeration tank is designed to provide the required detention time (depending on the specific modification) and ensure that the activated sludge and the influent wastewater are thoroughly mixed. Tank design normally attempts to ensure that no dead spots are created.

Aeration

Aeration can be mechanical or diffused. Mechanical aeration systems use agitators or mixers to mix air and mixed liquor. Some systems use *sparge rings* to release air directly into the mixer. Diffused aeration systems use pressurized air released through diffusers near the bottom of the tank. Efficiency is directly related to the size of the air bubbles produced. Fine bubble systems have a higher efficiency. The diffused air system has a blower to produce large volumes of low pressure air (5 to 10 psi), air lines to carry the air to the aeration tank, and headers to distribute the air to the diffusers, which release the air into the wastewater.

Settling Tank

Activated sludge systems are equipped with plain settling tanks designed to provide 2 to 4 hr of hydraulic detention time.

Return Sludge

The return sludge system includes pumps, a timer or variable speed drive to regulate pump delivery, and a flow measurement device to determine actual flow rates.

Waste Sludge

In some cases, the waste activated sludge withdrawal is accomplished by adjusting valves on the return system. When a separate system is used it includes pumps, a timer or variable speed drive, and a flow measurement device.

OVERVIEW OF ACTIVATED SLUDGE PROCESS

The activated sludge process is a treatment technique in which wastewater and reused biological sludge full of living microorganisms are mixed and aerated. The biological solids are then separated from the treated wastewater in a clarifier and are returned to the aeration process or wasted. The microorganisms are mixed thoroughly with the incoming organic material, and they grow and reproduce by using the organic material as food. As they grow and are mixed with air, the individual organisms cling together (flocculate). Once flocculated, they more readily settle in the secondary clarifiers.

The wastewater being treated flows continuously into an aeration tank where air is injected to mix the wastewater with the return activated sludge and to supply the oxygen required by the microbes to live and feed on the organics. Aeration can be supplied by injection through air diffusers in the bottom of the tank or by mechanical aerators located at the surface. The mixture of activated sludge and wastewater in the aeration tank is called the *mixed liquor*. The mixed liquor flows to a secondary clarifier where the activated sludge is allowed to settle.

The activated sludge is constantly growing, and more is produced than can be returned for use in the aeration basin. Some of this sludge must, therefore, be wasted to a sludge handling system for treatment and disposal. The volume of sludge returned to the aeration basins is normally 40 to 60% of the wastewater flow. The rest is wasted.

Factors Affecting Operation of the Activated Sludge Process

A number of factors affect the performance of an activated sludge system. These include the following:

- Temperature
- Return rates
- Amount of oxygen available
- Amount of organic matter available
- pH
- Waste rates
- Aeration time
- Wastewater toxicity

To obtain the desired level of performance in an activated sludge system, it is necessary to maintain a proper balance among the amounts of food (organic matter), organisms (activated sludge), and oxygen (dissolved oxygen). The majority of problems occurring with the activated sludge process result from an imbalance among these three items.

To fully appreciate and understand the biological process taking place in a normally functioning activated sludge process, the operator must have knowledge of the key players in the process: the organisms. This makes a certain amount of sense when we consider that the heart of the activated sludge process is the mass of settleable solids formed by aerating wastewater containing biological degradable compounds in the presence of microorganisms. Activated sludge consists of organic solids plus bacteria, fungi, protozoa, rotifers, and nematodes.

Activated Sludge Formation

The formation of activated sludge is dependent on three steps. The first step is the transfer of food from wastewater to organism. Second is the conversion of wastes to a usable form. Third is the flocculation step.

1. *Transfer*—Organic matter (food) is transferred from the water to the organisms. Soluble material is absorbed directly through the cell wall. Particulate and colloidal matter is adsorbed to the cell wall, where it is broken down into simpler soluble forms, then absorbed through the cell wall.
2. *Conversion*—Food matter is converted to cell matter by synthesis and oxidation into end products such as CO_2, H_2O, NH_3, stable organic waste, and new cells.
3. *Flocculation*—Flocculation is the gathering of fine particles into larger particles. This process begins in the aeration tank and is the basic mechanism for removal of suspended matter in the final clarifier. The concentrated *biofloc* that settles and forms the sludge blanket in the secondary clarifier is known as activated sludge.

OXIDATION DITCHES

An oxidation ditch is a modified extended aeration activated sludge biological treatment process that utilizes long sludge retention times (SRTs) to remove biodegradable organics. Oxidation ditches are typically complete-mix systems, but they can be modified to approach plug-flow conditions. (*Note:* As conditions approach plug flow, diffused air must be used to provide enough mixing. The system will also no longer operate as an oxidation ditch.) Typical oxidation ditch treatment systems consist of a single or multi-channel configuration within a ring, oval, or horseshoe-shaped basin. As a result, oxidation ditches are called "racetrack type" reactors. Horizontally or vertically mounted aerators provide circulation, oxygen transfer, and aeration in the ditch.

Preliminary treatment, such as bar screens and grit removal, normally precedes the oxidation ditch. Primary settling prior to an oxidation ditch is sometimes practiced but is not typical in this design. Tertiary filters may be required after clarification, depending on the effluent requirements. Disinfection is required and reaeration may be necessary prior to final discharge. Flow to the oxidation ditch is aerated and mixed with return sludge from a secondary clarifier. A typical process flow diagram for an activated sludge plant using an oxidation ditch is shown in Figure 7.13.

Surface aerators, such as brush rotors, disc aerators, draft tube aerators, or fine bubble diffusers, are used to circulate the mixed liquor. The mixing process entrains oxygen into the mixed liquor to foster microbial growth and the motive velocity ensures contact of microorganisms with the incoming wastewater. The aeration sharply increases the dissolved oxygen concentration but it decreases when the biomass uptakes oxygen as the mixed liquor travels through the ditch. Solids are maintained in suspension as the mixed liquor travels through the ditch. If design SRTs are selected for nitrification, a high degree of nitrification will occur. Oxidation ditch effluent is usually settled in a separate secondary clarifier. An anaerobic tank may be added prior to the ditch to enhance biological phosphorus removal.

An oxidation ditch may also be operated to achieve partial denitrification. One of the most common design modifications for enhanced nitrogen removal is known as the Modified Ludzack–Ettinger (MLE) process. In this process, illustrated in Figure 7.14, an anoxic tank is added upstream of the ditch along with mixed liquor

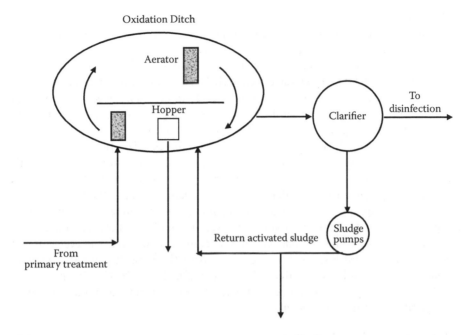

FIGURE 7.13 Typical oxidation ditch activated sludge system.

recirculation from the aerobic zone to the tank to achieve higher levels of deni-trification. In the aerobic basin, autotrophic bacteria (nitrifiers) convert ammonia nitrogen to nitrite nitrogen and then to nitrate nitrogen. In the anoxic zone, hetero-trophic bacteria convert nitrate nitrogen to nitrogen gas, which is released to the atmosphere. Some mixed liquor from the aerobic basin is recirculated to the anoxic zone to provide mixed liquor with a high concentration of nitrate nitrogen to the anoxic zone.

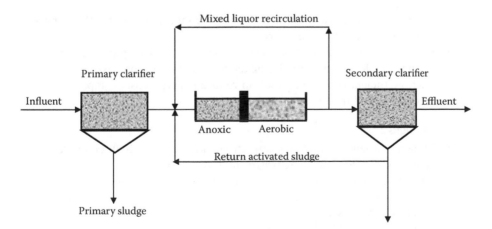

FIGURE 7.14 Modified Ludzack–Ettinger (MLE) process.

Several manufacturers have developed modifications to the oxidation ditch design to remove nutrients in conditions cycled or phased between the anoxic and aerobic states. Although the mechanics of operation differ by manufacturer, in general the process consists of two separate aeration basins, the first anoxic and the second aerobic. Wastewater and return activated sludge (RAS) are introduced into the first reactor, which operates under anoxic conditions. Mixed liquor then flows into the second reactor, which operates under aerobic conditions. The process is then reversed, and the second reactor begins to operate under anoxic conditions.

With regard to applicability, the oxidation ditch process is a fully demonstrated secondary wastewater treatment technology, applicable in any situation where activated sludge treatment (conventional or extend aeration) is appropriate. Oxidation ditches are applicable in plants that require nitrification because the basins can be sized using an appropriate SRT to achieve nitrification at the mixed liquor minimum temperature. This technology is very effective in small installations, small communities, and isolated institutions, because it requires more land than conventional treatment plants (USEPA, 2000a). There are currently more than 9000 municipal oxidation ditch installations in the United States (Spellman, 2007). Nitrification to less than 1 mg/L ammonia nitrogen consistently occurs when ditches are designed and operated for nitrogen removal.

Advantages and Disadvantages

Advantages

The main advantage of the oxidation ditch is the ability to achieve removal performance objectives with low operational requirements and low operation and maintenance (O&M) costs. Some specific advantages of oxidation ditches include the following:

- An added measure of reliability and performance is provided compared to other biological processes due to a constant water level and continuous discharge which lowers the weir overflow rate and eliminates the periodic effluent surge common to other biological processes, such as sequencing batch reactors.
- Long hydraulic detention time and complete mixing minimize the impact of a shock load or hydraulic surge.
- It produces less sludge than other biological treatment processes due to extended biological activity during the activated sludge process.
- Energy-efficient operations result in reduced energy costs compared with other biological treatment processes.

Disadvantages

- Effluent suspended solids concentrations are relatively high compared to other modifications of the activated sludge process.
- The process requires a larger land area than other activated sludge treatment options. This can prove costly, limiting the feasibility of oxidation ditches in urban, suburban, or other areas where land acquisition costs are relatively high (USEPA, 2000a).

ACTIVATED SLUDGE PROCESS CONTROL PARAMETERS

When operating an activated sludge process, the operator must be familiar with the many important process control parameters that must be monitored frequently and adjusted occasionally to maintain optimal performance.

Alkalinity

Monitoring alkalinity in the aeration tank is essential to control of the process. Insufficient alkalinity will reduce organism activity and may result in low effluent pH and, in some cases, extremely high chlorine demand in the disinfection process.

Dissolved Oxygen

The activated sludge process is an aerobic process that requires some dissolved oxygen (DO) to be present at all times. The amount of oxygen required is dependent on the influent food (BOD), the activity of the activated sludge, and the degree of treatment desired.

pH

Activated sludge microorganisms can be injured or destroyed by wide variations in pH. The pH of the aeration basin will normally be in the range of 6.5 to 9.0. Gradual variations within this range will not cause any major problems; however, rapid changes of one or more pH units can have a significant impact on performance. Industrial waste discharges, septic wastes, or significant amounts of stormwater flows may produce wide variations in pH. pH should be monitored as part of the routine process control testing schedule. Sudden changes or abnormal pH values may indicate an industrial discharge of strongly acidic or alkaline wastes. Because these wastes can upset the environmental balance of the activated sludge, the presence of wide pH variations can result in poor performance. Processes undergoing nitrification may show a significant decrease in effluent pH.

Mixed Liquor Suspended Solids, Volatile Suspended Solids, and Total Suspended Solids

The mixed liquor suspended solids (MLSS) or mixed liquor volatile suspended solids (MLVSS) can be used to represent the activated sludge or microorganisms present in the process. Process control calculations, such as sludge age and sludge volume index, cannot be calculated unless the MLSS is determined. The MLSS and MLVSS are adjusted by increasing or decreasing the waste sludge rates. The mixed liquor total suspended solids (MLTSS) are an important activated sludge control parameter. To increase the MLTSS, for example, the operator must decrease the waste rate or increase the MCRT. The MCRT must be decreased to prevent the MLTSS from changing when the number of aeration tanks in service is reduced.

Note: When performing the Gould sludge age test, assume that the source of the MLTSS in the aeration tank is influent solids.

Return Activated Sludge Rate and Concentration

The sludge rate is a critical control variable. The operator must maintain a continuous return of activated sludge to the aeration tank or the process will show a drastic decrease in performance. If the rate is too low, solids remain in the settling tank, resulting in solids loss and a septic return. If the rate is too high, the aeration tank can become hydraulically overloaded, causing reduced aeration time and poor performance. The return concentration is also important because it may be used to determine the return rate required to maintain the desired MLSS.

Waste Activated Sludge Flow Rate

Because the activated sludge contains living organisms that grow, reproduce, and produce waste matter, the amount of activated sludge is continuously increasing. If the activated sludge is allowed to remain in the system too long, the performance of the process will decrease. If too much activated sludge is removed from the system, the solids become very light and will not settle quickly enough to be removed in the secondary clarifier.

Temperature

Because temperature directly affects the activity of the microorganisms, accurate monitoring of temperature can be helpful in identifying the causes of significant changes in microorganism populations or process performance.

Sludge Blanket Depth

The separation of solids and liquid in the secondary clarifier results in a blanket of solids. If solids are not removed from the clarifier at the same rate they enter, the blanket will increase in depth. If this occurs, the solids may carryover into the process effluent. The sludge blanket depth may be affected by other conditions, such as temperature variation, toxic wastes, or sludge bulking. The best sludge blanket depth is dependent on such factors as hydraulic load, clarifier design, and sludge characteristics. The best blanket depth must be determined on an individual basis by experimentation.

> *Note:* When measuring sludge blanket depth, it is general practice to use a 15- to 20-ft long clear plastic pipe marked at 6-in. intervals; the pipe is equipped with a ball valve at the bottom.

Activated Sludge Operational Control Levels

The operator has two methods available to operate an activated sludge system. The operator can wait until the process performance deteriorates and make drastic changes, or the operator can establish *normal* operational levels and make minor adjustments to keep the process within the established operational levels.

> *Note:* Control levels can be defined as the upper and lower values for a process control variable that can be expected to produce the desired effluent quality.

Although the first method will guarantee that plant performance is always maintained within effluent limitations, the second method has a much higher probability of achieving this objective. This section discusses methods used to establish *normal* control levels for the activated sludge process. Several major factors should be considered when establishing control levels for the activated sludge system, including the following:

- Influent characteristics
- Industrial contributions
- Process sidestreams
- Seasonal variations
- Required effluent quality

Influent Characteristics

Influent characteristics were discussed earlier; however, major factors to consider when evaluating influent characteristics are the nature and volume of industrial contributions to the system. Waste characteristics (BOD, solids, pH, metals, toxicity, and temperature), volume, and discharge pattern (e.g., continuous, slug, daily, weekly) should be evaluated when determining if a waste will require pretreatment by the industry or adjustments to operational control levels.

Industrial Contributions

One or more industrial contributors produce a significant portion of the plant loading (in many systems). Identifying and characterizing all industrial contributors is important. Remember that the volume of waste generated may not be as important as the characteristics of the waste. Extremely high-strength wastes can result in organic overloading or poor performance because of insufficient nutrient availability. A second consideration is the presence of materials that even in small quantities are toxic to the process microorganisms or that create a toxic condition in the plant effluent or plant sludge. Industrial contributions to a biological treatment system should be thoroughly characterized prior to acceptance, monitored frequently, and controlled by either local ordinance or implementation of a pretreatment program.

Process Sidestreams

Process sidestreams are flows produced in other treatment processes that must be returned to the wastewater system for treatment prior to disposal. Examples of process sidestreams include the following:

- Thickener supernatant
- Aerobic and anaerobic digester supernatant
- Liquids removed by sludge dewatering processes (filtrate, centrate, and subnate)
- Supernatant from heat treatment and chlorine oxidation sludge treatment processes

Testing these flows periodically to determine both their quantity and strength is important. In many treatment systems, a significant part of the organic or hydraulic loading for the plant is generated by sidestream flows. The contribution of the plant sidestream flows can significantly change the operational control levels of the activated sludge system.

Seasonal Variations

Seasonal variations in temperature, oxygen solubility, organism activity, and waste characteristics may require several normal control levels for the activated sludge process; for example, during cold months of the year, aeration tank solids levels may have to be maintained at a significantly higher level than are required during warm weather. Likewise, the aeration rate may be controlled by the mixing requirements of the system during the colder months and by the oxygen demand of the system during the warm months.

Control Levels at Startup

Control levels for an activated sludge system during startup are usually based on design engineer recommendations or on information that is available from recognized reference sources. Although these levels provide a starting point, both the process control parameter sensitivity and control levels should be established on a plant-by-plant basis. During the first 12 months of operation, it is important to evaluate all of the potential process control options to determine the following:

- Sensitivity to effluent quality changes
- Seasonal variability
- Potential problems

DISINFECTION OF WASTEWATER

Like drinking water, liquid wastewater effluent is disinfected. Unlike drinking water, however, wastewater effluent is disinfected not to directly protect a drinking water supply but instead to protect public health in general. This is particularly important when the secondary effluent is discharged into a body of water used for swimming or as a downstream water supply. In the treatment of water for human consumption, treated water is typically chlorinated (although ozonation is also currently being applied in many cases). Chlorination is the preferred disinfection in potable water supplies because of chlorine's unique ability to provide a residual. This chlorine residual is important because when treated water leaves the waterworks facility and enters the distribution system the possibility of contamination is increased. The residual works to continuously disinfect water right up to the consumer's tap. In this section, we discuss basic chlorination and dechlorination. In addition, we describe the use of ultraviolet (UV) irradiation, ozonation, bromine chloride, and no disinfection. Keep in mind that much of the chlorination material presented in the following is similar to the chlorination information presented in the water treatment section of this chapter.

CHLORINE DISINFECTION

Chlorination for disinfection, as shown in Figure 7.4, follows all other steps in conventional wastewater treatment. The purpose of chlorination is to reduce the population of organisms in the wastewater to levels low enough to ensure that pathogenic organisms will not be present in sufficient quantities to cause disease when discharged. You might wonder why it is that chlorination of critical waters such as natural trout streams is not normal practice. This practice is strictly prohibited because chlorine and its byproducts (chloramines) are extremely toxic to aquatic organisms.

Note: Chlorine gas (vapor density of 2.5) is heavier than air; therefore, exhaust from a chlorinator room should be taken from floor level.

Note: The safest action to take in the event of a major chlorine container leak is to call the fire department.

Chlorine is a very reactive substance. Chlorine is added to wastewater to satisfy all of the chemical demands—in other words, to react with certain chemicals (such as sulfide, sulfite, or ferrous iron). When these initial chemical demands have been satisfied, chlorine will react with substances such as ammonia to produce chloramines and other substances that, although not as effective as chlorine, also have disinfecting capability. This produces a combined residual, which can be measured using residual chlorine test methods. If additional chlorine is added, free residual chlorine can be produced. Due to the chemicals normally found in wastewater, chlorine residuals are normally combined rather than free residuals. Control of the disinfection process is normally based on maintaining total chlorine residual of at least 1.0 mg/L for a contact time of at least 30 min at design flow.

Note: Residual level, contact time, and effluent quality affect disinfection. Failure to maintain the desired residual levels for the required contact time will result in lower efficiency and increased probability that disease organisms will be discharged.

Based on water quality standards, the total residual limitations on chlorine are as follows:

- *Freshwater*—Less than 11 ppb total residual chlorine
- *Estuaries*—Less that 7.5 ppb for halogen-produced oxidants
- *Endangered species*—Use of chlorine prohibited

CHLORINATION EQUIPMENT

Hypochlorite Systems

Depending on the form of hypochlorite selected for use, special equipment to control the addition of hypochlorite to the wastewater is required. Liquid forms require the use of metering pumps, which can deliver varying flows of hypochlorite solution. Dry chemicals require the use of a feed system designed to provide variable doses of the form used. The tablet form of hypochlorite requires the use of a tablet chlorinator designed specifically to provide the desired dose of chlorine. The hypochlorite solution or dry feed systems dispense the hypochlorite, which is then mixed with the flow. The treated wastewater then enters the contact tank to provide the required contact time.

Chlorine Systems

Because of the potential hazards associated with the use of chlorine, the equipment requirements are significantly greater than those associated with hypochlorite use. The system most widely used is a solution feed system. In this system, chlorine is removed from the container at a flow rate controlled by a variable orifice. Water moving through the chlorine injector creates a vacuum, which draws the chlorine gas to the injector and mixes it with the water. The chlorine gas reacts with the water to form hypochlorous and hydrochloric acid. The solution is then piped to the chlorine contact tank and dispersed into the wastewater through a diffuser. Larger facilities may withdraw the liquid form of chlorine and use evaporators (heaters) to convert to the gas form. Small facilities will normally draw the gas form of chlorine from the cylinder. As gas is withdrawn, liquid will be converted to the gas form. This requires heat energy and may result in chlorine line freeze-up if the withdrawal rate exceeds the available energy levels.

Normal Operation

In either type of system, normal operation requires adjustment of feed rates to ensure that required residual levels are maintained. This normally requires chlorine residual testing and adjustment based on the results of the test. Other activities include removal of accumulated solids from the contact tank, collection of bacteriological samples to evaluate process performance, and maintenance of safety equipment (respirator/air pack, safety lines, etc.). Hypochlorite operation may also include make-up solution (solution feed systems) or adding powder or pellets to the dry chemical feeder or tablets to the tablet chlorinator.

Chlorine operations include adjustment of chlorinator feed rates, inspection of mechanical equipment, testing for leaks using ammonia swabs (white smoke indicates the presence of leaks), changing containers (which requires more than one person for safety), and adjusting the injector water feed rate when required. Chlorination requires routine testing of plant effluent for total residual chlorine and may also require collection and analysis of samples to determine the fecal coliform concentration in the effluent.

ULTRAVIOLET IRRADIATION

Although ultraviolet (UV) disinfection was recognized as a method for achieving disinfection in the late 19th century, its application virtually disappeared with the evolution of chlorination technologies. In recent years, however, there has been a resurgence in its use in the wastewater field, largely as a consequence of concern for discharge of toxic chlorine residual. Even more recently, UV has gained more attention because of the tough new regulations on chlorine use imposed by both OSHA and the USEPA. Because of this relatively recent increased regulatory pressure, many facilities are actively engaged in substituting chlorine for other disinfection alternatives. Moreover, UV technology itself has made many improvements, which now makes UV attractive as a disinfection alternative. Ultraviolet light has very good germicidal qualities and is very effective in destroying microorganisms. It is used in hospitals, biological testing facilities, and many other similar locations. In wastewater treatment, the plant effluent is exposed to ultraviolet light of a specified wavelength and intensity for a specified contact period. The effectiveness of the process is dependent on

- UV light intensity
- Contact time
- Wastewater quality (turbidity)

For any one treatment plant, disinfection success is directly related to the concentration of colloidal and particulate constituents in the wastewater. The Achilles' heel of UV for disinfecting wastewater is turbidity. If the wastewater quality is poor, the ultraviolet light will be unable to penetrate the solids, and the effectiveness of the process decreases dramatically. For this reason, many states limit the use of UV disinfection to facilities that can reasonably be expected to produce an effluent containing ≤ 30 mg/L BOD_5 and total suspended solids. The main components of a UV disinfection system are mercury arc lamps, a reactor, and ballasts. The source of UV radiation is either the low-pressure or medium-pressure mercury arc lamp with low or high intensities.

In the operation of UV systems, UV lamps must be readily available when replacements are required. The best lamps are those with a stated operating life of at least 7500 hr and those that do not produce significant amounts of ozone or hydrogen peroxide. The lamps must also meet technical specifications for intensity, output, and arc length. If the UV light tubes are submerged in the waste stream, they must be protected inside quartz tubes, which not only protect the lights but also make cleaning and replacement easier.

Contact tanks must be used with UV disinfection. They are designed with the banks of UV lights in a horizontal position, either parallel or perpendicular to the flow, or with the banks of lights placed in a vertical position perpendicular to the flow.

Note: The contact tank must provide, at a minimum, a 10-second exposure time.

We stated earlier that turbidity has been a problem with UV wastewater treatment—and this is the case. However, if turbidity is its Achilles' heel, then the need for increased maintenance (as compared to other disinfection alternatives) is the toe of the same foot. UV maintenance requires that the tubes be cleaned on a regular basis or as needed. In addition, periodic acid washing is also required to remove chemical buildup.

Routine monitoring of UV disinfection systems is required. Checking on bulb burnout, buildup of solids on quartz tubes, and UV light intensity is necessary.

Note: UV light is extremely hazardous to the eyes. Never enter an area where UV lights are in operation without proper eye protection. Never look directly into the ultraviolet light.

OZONATION

Ozone is a strong oxidizing gas that reacts with most organic and many inorganic molecules. It is produced when oxygen molecules separate, collide with other oxygen atoms, and form a molecule consisting of three oxygen atoms. For high-quality effluents, ozone is a very effective disinfectant. Current regulations for domestic treatment systems limit the use of ozonation to filtered effluents unless the effectiveness of the system can be demonstrated prior to installation.

Note: Effluent quality is the key performance factor for ozonation.

For ozonation of wastewater, the facility must have the capability to generate pure oxygen along with an ozone generator. A contact tank with a ≥10-minute contact time at design average daily flow is required. Off-gas monitoring for process control is also required. In addition, safety equipment capable of monitoring ozone in the atmosphere and a ventilation system to prevent ozone levels exceeding 0.1 ppm are necessary.

The actual operation of the ozonation process consists of monitoring and adjusting the ozone generator and monitoring the control system to maintain the required ozone concentration in the off-gas. The process must also be evaluated periodically using biological testing to assess its effectiveness.

Note: Ozone is an extremely toxic substance. Concentrations in air should not exceed 0.1 ppm. It also has the potential to create an explosive atmosphere. Sufficient ventilation and purging capabilities should be provided.

Note: Ozone has certain advantages over chlorine for disinfection of wastewater, in that: (1) ozone increases DO in the effluent, (2) ozone has a briefer contact time, (3) ozone has no undesirable effects on marine organisms, and (4) ozone decreases turbidity and odor.

Ozone disinfection is generally used at medium- to large-sized plants after at least secondary treatment. In addition to disinfection, another common use for ozone in wastewater treatment is odor control. Ozone disinfection is the least used method in the United States, although this technology has been widely accepted in Europe for decades. Ozone treatment has the ability to achieve higher levels of disinfection than either chlorine or UV; however, the capital costs as well as maintenance expenditures are not competitive with available alternatives. Ozone is therefore used only sparingly, primarily in special cases where alternatives are not effective (USEPA, 1999b).

BROMINE CHLORIDE

Bromine chloride is a mixture of bromine and chlorine. It forms hydrocarbons and hydrochloric acid when mixed with water. Bromine chloride is an excellent disinfectant that reacts quickly and normally does not produce any long-term residuals.

Note: Bromine chloride is an extremely corrosive compound in the presence of low concentrations of moisture.

The reactions occurring when bromine chloride is added to the wastewater are similar to those occurring when chlorine is added. The major difference is the production of bromamine compounds rather than chloramines. The bromamine compounds are excellent disinfectants but are less stable and dissipate quickly. In most cases, the bromamines decay into other, less toxic compounds rapidly and are undetectable in the plant effluent. The factors that affect performance are similar to those affecting the performance of the chlorine disinfection process. Effluent quality, contact time, etc. have a direct impact on the performance of the process.

No Disinfection

In a very limited number of cases, treated wastewater discharges without disinfection are permitted. These are approved on a case-by-case basis. Each request must be evaluated based on the point of discharge, the quality of the discharge, the potential for human contact, and many other factors.

ADVANCED WASTEWATER TREATMENT

Advanced wastewater treatment is defined as the methods and processes that remove more contaminants (suspended and dissolved substances) from wastewater than are taken out by conventional biological treatment. Put another way, advanced wastewater treatment is the application of a process or system that follows secondary treatment or that includes phosphorus removal or nitrification in conventional secondary treatment.

Advanced wastewater treatment is used to augment conventional secondary treatment because secondary treatment typically removes only between 85 and 95% of the biochemical oxygen demand (BOD) and total suspended solids (TSS) in raw sanitary sewage. Generally, this leaves 30 mg/L or less of BOD and TSS in the secondary effluent. To meet stringent water quality standards, this level of BOD and TSS in secondary effluent may not prevent violation of water quality standards—the plant may not make permit. Thus, advanced wastewater treatment is often used to remove additional pollutants from treated wastewater.

In addition to meeting or exceeding the requirements of water quality standards, treatment facilities use advanced wastewater treatment for other reasons, as well; for example, conventional secondary wastewater treatment is sometimes not sufficient to protect the aquatic environment. In a stream, for example, when periodic flow events occur, the stream may not provide the amount of dilution of effluent required to maintain the necessary dissolved oxygen (DO) levels for aquatic organism survival.

Secondary treatment has other limitations. It does not significantly reduce the effluent concentration of nitrogen and phosphorus (important plant nutrients) in sewage. An overabundance of these nutrients can overstimulate plant and algae growth such that they create water quality problems. If they are discharged into lakes, for example, these nutrients contribute to algal blooms and accelerated eutrophication (lake aging). Also, the nitrogen in the sewage effluent may be present mostly in the form of ammonia compounds. At high enough concentrations, ammonia compounds can be toxic to aquatic organisms. Yet another problem with these compounds is that they exert a *nitrogenous oxygen* demand in the receiving water, as they convert to nitrates. This process is called *nitrification*.

> **Note:** The term *tertiary treatment* is commonly used as a synonym for advanced wastewater treatment; however, these two terms do not have precisely the same meaning. Tertiary suggests a third step that is applied after primary and secondary treatment.

Advanced wastewater treatment can remove more than 99% of the pollutants from raw sewage and can produce an effluent of almost potable (drinking) water quality. Obviously, however, advanced treatment is not cost free. The cost of advanced treatment—costs of operation and maintenance, as well as retrofit of existing conventional processes—is very high (sometimes doubling the cost of secondary treatment). A plan to install advanced treatment technology calls for careful study—the benefit-to-cost ratio is not always significant enough to justify the additional expense.

Even considering the expense, application of some form of advanced treatment is not uncommon. These treatment processes can be physical, chemical, or biological. The specific process used is based on the purpose of the treatment and the quality of the effluent desired.

MICROSCREENING

Microscreening (also called *microstraining*) is an advanced treatment process used to reduce suspended solids. The microscreens are composed of specially woven steel wire fabric mounted around the perimeter of a large revolving drum. The steel wire cloth acts as a fine screen, with openings as small as 20 μm—small enough to remove microscopic organisms and debris. The rotating drum is partially submerged in the secondary effluent, which must flow into the drum then outward through the microscreen. As the drum rotates, captured solids are carried to the top, where a high-velocity water spray flushes them into a hopper or backwash tray mounted on the hollow axle of the drum. Backwash solids are recycled to plant influent for treatment. These units have found greatest application in treatment of industrial waters and final polishing filtration of wastewater effluents. Expected performance for suspended solids removal is 95 to 99%, but the typical suspended solids removal achieved with these units is about 55%. The normal range is from 10 to 80%.

The functional design of the microscreen unit involves the following considerations: (1) characterization of the suspended solids with respect to the concentration and degree of flocculation, (2) selection of unit design parameter values that will not only ensure capacity to meet maximum hydraulic loadings with critical solids characteristics but also provide desired design performance over the expected range of hydraulic and solids loadings, and (3) provision of backwash and cleaning facilities to maintain the capacity of the screen (Metcalf & Eddy, 2003).

FILTRATION

The purpose of filtration processes used in advanced treatment is to remove suspended solids. The specific operations associated with a filtration system are dependent on the equipment used. A general description of the process follows. Wastewater flows to a filter (gravity or pressurized). The filter contains single media, dual media, or multimedia. Wastewater flows through the media, which remove solids. The solids remain in the filter. Backwashing the filter as needed removes trapped solids. Backwashed solids are returned to the plant for treatment. Processes typically remove 95 to 99% of the suspended matter.

MEMBRANE FILTRATION

Earlier we briefly discussed membrane filtration. In this section, we present a more in-depth discussion of membrane filtration because the technologies most commonly used for performing secondary treatment of municipal wastewater rely on microorganisms suspended in the wastewater to treat it. Although these technologies work well in many situations, they have several drawbacks, including the difficulty of growing the right types of microorganisms and the physical requirement of a large site. The use of microfiltration membrane bioreactors (MBRs), a technology that has become increasingly used in the past 10 years, overcomes many of the limitations of conventional systems. These systems have the advantage of combining a suspended-growth biological reactor with solids removal via filtration. The membranes can be designed for and operated in small spaces and with high removal efficiency of contaminants such as nitrogen, phosphorus, bacteria, biochemical oxygen demand, and total suspended solids. The membrane filtration system in effect can replace the secondary clarifier and sand filters in a typical activated sludge treatment system. Membrane filtration allows a higher biomass concentration to be maintained, thereby allowing smaller bioreactors to be used (USEPA, 2007a).

Membrane filtration involves the flow of water-containing pollutants across a membrane. Water permeates through the membrane into a separate channel for recovery. Because of the cross-flow movement of water and the waste constituents, materials left behind do not accumulate at the membrane surface but are carried out of the system for later recovery or disposal. The water passing through the membrane is called the *permeate*, and the water with the more concentrated materials is called the *concentrate* or *retentate*.

Membranes are constructed of cellulose or other polymer material, with a maximum pore size set during the manufacturing process. The requirement is that the membranes prevent passage of particles the size of microorganisms, or about 1 μm (0.001 mm), so that they remain in the system. This means that MBR systems are good for removing solid material, but the removal of dissolved wastewater components must be facilitated by using additional treatment steps. Membranes can be configured in a number of ways. For MBR applications, the two configurations most often used are hollow fibers grouped in bundles or as flat plates. The hollow-fiber bundles are connected by manifolds in units that are designed for easy changing and servicing.

Designers of MBR systems require only basic information about the wastewater characteristics (e.g., influent characteristics, effluent requirements, flow data) to design an MBR system. Depending on effluent requirements, certain supplementary options can be included with the MBR system. For example, chemical addition (at various places in the treatment chain, including before the primary settling tank, before the secondary settling tank [clarifier], and before the MBR or final filters) for phosphorus removal can be included in an MBR system if needed to achieve low phosphorus concentrations in the effluent.

Membrane bioreactor systems historically have been used for small-scale treatment applications when portions of the treatment system were shut down and the wastewater was routed around (or bypassed) during maintenance periods. However, MBR systems are now often used in full-treatment applications. In these instances, it is recommended that the installation include one additional membrane tank/ unit beyond what the design would nominally call for. This "N plus 1" concept is a blend between conventional activated sludge and membrane process design. It is especially important to consider both operation and maintenance requirements when selecting the number of units for MBRs. The inclusion of an extra unit gives operators flexibility and ensures that sufficient operating capacity will be available (Wallis-Lage et al., 2006). For example, bioreactor sizing is often limited by oxygen transfer, rather than the volume required to achieve the required sludge retention time, a factor that significantly affects bioreactor numbers and sizing (Crawford et al., 2000).

Although MBR systems provide operational flexibility with respect to flow rates, as well as the ability to readily add or subtract units as conditions dictate, that flexibility has limits. Membranes typically require that the water surface be maintained above a minimum elevation so that the membranes remain wet during operation. Throughput limitations are dictated by the physical properties of the membrane, and the result is that peak design flows should be no more than 1.5 to 2 times the average design flow. If peak flows exceed that limit, either additional membranes are needed simply to process the peak flow or equalization should be included in the overall design. The equalization is done by including a separate basin (external equalization) or by maintaining water in the aeration and membrane tanks at depths higher than those required and then removing that water to accommodate high flows when necessary (internal equalization) (USEPA, 2007a).

BIOLOGICAL NITRIFICATION

Biological nitrification is the first basic step in the process of *biological nitrification–denitrification*. In nitrification, the secondary effluent is introduced into another aeration tank, trickling filter, or biodisc. Because most of the carbonaceous BOD has already been removed, the microorganisms that drive in this advanced step are the nitrifying bacteria *Nitrosomonas* and *Nitrobacter*. In nitrification, the ammonia nitrogen is converted to nitrate nitrogen, producing a *nitrified effluent*. At this point, the nitrogen has not actually been removed, only converted to a form that is not toxic to aquatic life and that does not cause an additional oxygen demand. The nitrification process can be limited (performance affected) by alkalinity (requires 7.3 parts alkalinity to 1.0 part ammonia nitrogen), pH, dissolved oxygen availability, toxicity (ammonia or other toxic materials), and process mean cell residence time (sludge retention time). As a general rule, biological nitrification is more effective and achieves higher levels of removal during the warmer times of the year.

BIOLOGICAL DENITRIFICATION

Biological denitrification removes nitrogen from the wastewater. When bacteria come in contact with a nitrified element in the absence of oxygen, they reduce the nitrates to nitrogen gas, which escapes the wastewater. The denitrification process can be carried out in either an anoxic activated sludge system (suspended-growth) or a column system (fixed-growth). The denitrification process can remove up to 85% or more of nitrogen. After effective biological treatment, little oxygen demanding material is left in the wastewater when it reaches the denitrification process. The denitrification reaction will only occur if an oxygen demand source exists when no dissolved oxygen is present in the wastewater. An oxygen demand source is usually added to reduce the nitrates quickly. The most common demand source added is soluble BOD or methanol. Approximately 3 mg/L of methanol are added for every 1 mg/L of nitrate-nitrogen. Suspended-growth denitrification reactors are mixed mechanically but only enough to keep the biomass from settling without adding unwanted oxygen. Submerged filters of different types of media may also be used to provide denitrification. A fine media downflow filter is sometimes used to provide both denitrification and effluent filtration. A fluidized sand bed—where wastewater flows upward through a media of sand or activated carbon at a rate to fluidize the bed—may also be utilized. Denitrification bacteria grow on the media.

CARBON ADSORPTION

The main purpose of carbon adsorption used in advanced treatment processes is the removal of refractory organic compounds (non-BOD_5) and soluble organic materials that are difficult to eliminate by biological or physicochemical treatment. In the carbon adsorption process, wastewater passes through a container filled either with carbon powder or carbon slurry. Organics adsorb onto the carbon (i.e., organic molecules are attracted to the activated carbon surface and are held there) with sufficient contact time. A carbon system usually has several columns or basins used as contactors. Most contact chambers are either open concrete gravity-type systems or steel pressure containers applicable to upflow or downflow operation. With use, carbon loses its adsorptive capacity. The carbon must then be regenerated or replaced with fresh carbon. As head loss develops in carbon contactors, they are backwashed with clean effluent in much the same way the effluent filters are backwashed. Carbon used for adsorption may be in a granular form or in a powdered form.

Note: Powdered carbon is too fine for use in columns and is usually added to the wastewater, then later removed by coagulation and settling.

LAND APPLICATION

The application of secondary effluent onto a land surface can provide an effective alternative to the expensive and complicated advanced treatment methods discussed previously and the biological nutrient removal (BNR) system discussed later. A high-quality polished effluent (i.e., effluent with high levels of TSS, BOD, phosphorus, and nitrogen compounds as well as reduced refractory organics) can be obtained by the natural processes that occur as the effluent flows over the vegetated ground surface and percolates through the soil. Limitations are involved with land application of wastewater effluent. For example, the process needs large land areas. Soil type and climate are also critical factors in controlling the design and feasibility of a land treatment process.

Type and Modes of Land Application

Three basic types of land application or treatment are commonly used: *irrigation* (slow rate), *overland flow*, and *infiltration–percolation* (rapid rate). The basic objectives of these types of land applications and the conditions under which they can function vary. In irrigation (also called *slow rate*), wastewater is sprayed or applied (usually by ridge-and-furrow surface spreading or by sprinkler systems) to the surface of the land. Wastewater enters the soil. Crops growing on the irrigation area utilize available nutrients. Soil organisms stabilize the organic content of the flow. Water returns to the hydrologic cycle through evaporation or by entering the surface water or groundwater.

The irrigation land application method provides the best results (compared with the other two types of land application systems) with respect to advanced treatment levels of pollutant removal. Not only are suspended solids and BOD significantly reduced by filtration of the wastewater, but also biological oxidation of the organics occurs in the top few inches of soil. Nitrogen is removed primarily by crop uptake, and phosphorus is removed by adsorption within the soil. Expected performance levels for irrigation include the following:

- BOD_5, 98%
- Suspended solids, 98%
- Nitrogen, 85%
- Phosphorus, 95%
- Metals, 95%

The overland flow application method utilizes physical, chemical, and biological processes as the wastewater flows in a thin film down the relatively impermeable surface. In the process, wastewater sprayed over sloped terraces flows slowly over the surface. Soil and vegetation remove suspended solids, nutrients, and organics. A small portion of the wastewater evaporates. The remainder flows to collection channels. Collected effluent is discharged to surface waters. Expected performance levels for overland flow include the following:

- BOD_5, 92%
- Suspended solids, 92%
- Nitrogen, 70–90%

- Phosphorus, 40–80%
- Metals, 50%

In the infiltration–percolation application method, wastewater is sprayed or pumped to spreading basins (also known as recharge basins or large ponds). Some wastewater evaporates. The remainder percolates or infiltrates into the soil. Solids are removed by filtration. Water recharges the groundwater system. Most of the effluent percolates to the groundwater; very little of it is absorbed by vegetation. The filtering and adsorption action of the soil removes most of the BOD, TSS, and phosphorus from the effluent; however, nitrogen removal is relatively poor. Expected performance levels for infiltration–percolation include the following:

- BOD_5, 85–99%
- Suspended solids, 98%
- Nitrogen, 0–50%
- Phosphorus, 60–95%
- Metals, 50–95%

BIOLOGICAL NUTRIENT REMOVAL

Nitrogen and phosphorus are the primary causes of cultural eutrophication (i.e., nutrient enrichment due to human activities) in surface waters. The most recognizable manifestations of this eutrophication are algal blooms that occur during the summer. Chronic symptoms of over-enrichment include low dissolved oxygen, fish kills, murky water, and depletion of desirable flora and fauna. In addition, the increase in algae and turbidity increases the need to chlorinate drinking water, which, in turn, leads to higher levels of disinfection byproducts that have been shown to increase the risk of cancer (USEPA, 2007c). Excessive amounts of nutrients can also stimulate the activity of microbes, such as *Pfisteria*, which may be harmful to human health (USEPA, 2001).

Approximately 25% of all water body impairments are due to nutrient-related causes (e.g., nutrients, oxygen depletion, algal growth, ammonia, harmful algal blooms, biological integrity, turbidity) (USEPA, 2007d). In efforts to reduce the number of nutrient impairments, many point-source dischargers have received more stringent effluent limits for nitrogen and phosphorus. To achieve these new, lower effluent limits, facilities have begun to look beyond traditional treatment technologies.

Recent experience has reinforced the concept that biological nutrient removal (BNR) systems are reliable and effective in removing nitrogen and phosphorus. The process is based on the principle that, under specific conditions, microorganisms will remove more phosphorus and nitrogen than is required for biological activity; thus, treatment can be accomplished without the use of chemicals. Not having to use and therefore having to purchase chemicals to remove nitrogen and phosphorus potentially has numerous cost–benefit implications. In addition, because chemicals are not required to be used, chemical waste products are not produced, reducing the need to handle and dispose of waste. Several patented processes are available for this purpose. Performance depends on the biological activity and the process employed.

Biological nutrient removal removes total nitrogen (TN) and total phosphorus (TP) from wastewater through the use of microorganisms under different environmental conditions in the treatment process (Metcalf & Eddy, 2003).

Enhanced Biological Nutrient Removal

Removing phosphorus from wastewater in secondary treatment processes has evolved into innovative *enhanced biological nutrient removal* (EBNR) technologies. An EBNR treatment process promotes the production of phosphorus-accumulating organisms, which utilize more phosphorus in their metabolic processes than a conventional secondary biological treatment process (USEPA, 2007b). The average total phosphorus concentrations in raw domestic wastewater are usually between 6 and 8 mg/L, and the total phosphorus concentration in municipal wastewater after conventional secondary treatment is routinely reduced to 3 or 4 mg/L. EBNR incorporated into the secondary treatment system can often reduce total phosphorus concentrations to 0.3 mg/L and less.

Facilities using EBNR have significantly reduced the amount of phosphorus to be removed through the subsequent chemical addition and tertiary filtration process. This improved the efficiency of the tertiary process and significantly reduced the costs of chemicals used to remove phosphorus. Facilities using EBNR reported that their chemical dosing was cut in half after EBNR was installed to remove phosphorus (USEPA, 2007b). Treatment provided by these EBNR processes also reduces other pollutants that commonly affect water quality to very low levels (USEPA, 2007b). Biochemical oxygen demand (BOD) and total suspended solids (TSS) are routinely less than 2 mg/L and fecal coliform bacteria less than 10 cfu/100 mL. Turbidity of the final effluent is very low, which allows for effective disinfection using ultraviolet light, rather than chlorination. Recent studies report finding that wastewater treatment facilities using EBNR also significantly reduced the amount of pharmaceuticals and healthcare products from municipal wastewater, as compared to removal accomplished by conventional secondary treatment.

SOLIDS (SLUDGE/BIOSOLIDS) HANDLING

The wastewater treatment unit processes described to this point remove solids and BOD from the waste stream before the liquid effluent is discharged to its receiving waters. What remains to be disposed of is a mixture of solids and wastes, called *process residuals*, more commonly referred to as *sludge* or *biosolids*.

> **Note:** *Sludge* is the commonly accepted name for wastewater solids; however, if wastewater sludge is used for beneficial reuse (e.g., as a soil amendment or fertilizer), it is commonly referred to as *biosolids*.

Because sludge can be as much as 97% water content and because the cost of disposal will be related to the volume of sludge being processed, one of the primary purposes or goals of sludge treatment (along with stabilizing it so it is no longer objectionable or environmentally damaging) is to separate as much of the water from the solids as possible. Sludge treatment methods may be designed to accomplish both of these purposes.

Note: Sludge treatment methods are generally divided into three major categories: *thickening, stabilization,* and *dewatering.* Many of these processes include complex sludge treatment methods such as heat treatment, vacuum filtration, and incineration.

BACKGROUND INFORMATION ON SLUDGE/BIOSOLIDS

When we speak of *sludge* or *biosolids,* we are speaking of the same substance or material; each is defined as the suspended solids removed from wastewater during sedimentation and then concentrated for further treatment and disposal or reuse. The difference between the terms *sludge* and *biosolids* is determined by the way they are managed.

Note: The task of disposing of, treating, or reusing wastewater solids is *sludge* or *biosolids management.*

Sludge is typically seen as wastewater solids that are disposed of. Biosolids are the same substance but managed for reuse, commonly called *beneficial reuse* (e.g., for land application as a soil amendment, such as biosolids compost). Note that even as wastewater treatment standards have become more stringent because of increasing environmental regulations, so has the volume of wastewater sludge increased. Note also that, before sludge can be disposed of or reused, it requires some form of treatment to reduce its volume, to stabilize it, and to inactivate pathogenic organisms.

Sludge initially forms as a 3 to 7% suspension of solids; with each person typically generating about 4 gal of sludge per week, the total quantity generated each day, week, month, and year is significant. Because of the volume and nature of the material, sludge management is a major factor in the design and operation of all water pollution control plants.

Note: Wastewater solids account for more than half of the total costs in a typical secondary treatment plant.

Sources of Sludge

Wastewater sludge is generated in primary, secondary, and chemical treatment processes. In primary treatment, the solids that float or settle are removed. The floatable material makes up a portion of the solid waste known as *scum.* Scum is not normally considered sludge; however, it should be disposed of in an environmentally sound way. The settleable material that collects on the bottom of the clarifier is known as *primary sludge.* Primary sludge can also be referred to as *raw sludge* because it has not undergone decomposition. Raw primary sludge from a typical domestic facility is quite objectionable and has a high percentage of water, two characteristics that make handling difficult.

Solids not removed in the primary clarifier are carried out of the primary unit. These solids are known as *colloidal suspended solids.* The secondary treatment system (e.g., trickling filter, activated sludge) is designed to change those colloidal solids into settleable solids that can be removed. Once in the settleable form, these solids are removed in the secondary clarifier. The sludge at the bottom of the secondary clarifier is called *secondary sludge.* Secondary sludges are light and fluffy and more difficult to process than primary sludges—in short, secondary sludges do not dewater well.

The addition of chemicals and various organic and inorganic substances prior to sedimentation and clarification may increase the solids capture and reduce the amount of solids lost in the effluent. This *chemical addition* results in the formation of heavier solids, which trap the colloidal solids or convert dissolved solids to settleable solids. The resultant solids are known as *chemical sludges*. As chemical usage increases, so does the quantity of sludge that must be handled and disposed of. Chemical sludges can be very difficult to process; they do not dewater well and contain lower percentages of solids.

Sludge Characteristics

The composition and characteristics of sewage sludge vary widely and can change considerably with time. Notwithstanding these facts, the basic components of wastewater sludge remain the same. The only variations occur in the quantity of the various components as the type of sludge and the process from which it originated changes. The main component of all sludges is *water.* Prior to treatment, most sludge contains 95 to 99% water. This high water content makes sludge handling and processing extremely costly in terms of both money and time. Sludge handling may represent up to 40% of the capital costs and 50% of the operating costs of a treatment plant. As a result, the importance of optimum design for handling and disposal of sludge cannot be overemphasized. The water content of the sludge is present in a number of different forms. Some forms can be removed by several sludge treatment processes, thus allowing the same flexibility in choosing the optimum sludge treatment and disposal method. The forms of water associated with sludges include the following:

- *Free water*—Water that is not attached to sludge solids in any way and can be removed by simple gravitational settling.
- *Floc water*—Water that is trapped within the floc and travels with them; it can be removed by mechanical dewatering.
- *Capillary water*—Water that adheres to the individual particles and can be squeezed out of shape and compacted.
- *Particle water*—Water that is chemically bound to the individual particles and cannot be removed without inclination.

From a public health view, the second and probably more important component of sludge is the *solids matter.* Representing from 1 to 8% of the total mixture, these solids are extremely unstable. Wastewater solids can be classified into two categories based on their origin—organic and inorganic. *Organic solids* in wastewater, simply put, are materials that were at one time alive and will burn or volatilize at 550°C after 15 min in a muffle furnace. The percent organic material within a sludge will determine how unstable it is. The inorganic material within sludge will determine how stable it is. The *inorganic solids* are those solids that were never alive and will not burn or volatilize at 550°C after 15 min in a muffle furnace. Inorganic solids are generally not subject to breakdown by biological action and are considered stable. Certain inorganic solids, however, can create problems when related to the environment—for example, heavy metals such as copper, lead, zinc, mercury, and others. These can be extremely harmful if discharged.

DID YOU KNOW?

Subpart D (pathogen and vector attraction reduction) requirements of the 40 CFR Part 503 regulation apply to sewage sludge (both bulk sewage sludge and sewage sludge that is sold or given away in a bag or other container for application to the land) and domestic septage applied to the land or placed on a surface disposal site. The regulated community includes persons who generate or prepare sewage sludge for application to the land, as well as those who apply it to the land. Included is anyone who

- Generates treated sewage sludge (biosolids) that is land applied or placed on a surface disposal site
- Derives a material from treated sewage sludge (biosolids)
- Applies biosolids to the land
- Owns or operates a surface disposal site

Organic solids may be subject to biological decomposition in either an aerobic or anaerobic environment. Decomposition of organic matter (with its production of objectionable byproducts) and the possibility of toxic organic solids within the sludge compound the problems of sludge disposal. The pathogens in domestic sewage are primarily associated with insoluble solids. Primary wastewater treatment processes concentrate these solids into sewage sludge, so untreated or raw primary sewage sludges have higher quantities of pathogens than the incoming wastewater. Biological wastewater treatment processes such as lagoons, trickling filters, and activated sludge treatment may substantially reduce the number of pathogens in wastewater (USEPA, 1989). These processes may also reduce the number of pathogens in sewage sludge by creating adverse conditions for pathogen survival. Nevertheless, the resulting biological sewage sludges may still contain sufficient levels of pathogens to pose a public health and environmental concern. Moreover, insects, birds, rodents, and domestic animals may transport sewage sludge and pathogens from sewage sludge to humans and to animals. Vectors are attracted to sewage sludge as a food source, and reducing the attraction of vectors to sewage sludge to prevent the spread of pathogens is a focus of current regulations. Sludge-borne pathogens and vector attraction are discussed in the following section.

Sludge Pathogens and Vector Attraction

A pathogen is an organism capable of causing disease. Pathogens infect humans through several different pathways including ingestion, inhalation, and dermal contact. The infective dose, or the number of pathogenic organisms to which a human must be exposed to become infected, varies depending on the organism and on the health status of the exposed individual. Pathogens that propagate in the enteric or urinary systems of humans and are discharged in feces or urine pose the greatest risk to public health with regard to the use and disposal of sewage sludge. Pathogens are also found in the urinary and enteric systems of other animals and may propagate in non-enteric settings.

The three major types of human pathogenic (disease-causing) organisms—bacteria, viruses, and protozoa—all may be present in domestic sewage. The actual species and quantity of pathogens present in the domestic sewage from a particular municipality (and the sewage sludge produced when treating the domestic sewage) depend on the health status of the local community and may vary substantially at different times. The level of pathogens present in treated sewage sludge (biosolids) also depends on the reductions achieved by the wastewater and sewage sludge treatment processes.

If improperly treated sewage sludge is illegally applied to land or placed on a surface disposal site, humans and animals could be exposed to pathogens directly by coming into contact with the sewage sludge or indirectly by consuming drinking water or food contaminated by sewage sludge pathogens. Insects, birds, rodents, and even farm workers could contribute to these exposure routes by transporting sewage sludge and sewage sludge pathogens away from the site. Potential routes of exposure include the following:

Direct contact
- Touching the sewage sludge
- Walking through an area (e.g., field, forest, reclamation area) shortly after sewage sludge application
- Handling soil from fields where sewage sludge has been applied
- Inhaling microbes that become airborne (via aerosols, dust, etc.) during sewage sludge spreading or by strong winds, plowing, or cultivating the soils after application

Indirect contact
- Consumption of pathogen-contaminated crops grown on sewage sludge-amended soil or of other food products that have been contaminated by contact with these crops or field workers, etc.
- Consumption of pathogen-contaminated milk or other food products from animal contaminated by grazing in pastures or fed crops grown on sewage sludge-amended fields
- Ingestion of drinking water or recreational waters contaminated by runoff from nearby land application sites or by organisms from sewage sludge migrating into groundwater aquifers
- Consumption of inadequately cooked or uncooked pathogen-contaminated fish from water contaminated by runoff from a nearby sewage sludge application site
- Contact with sewage sludge or pathogens transported away from the land application or surface disposal site by rodents, insects, or other vectors, including grazing animals or pets

One of the lesser impacts to public health can be from inhalation of airborne pathogens. Pathogens may become airborne via the spray of liquid biosolids from a splash plate or high-pressure hose, or in fine particulate dissemination as dewatered biosolids are applied or incorporated. While high-pressure spray applications may result in some aerosolization of pathogens, this type of equipment is generally used on large, remote sites such as forests, where the impact on the public is minimal. Fine particulates created by the application of dewatered biosolids or the incorporation of biosolids into soil may cause very localized fine particulate/dusty conditions, but particles in dewatered biosolids are too large to travel far, and the fine particulates do not spread beyond the immediate area. The activity of applying and incorporating biosolids may create dusty conditions. However, the biosolids are moist materials and do not add to the dusty condition, and by the time biosolids have dried sufficiently to create fine particulates, the pathogens have been reduced (Yeager and Ward, 1981). With regard to vector attraction reduction, it can be accomplished in two ways: by treating the sewage sludge to the point at which vectors will no longer be attracted to the sewage sludge and by placing a barrier between the sewage sludge and vectors.

> *Note:* Before moving on to a discussion of the fundamentals of sludge treatment methods, it is necessary to begin by covering sludge pumping calculations. It is important to point out that it is difficult (if not impossible) to treat the sludge unless it is pumped to the specific sludge treatment process.

SLUDGE THICKENING

The solids content of primary, activated, trickling-filter, or even mixed sludge (i.e., primary plus activated sludge) varies considerably, depending on the characteristics of the sludge. Note that the sludge removal and pumping facilities and the method of operation also affect the solids content. *Sludge thickening* (or *concentration*) is a unit process used to increase the solids content of the sludge by removing a portion of the liquid fraction. By increasing the solids content, more economical treatment of the sludge can be achieved. Sludge thickening processes include

- Gravity thickeners
- Flotation thickeners
- Solids concentrators

Gravity Thickening

Gravity thickening is most effective on primary sludge. In operation, solids are withdrawn from primary treatment (and sometimes secondary treatment) and pumped to the thickener. The solids buildup in the thickener forms a solids blanket on the bottom. The weight of the blanket compresses the solids on the bottom and squeezes the water out. By adjusting the blanket thickness, the percent solids in the underflow (solids withdrawn from the bottom of the thickener) can be increased or decreased. The supernatant (clear water) that rises to the surface is returned to the wastewater flow for treatment. Daily operations of the thickening process include pumping, observation, sampling and testing, process control calculations, maintenance, and housekeeping.

> *Note:* The equipment employed in thickening depends on the specific thickening processes used.

Equipment used for gravity thickening consists of a *thickening tank* that is similar in design to the settling tank used in primary treatment. Generally, the tank is circular and provides equipment for continuous solids collection. The collector mechanism uses heavier construction than that in a settling tank because the solids being moved are more concentrated. The gravity thickener pumping facilities (i.e., pump and flow measurement) are used for withdrawal of thickened solids.

Solids concentrations achieved by gravity thickeners are typically 8 to 10% solids from primary underflow, 2 to 4% solids from waste activated sludge, 7 to 9% solids from trickling filter residuals, and 4 to 9% from combined primary and secondary residuals. The performance of gravity thickening processes depends on

- Type of sludge
- Condition of influent sludge
- Temperature
- Blanket depth
- Solids loading
- Hydraulic loading
- Solids retention time
- Hydraulic detention time

Flotation Thickening

Flotation thickening is used most efficiently for waste sludges from suspended-growth biological treatment process, such as the activated sludge process. In operation, recycled water from the flotation thickener is aerated under pressure. During this time, the water absorbs more air than it would under normal pressure. The recycled flow together with chemical additives (if used) are mixed with the flow. When the mixture enters the flotation thickener, the excess air is released in the form of fine bubbles. These bubbles become attached to the solids and lift them toward the surface. The accumulation of solids on the surface is called the *float cake*. As more solids are added to the bottom of the float cake, it becomes thicker and water drains from the upper levels of the cake. The solids are then moved up an inclined plane by a scraper and discharged. The supernatant leaves the tank below the surface of the float solids and is recycled or returned to the waste stream for treatment. Typically, flotation thickener performance is 3 to 5% solids for waste activated sludge with polymer addition and 2 to 4% solids without polymer addition.

The flotation thickening process requires pressurized air, a vessel for mixing the air with all or part of the process residual flow, a tank in which the flotation process can occur, and solids collector mechanisms to remove the float cake (solids) from the top of the tank and accumulated heavy solids from the bottom of the tank. Because the process normally requires chemicals to be added to improve separation, chemical mixing equipment, storage tanks, and metering equipment to dispense the chemicals at the desired dose are required. The performance of a dissolved air-thickening process depends on various factors:

- Bubble size
- Solids loading
- Sludge characteristics
- Chemical selection
- Chemical dose

Solids Concentrators

Solids concentrators (belt thickeners) usually consist of a mixing tank, chemical storage and metering equipment, and a moving porous belt. In operation, the process residual flow is chemically treated and then spread evenly over the surface of the moving porous belt. As the flow is carried down the belt (similar to a conveyor belt), the solids are mechanically turned or agitated and water drains through the belt. This process is primarily used in facilities where space is limited.

SLUDGE STABILIZATION

The purpose of sludge stabilization is to reduce volume, stabilize the organic matter, and eliminate pathogenic organisms to permit reuse or disposal. The equipment required for stabilization depends on the specific process used. Sludge stabilization processes include the following:

- Aerobic digestion
- Anaerobic digestion
- Composting
- Lime stabilization
- Wet air oxidation (heat treatment)
- Chemical oxidation (chlorine oxidation)
- Incineration

Aerobic Digestion

Equipment used for aerobic digestion includes an aeration tank (digester), which is similar in design to the aeration tank used for the activated sludge process. Either diffused or mechanical aeration equipment is necessary to maintain the aerobic conditions in the tank. Solids and supernatant removal equipment is also required. In operation, process residuals (sludge) are added to the digester and aerated to maintain a dissolved oxygen (DO) concentration of 1 mg/L. Aeration also ensures that the tank contents are well mixed. Generally, aeration continues for approximately 20 days of retention time. Periodically, aeration is stopped and the solids are allowed to settle. Sludge and the clear liquid supernatant are withdrawn as needed to provide more room in the digester. When no additional volume is available, mixing is stopped for 12 to 24 hours before solids are withdrawn for disposal. Process control testing should include alkalinity, pH, percent solids, percent volatile solids for influent sludge, supernatant, digested sludge, and digester contents. A typical operational problem associated with an aerobic digester is pH control. When pH drops, for example, this may indicate normal biological activity or low influent alkalinity. This problem is corrected by adding alkalinity (e.g., lime, bicarbonate).

Anaerobic Digestion

Anaerobic digestion, the traditional method of sludge stabilization, uses bacteria that thrive in the absence of oxygen. It is slower than aerobic digestion but has the advantage that only a small percentage of the wastes are converted into new bacterial cells. Instead, most of the organics are converted into carbon dioxide and methane gas.

Note: In an anaerobic digester, the entrance of air should be prevented because of the potential for an explosive mixture resulting from air mixing with gas produced in the digester.

Equipment used in anaerobic digestion includes a sealed digestion tank with either a fixed or a floating cover, heating and mixing equipment, gas storage tanks, solids and supernatant withdrawal equipment, and safety equipment (e.g., vacuum relief, pressure relief, flame traps, explosion proof electrical equipment).

In operation, process residual (thickened or unthickened sludge) is pumped into the sealed digester. The organic matter digests anaerobically by a two-stage process. Sugars, starches, and carbohydrates are converted to volatile acids, carbon dioxide, and hydrogen sulfide. The volatile acids are then converted to methane gas. This operation can occur in a single tank (single stage) or in two tanks (two stages). In a single-stage system, supernatant and digested solids must be removed whenever flow is added. In a two-stage operation, solids and liquids from the first stage flow into the second stage each time fresh solids are added. Supernatant is withdrawn from the second stage to provide additional treatment space. Periodically, solids are withdrawn for dewatering or disposal. The methane gas produced in the process may be used for many plant activities.

Note: The primary purpose of a secondary digester is to allow for solids separation.

Various performance factors affect the operation of the anaerobic digester; for example, the percent volatile matter in raw sludge, digester temperature, mixing, volatile acids/alkalinity ratio, feed rate, percent solids in raw sludge, and pH are all important operational parameters that the operator must monitor.

OTHER SLUDGE STABILIZATION PROCESSES

In addition to aerobic and anaerobic digestion, other sludge stabilization processes include composting, lime stabilization, wet air oxidation, and chemical (chlorine) oxidation. These other stabilization processes are briefly described in this section.

Composting

The purpose of composting sludge is to stabilize the organic matter, reduce volume, and eliminate pathogenic organisms. In a composting operation, dewatered solids are usually mixed with a bulking agent (e.g., hardwood chips) and stored until biological stabilization occurs. The composting mixture is ventilated during storage to provide sufficient oxygen for oxidation and to prevent odors. After the solids are stabilized, they are separated from the bulking agent. The composted solids are then stored for curing and applied to farmlands or other beneficial uses. Expected performance of the composting operation for both percent volatile matter reduction and percent moisture reduction ranges from 40 to 60%.

Definitions of Key Terms

Aerated static pile—Composting system using controlled aeration from a series of perforated pipes running underneath each pile and connected to a pump that draws or blows air through the piles.

Aeration (for composting)—Bringing about contact of air with composted solid organic matter by means of turning or ventilating to allow microbial aerobic metabolism (bio-oxidation).

Aerobic—Composting environment characterized by bacteria active in the presence of oxygen (aerobes); generates more heat and is a faster process than anaerobic composting.

Anaerobic—Composting environment characterized by bacteria active in the absence of oxygen (anaerobes).

Bagged biosolids—Biosolids that are sold or given away in a bag or other container (i.e., either an open or a closed vessel containing 1 metric ton or less of biosolids).

Bioaerosols—Biological aerosols that can pose potential health risks during the composting and handling of organic materials. Bioaerosols are suspensions of particles in the air consisting partially or wholly of microorganisms. The bioaerosols of concern during composing include actinomycetes, bacteria, viruses, molds, and fungi.

Biosolids composting—Aerobic biological degradation or bacterial conversion of dewatered biosolids, which works to produce compost that can be used as a soil amendment or conditioner.

Biosolids quality parameters—The USEPA determined that three main parameters of concern should be used in gauging biosolids quality: (1) the relevant presence or absence of pathogenic organisms, (2) pollutants, and (3) the degree of attractiveness of the biosolids to vectors. There can be a number of possible biosolids qualities. In order to express or describe those biosolids meeting the highest quality for all three of these biosolids quality parameters the term *exceptional quality* (EQ) has come into common use.

Bulk biosolids—Biosolids that are not sold or given away in a bag or other container for application to the land.

Bulking agents—Materials, usually carbonaceous such as sawdust or woodchips, added to a compost system to maintain airflow by preventing settlement and compaction of the compost.

Carbon-to-nitrogen ratio (C:N ratio)—Ratio representing the quantity of carbon (C) in relation to the quantity of nitrogen (N) in a soil or organic material; determines the composting potential of a material and serves to indicate product quality.

Compost—The end product (innocuous humus) remaining after the composting process is completed.

Curing—Late stage of composting, after much of the readily metabolized material has been decomposed, which provides additional stabilization and allows further decomposition of cellulose and lignin (found in woody-like substances).

Curing air—Curing piles are aerated primarily for moisture removal to meet final product moisture requirements and to keep odors from building up in the compost pile as biological activity is dissipating. Final product moisture requirements and summer ambient conditions are used to determine air requirements for moisture removal for the curing process.

Endotoxins—A toxin produced within a microorganism and released upon destruction of the cell in which it is produced. Endotoxins can be carried by airborne dust particles at composting facilities.

Exceptional quality (EQ) sludge (biosolids)—Although this term is not used in 40 CFR Part 503, it has become a shorthand term for biosolids that meet the pollutant concentrations in Table 3 of Part 503.13(b)(3); one of the six Class A pathogen reduction alternatives in 503.32(a); and one of the vector attraction reduction options in 503.33(b)(1)–(8) (Spellman, 1997).

Feedstock—Decomposable organic material used for the manufacture of compost.

Heat removal and temperature control—The biological oxidation process for composting biosolids is an exothermic reaction. The heat given off by the composting process can raise the temperature of the compost pile high enough to destroy the organisms responsible for biodegradation; therefore, the compost pile cells are aerated to control the temperature of the compost process by removing excess heat to maintain optimum temperature for organic solids degradation and pathogen reduction. Optimum temperatures are typically between 50 and 60°C (122 and 140°F). Using summer ambient air conditions, aeration requirements for heat removal can be calculated.

Metric ton—One metric ton, or 1000 kg, equals about 2205 lb, which is larger than the short ton (2000 lb) usually referred to in the British system of units. The metric ton unit is used throughout this text.

Moisture removal—When temperature increases, the quantity of moisture in saturated air increases. Air is required for the composting process to remove water that is present in the mix and produced by the oxidation of organic solids. The quantity of air required for moisture removal is calculated based on the desired moisture content for the compost product and the psychometric properties of the ambient air supply. Air requirements for moisture removal are calculated from summer ambient air conditions and required final compost characteristics.

Oxidation air—The composting process requires oxygen to support aerobic biological oxidation of degradable organics in the biosolids and wood chips. Stoichiometric requirements for oxygen are related to the extent of organic solids degradation expected during the composting cycle time.

Pathogenic organisms—Specifically, *Salmonella* and *Escherichia coli* bacteria, enteric viruses, or visible helminth ova.

Peaking air—The rate of organic oxidation and, therefore, heat release can vary greatly for the composting process. If sufficient aeration capacity is not provided to meet peak requirements for heat or moisture removal, temperature limits for the process may be exceeded. Peaking air rates are typically 1.9 times the average aeration rate for heat removal.

Pollutant—An organic substance, an inorganic substance, a combination of organic and inorganic substances, or a pathogenic organism that, after discharge and upon exposure, ingestion, inhalation, or assimilation into an organism either directly from the environment or indirectly by ingestion through the food chain, could, on the basis of information available to the

USEPA, cause death, disease, behavioral abnormalities, cancer, genetic mutations, physiological malfunctions, or physical deformations in either organisms or offspring of the organisms.

Stability—State or condition in which the composted material can be stored without giving rise to nuisances or can be applied to the soil without causing problems there; the desired degree of stability for finished compost is one in which the readily decomposed compounds are broken down and only the decomposition of the more resistant biologically decomposable compounds remains to be accomplished.

USEPA's 503 regulation—In order to ensure that sewage sludge (biosolids) is used or disposed of in a way that protects both human health and the environment, under the authority of the Clean Water Act as amended, the U.S. Environmental Protection Agency promulgated, at 40 CFR Part 503, Phase I of the risk-based regulation that governs the final use or disposal of sewage sludge (biosolids).

Vectors—Refers to the degree of attractiveness of biosolids to flies, rats, and mosquitoes that could come into contact with pathogenic organisms and spread disease.

Aerated Static Pile

Three methods of composting wastewater biosolids are common. Each method involves mixing dewatered wastewater solids with a bulking agent to provide carbon and increase porosity. The resulting mixture is piled or placed in a vessel where microbial activity causes the temperatures of the mixture to rise during the active composing period. The specific temperatures that must be achieved and maintained for successful composing vary based on the method and use of the end product. After active composting the material is cured and distributed. Again, there are three commonly employed composting methods but we only describe the aerated static pile (ASP) method because it is commonly used. For an in-depth treatment of the other two methods, windrow and in-vessel, refer to Spellman (1997).

In the aerated static pile type of composting facility, the homogenized mixture of a bulking agent (coarse hardwood wood chips) and dewatered biosolids is piled by front-end loaders onto a large concrete composting pad where it is mechanically aerated via PVC plastic pipe embedded within the concrete slab. This ventilation procedure is part of the 26-day period of active composting when adequate air and oxygen are necessary to support aerobic biological activity in the compost mass and to reduce the heat and moisture content of the compost mixture. Keep in mind that a compost pile without a properly sized air distribution system can lead to the onset of anaerobic conditions and can give rise to putrefactive odors.

For illustration and discussion purposes, we assume that a typical overall composting pad area is approximately 200 feet by 240 feet and consists of 11 blowers and 24 pipe troughs. Three blowers are 20-hp, 2400-cfm, variable-speed-drive units capable of operating in either the positive or negative aeration mode. Blowers A, B, and C are each connected to two piping troughs that run the full length of the pad. The two troughs are connected at the opposite end of the composting pad to create an *aeration pipe loop*. The other eight blowers are rated at 3 hp and 1200 cfm and are arranged with one blower per six troughs at half length feeding 200 cfm per trough. These blowers can be operated in the positive or negative aeration mode. Aeration piping within the six pipe troughs is perforated PVC plastic pipe (6-inch inside diameter and 1/4-inch wall thickness). Perforation holes or orifices vary in size from 7/32 inch to 1/2 inch, increasing in diameter as the distance from the blower increases.

The variable-speed motor drives installed with blowers A, B, and C are controlled by five thermal probes mounted at various depths in the compost pile, and various parameters are fed back to the recorder; the other eight blowers are constant speed, controlled by a timer that cycles them on and off. To ensure optimum composting operations it is important to verify that these thermal probes are calibrated on a regular basis. In the constant-speed system, thermal probes are installed but all readings are taken and recorded manually. For water and leachate drainage purposes, all aeration piping within the troughs slopes downward with the highest point at the center of the composting pad. Drain caps located at each end of the pipe length are manually removed on a regular basis so that any buildup of debris or moisture will not interfere with the airflow.

The actual construction process involved in building the compost pile will be covered in detail later but for now a few key points should be made. Prior to piling the mixture onto the composting pad, an 18-inch layer of wood chips is laid down and serves as a base material. The primary purpose of the wood chips base is to keep the composting mixture clear of the aeration pipes, thus reducing clogging of the air distribution openings in the pipes and allowing free air circulation. A secondary benefit is that the wood chips insulate the composting mixture from the pad. The compost pad is like a heat sink, and this insulating barrier improves the uniformity of heat distribution within the composting mixture.

Lime Stabilization

Lime or alkaline stabilization can achieve the minimum requirements for both Class A (no detectable pathogens) and Class B (a reduced level of pathogens) biosolids with respect to pathogens, depending on the amount of alkaline material added and other processes employed. Generally, alkaline stabilization meets the Class B requirements when the pH of the mixture of wastewater solids and alkaline material is at 12 or above after 2 hours of contact.

Class A requirements can be achieved when the pH of the mixture is maintained at or above 12 for at least 72 hours, with a temperature of 52°C being maintained for at least 12 hours during this time. In one process, the mixture is air dried to over 50% solids after the 72-hour period of elevated pH. Alternatively, the process may be manipulated to maintain temperatures at or above 70°F for 30 or more minutes, while maintaining the pH requirement of 12. This higher temperature can be achieved by overdosing with lime (that is, by adding more than is needed to reach a pH of 12), by using a supplemental heat source, or by using a combination of the two. Monitoring for fecal coliforms or *Salmonella* sp. is required prior to release by the generator for use.

Materials that may be used for alkaline stabilization include hydrated lime, quicklime (calcium oxide), fly ash, lime and cement kiln dust, and carbide lime. Quicklime is commonly used because it has a high heat of hydrolysis (491 British thermal units) and can significantly enhance pathogen destruction. Fly ash, lime kiln dust, or cement kiln dust are often used for alkaline stabilization because of their availability and relatively low cost.

The alkaline-stabilized product is suitable for application in many situations, such as landscaping, agriculture, and mine reclamation. The product serves as a lime substitute, source of organic matter, and a specialty fertilizer. The addition of alkaline-stabilized biosolids results in more favorable conditions for vegetative growth by improving soil properties such as pH, texture, and water holding capacity. Appropriate applications depend on the needs of the soil and crops that will be grown and the pathogen classification. For example, a Class B material would not be suitable for blending in a top soil mix intended for use in home landscaping but is suitable for agriculture, mine reclamation, and landfill cover where the potential for contact with the public is lower and access can be restricted. Class A alkaline-stabilized biosolids are useful in agriculture and as a topsoil blend ingredient. Alkaline-stabilized biosolids provide pH adjustment, nutrients, and organic matter, reducing reliance on other fertilizers.

Alkaline-stabilized biosolids are also useful as daily landfill cover. They satisfy the federal requirement that landfills must be covered with soil or soil-like material at the end of each day (40 CFR 258). In most cases, lime stabilized biosolids are blended with other soil to achieve the proper consistency for daily cover.

As previously mentioned, alkaline-stabilized biosolids are excellent for land reclamation in degraded areas, including acid mine spills or mine tailings. Soil conditions at such sites are very unfavorable for vegetative growth often due to acid content, lack of nutrients, elevated levels of heavy metals, and poor soil texture. Alkaline-stabilized biosolids help to remedy these problems, making conditions more favorable for plant growth and reducing erosion potential. In addition, once a vegetative cover is established, the quality of mine drainage improves.

Thermal Treatment

Thermal treatment (or wet air oxidation) subjects sludge to high temperature and pressure in a closed reactor vessel. The high temperature and pressure rupture the cell walls of any microorganisms present in the solids and causes chemical oxidation of the organic matter. This process substantially improves dewatering and reduces the volume of material for disposal. It also produces a very high-strength waste, which must be returned to the wastewater treatment system for further treatment.

Chlorine Oxidation

Chlorine oxidation also occurs in a closed vessel. In this process, chlorine (100 to 1000 mg/L) is mixed with a recycled solids flow. The recycled flow and process residual flow are mixed in the reactor. The solids and water are separated after leaving the reactor vessel. The water is returned to the wastewater treatment system, and the treated solids are dewatered for disposal. The main advantage of chlorine oxidation is that it can be operated intermittently. The main disadvantage is production of extremely low pH and high chlorine content in the supernatant.

SLUDGE DEWATERING

Digested sludge removed from the digester is still mostly liquid. The primary objective of dewatering biosolids is to reduce moisture and consequently volume to a degree that will allow for economical disposal or reuse. Dewatering biosolids is important because it has a significant impact on the economics, functioning, and required capacity of downstream operations. As an example of the economic importance of dewatering biosolids to achieve a higher solids content, consider the example provided by Padmanabha et al. (1994). A plant where more than 1800 wet ton/day of cake is produced must haul it approximately 50 miles to the land application site. If the cake produced and hauled such long distances contained a higher water content, the overall hauling costs involved would be higher than for cargo that is low in moisture and high in solids content. A biosolids cake that is higher in solids content reduces the need for space, fuel, labor, equipment, and size of the receiving facility (e.g., a composting facility) (Epstein and Alpert, 1984).

Probably one of the best summarizations of the various reasons why it is important to dewater biosolids was given by Metcalf & Eddy (1991): (1) the costs of transporting biosolids to the ultimate disposal site are greatly reduced when biosolids volume is reduced; (2) dewatered biosolids allow for easier handling; (3) dewatering biosolids allows for more efficient incineration; (4) if composting is the beneficial reuse choice, dewatered biosolids decrease the amount and therefore the cost of bulking agents; (5) with the USEPA's 503 Rule, dewatering biosolids may be required to render the biosolids less offensive; and (6) when landfilling is the ultimate disposal option, dewatering biosolids is required to reduce leachate production. Again, the importance of adequately dewatering biosolids for proper disposal or reuse cannot be overstated.

The unit processes that are most often used for dewatering biosolids are (1) vacuum filtration, (2) pressure filtration, (3) centrifugation, and (4) drying beds. The biosolids cake produced by common dewatering processes has a consistency similar to dry, crumbly, bread pudding (Spellman, 1996). This dry, non-fluid dewatered, crumbly cake product is easily handled and non-offensive, and it can be land applied manually and by conventional agricultural spreaders (Outwater, 1994).

Dewatering processes are usually divided into natural air drying and mechanical methods. Natural dewatering methods include removing moisture by evaporation and gravity or induced drainage such as sand beds, biosolids lagoons, paved beds, *Phragmites* reed beds, vacuum-assisted beds, Wedgewater™ beds, and dewatering via freezing. These natural dewatering methods are less controllable than mechanical dewatering methods but are typically less expensive. Moreover, these natural dewatering methods require less power because they rely on solar energy, gravity, and biological processes as the source of energy for dewatering. Mechanical dewatering processes include pressure filters, vacuum filters, belt filters, and centrifuges.

Sand Drying Beds

Sand beds have been used successfully for years to dewater sludge. Composed of a sand bed (consisting of a gravel base, underdrains, and 8 to 12 inches of filter-grade sand), a drying bed includes an inlet pipe, splash pad containment walls, and a system to return filtrate (water) for treatment. In some cases, the sand beds are covered to protect drying solids from the elements. In operation, solids are pumped to the sand bed and allowed to dry by first draining off excess water through the sand and then by evaporation. This is the simplest and cheapest method for dewatering sludge. Moreover, no special training or expertise is required. There is a downside, however, in that drying beds require a great deal of manpower to clean the beds, they can create odor and insect problems, and they can cause sludge buildup during inclement weather.

Four types of drying beds are commonly used to dewater biosolids: (1) sand, (2) paved, (3) artificial media, and (4) vacuum-assisted (Metcalf & Eddy, 1991). In addition to these commonly used dewatering methods, a few of the innovative methods of natural dewatering will also be discussed in this section. The innovative natural dewatering methods to be discussed include experimental work on biosolids dewatering via freezing. Moreover, dewatering biosolids with aquatic plants, which has been tested and installed in several sites throughout the United States, is also discussed.

Drying beds are generally used for dewatering well-digested biosolids. Attempting to air dry raw biosolids is generally unsuccessful and may result in odor and vector control problems. Biosolids drying beds consist of a perforated or open-joint drainage system in support media (usually gravel), covered with a filter media (usually sand but can consist of extruded plastic or wire mesh). Drying beds are usually separated into workable sections by wood, concrete, or other materials. Drying beds may be enclosed or open to the weather. They may rely entirely on natural drainage and evaporation processes or may use a vacuum to assist the operation (both types are discussed in the following sections).

Rotary Vacuum Filtration

Rotary vacuum filters have also been used for many years to dewater sludge. The vacuum filter includes filter media (belt, cloth, or metal coils), media support (drum), vacuum system, chemical feed equipment, and conveyor belts to transport the dewatered solids. In operation, chemically treated solids are pumped to a vat or tank in which a rotating drum is submerged. As the drum rotates, a vacuum is applied to the drum. Solids collect on the media and are held there by the vacuum as the drum rotates out of the tank. The vacuum removes additional water from the captured solids. When solids reach the discharge zone, the vacuum is released and the dewatered solids are discharged onto a conveyor belt for disposal. The media are then washed prior to returning to the start of the cycle.

Types of Rotary Vacuum Filters

The three principal types of rotary vacuum filters are rotary drum, coil, and belt. The *rotary drum* filter consists of a cylindrical drum rotating partially submerged in a vat or pan of conditioned sludge. The drum is divided lengthwise into a number of sections that are connected through internal piping to ports in the valve body (plant) at the hub. This plate rotates in contact with a fixed valve plate with similar parts, which are connected to a vacuum supply, a compressed air supply, and an atmosphere vent. As the drum rotates, each section is thus connected to the appropriate service.

The *coil type* of vacuum filter uses two layers of stainless steel coils arranged in corduroy fashion around the drum. After a dewatering cycle, the two layers of springs leave the drum bed and are separated from each other so the cake is lifted off the lower layer and is discharged from the upper layer. The coils are then washed and reapplied to the drum. The coil filter is used successfully for all types of sludges; however, sludges with extremely fine particles or ones that are resistant to flocculation dewater poorly with this system. The media on a *belt filter* leave the drum surface at the end of the drying zone and pass over a small diameter discharge roll to aid in cake discharge. Washing of the media occurs next. The media are then returned to the drum and to the vat for another cycle. This type of filter normally has a small-diameter curved bar between the point where the belt leaves the drum and the discharge roll. This bar primarily aids in maintaining belt dimensional stability.

Pressure Filtration

Pressure filtration differs from vacuum filtration in that the liquid is forced through the filter media by a positive pressure instead of a vacuum. Several types of presses are available, but the most commonly used types are plate-and-frame presses and belt presses. Filter presses include the belt or plate-and-frame types. The belt filter includes two or more porous belts, rollers, and related handling systems for chemical makeup and feed, as well as supernatant and solids collection and transport.

The plate-and-frame filter has a support frame, filter plates covered with porous material, a hydraulic or mechanical mechanism for pressing plates together, and related handling systems for chemical makeup and feed, as well as supernatant and solids collection and transport. Solids are pumped (sandwiched) between plates. Pressure (200 to 250 psi) is applied to the plates and water is squeezed from the solids. At the end of the cycle, the pressure is released; as the plates separate, the solids drop out onto a conveyor belt for transport to storage or disposal. Performance factors for plate-and-frame presses include feed sludge characteristics, type and amount of chemical conditioning, operating pressures, and the type and amount of precoat.

The belt filter uses a coagulant (polymer) mixed with the influent solids. The chemically treated solids are discharged between two moving belts. First, water drains from the solids by gravity. Then, as the two belts move between a series of rollers, pressure squeezes additional water out of the solids. The solids are then discharged onto a conveyor belt for transport to storage or disposal. Performance factors for the belt press include sludge feed rate, belt speed, belt tension, belt permeability, chemical dosage, and chemical selection. Filter presses have lower operation and maintenance (O&M) costs than those of vacuum filters or centrifuges. They typically produce a good-quality cake and can be batch operated; however, construction and installation costs are high. Moreover, chemical addition is required and the presses must be operated by skilled personnel.

Centrifugation

Centrifuges of various types have been utilized in dewatering operations for at least 30 years, and their use appears to be continuing to gain in popularity. Depending on the type of centrifuge that is used, in addition to centrifuge pumping equipment for solids removal, support systems for the removal of dewatered solids are also required.

LAND APPLICATION OF BIOSOLIDS

The purpose of land application of biosolids is to dispose of the treated biosolids in an environmentally sound manner by recycling nutrients and soil conditioners. To be land applied, wastewater biosolids must comply with state and federal biosolids management and disposal regulations. Biosolids must not contain materials that are dangerous to human health (e.g., toxicity, pathogenic organisms) or dangerous to the environment (e.g., toxicity, pesticides, heavy metals). Treated biosolids are land applied by either direct injection or application and plowing in (incorporation).

REFERENCES AND RECOMMENDED READING

Albrecht, R. (1987). How to succeed in compost marketing. *BioCycle*, 28(9): 26–27.

Alexander, R. (1991). Sludge compost use on athletic fields. *BioCycle*, 32(7): 69–71.

APHA. (1992). *Standard Methods for the Examination of Water and Wastewater*, 18th ed. Washington, DC: American Public Health Association.

Anderson, J.B. and Zwieg, H.P. (1962). Biology of waste stabilization ponds. *Southwest Water Works J.*, 44(2): 15–18.

Aptel, P. and Buckley, C.A. (1996). Categories of membrane operations. In: *Water Treatment Membrane Processes*, Chap. 2. New York: McGraw-Hill.

Assenzo, J.R. and Reid, G.W. (1966). Removing nitrogen and phosphorus by bio-oxidation ponds in central Oklahoma. *Water Sewage Works*, 13(8): 294–299.

Benedict, A.H., Epstein, E., and English, J.N. (1986). Municipal sludge composting technology evaluation. *J. WPCF*, 58(4): 279–289.

Brady, T.J. et al. (1996). Chlorination effectiveness for zebra and quagga mussels. *J. AWWA*. 88(1): 107–110.

Britton, J.C. and Morton, B.A. (1982). Dissection guide, field and laboratory manual for the introduced bivalve *Corbicula fluminea*. *Malacol. Rev.*, 3(1): 1–82.

Brockett, O.D. (1976). Microbial reactions in facultative ponds. 1. The anaerobic nature of oxidation pond sediments. *Water Res.*, 10(1): 45–49.

Burnett, C.H. (1992). Small Cities + Warm Climates = Windrow Composting, paper presented at the Water Environment Federation 65th Annual Conference & Exposition, New Orleans, LA, Sept. 20–24.

Burnett, G.W. and Schuster, G.S. (1973). *Pathogenic Microbiology*. St. Louis, MO: Mosby.

Butterfield, C.T. et al. (1943). Chlorine vs. hypochlorite. *Public Health Rep.*, 58: 1837.

Cameron, G.N., Symons, J.M., Spencer, S.R., and Ja, J.Y. (1989). Minimizing THM formation during control of the Asiatic clam: a comparison of biocides. *J. AWWA*, 81(10): 53–62.

Cheng, R.C. et al. (1994). Enhanced coagulation for arsenic removal. *J. AWWA*, 9: 79–90.

Cheremisinoff, P.N. (1995). Gravity separation for efficient solids removal. *Natl. Environ. J.*, 5(6): 29–32.

Cheremisinoff, P.N. and Young, R.A. (1981). *Pollution Engineering Practice Handbook*. Ann Arbor, MI: Ann Arbor Science Publishers.

Chick, H. (1908). Investigation of the laws of disinfection. *J. Hyg.*, 8: 92.

Clifford, D.A., and Lin, C.C., 1985). *Arsenic (Arsenite) and Arsenic (Arsenate) Removal from Drinking Water in San Ysidro, New Mexico*. Houston, TX: University of Houston.

Clifford, D.A. et al. (1997). *Final Report: Phases 1 and 2 City of Albuquerque Arsenic Study Field Studies on Arsenic Removal in Albuquerque, New Mexico Using the University of Houston/EPA Mobile Drinking Water Treatment Research Facility*. Houston, TX: University of Houston.

Connell, G.F. (1996). *The Chlorination/Chloramination Handbook*. Denver, CO: American Water Works Association.

Corbitt, R.A. (1990). *Standard Handbook of Environmental Engineering*. New York: McGraw-Hill.

Craggs, R. (2005). Nutrients. In: *Pond Treatment Technology* (Hilton, A., Ed.), pp. 282–310. London: IWA Publishing.

Craun, G.F., 1981). Outbreaks of waterborne disease in the United States, *J. AWWA*, 73: 360.

Craun, G.F., and Jakubowski, W. (1996). Status of Waterborne Giardiasis Outbreaks and Monitoring Methods, paper presented at American Water Resources Association Water Related Health Issue Symposium, Atlanta, GA.

Crawford, G., Daigger, G., Fisher, J., Blair, S., and Lewis, R. (2005). Parallel operation of large membrane bioreactors at Traverse City. In: *Proceedings of the Water Environment Federation 78th Annual Conference & Exposition*, Washington, DC, Oct. 29–Nov. 2.

Crawford, G., Fernandez, A., Shawwa, A., and Daigger, G. (2002). Competitive bidding and evaluation of membrane bioreactor equipment: three large plant case studies. In: *Proceedings of the Water Environment Federation 75th Annual Conference & Exposition*, Chicago, IL, Sept. 28–Oct. 2.

Crawford, G., Thompson, D., Lozier, J., Daigger, G., and Fleischer, E. (2000). Membrane bioreactors: a designer's perspective. In: *Proceedings of the Water Environment Federation 73rd Annual Conference & Exposition on Water Quality and Wastewater Treatment*, Anaheim, CA, Oct. 14–18.

Crites, R. and Tchobanoglous, G. (1998). *Small and Decentralized Wastewater Management Systems*. Boston, MA: WCB McGraw-Hill.

Crites, R.W., Middlebrooks, E.J., and Reed S.C. (2006). *Natural Wastewater Treatment Systems*. Boca Raton, FL: Taylor & Francis.

Culp, G.L. and Culp, R.L. (1974). Outbreaks of waterborne disease in the United States, *J. AWWA*, 73: 360.

Culp, G.L. et al. (1986). *Handbook of Public Water Systems*. New York: Van Nostrand Reinhold.

DeMers, L.D. and Renner, R.C. (1992). *Alternative Disinfection Technologies for Small Drinking Water Systems*. Denver, CO: American Water Works Association.

Edwards, M.A. (1994). Chemistry of arsenic removal during coagulation and Fe–Mn oxidation. *J. AWWA*, 86(9): 64–77.

Epstein, E. (1994). Composting and bioaerosols. *BioCycle*, 35(1): 51–58.

Epstein, E. (1998). *Design and Operations of Composting Facilities: Public Health Aspect*, http://www.rdptech.com/tch15.htm.

Epstein, E. and Alpert, J.E. (1984). Sludge dewatering and compost economics. *BioCycle*, 25(10): 31–34.

Epstein, E. and Epstein, J. (1989). Public health issues and composting. *BioCycle*, 30(8): 50–53.

Emrick, J. and Abraham, K. (2002). Long-term BNR operations—cold in Montana! In: *Proceedings of the Water Environment Federation 75th Annual Technical Exhibition & Conference*, Chicago, IL, Sept. 28–Oct. 2.

Finstein, M.S., Miller, F.C., Hogan, J.A., and Strom, P.F. (1987). Analysis of EPA guidance on composting sludge. *BioCycle*, 28(1): 20–26.

Fleischer, E.J., Broderick, T.A., Daigger, G.T., Lozier, J.C., Wollmann, A.M., and Fonseca, A.D. (2001). Evaluating the next generation of water reclamation processes. In: *Proceedings of the Water Environment Federation 74th Annual Technical Exhibition & Conference*, Atlanta, GA, Oct. 13–17.

Fleischer, E.J., Broderick, T.A., Daigger, G.T., Fonseca, A.D., Holbrook, R.D., and Murthy, S.N. (2005). Evaluation of membrane bioreactor process capabilities to meet stringent effluent nutrient discharge requirements. *Water Environ. Res.*, 77: 162–178.

Gallert, C. and Winter, J. (2005). Bacterial metabolism in wastewater treatment systems. In: *Environmental Biotechnology: Concepts and Applications* (Jördening, H.-J. and Winter, J., Eds.), pp. 1–48. Weinheim: Wiley-VCH.

Gannett, F. (2012) Refinement of Nitrogen Removal from Municipal Wastewater Treatment Plants, paper prepared for the Maryland Department of the Environment.

Gaudy, Jr., A.F. and Gaudy, E.T. (1980). *Microbiology for Environmental Scientists and Engineers*. New York: McGraw Hill.

Gloyna, E.F. (1976). Facultative waste stabilization pond design. In: *Ponds As a Waste Treatment Alternative*, Water Resources Symposium No. 9 (Gloyna, E.F., Malina, Jr., J.F., and Davis, E.M., Eds.), p. 143. Austin: University of Texas Press.

Gordon, G. et al. (1995). *Minimizing Chlorate Ion Formation in Drinking Water when Hypochlorite Ion Is the Chlorinating Agent*. Denver, CO: American Water Works Association.

Grady, Jr., C.P.L., Daigger, G.T., Lover, N.G., and Filipe, C.D.M. (2011). *Biological Wastewater Treatment*, 3rd ed. Boca Raton, FL: CRC Press.

Grönlund, E. (2002). Microalgae at Wastewater Treatment in Cold Climates, licentiate thesis, Department of Environmental Engineering, Luleå University of Technology, Luleå, Sweden.

Gurol, M.D. and Pidatella, M.A. (1983). Study of ozone-induced coagulation. In: *Environmental Engineering* (Medine, A. and Anderson, M., Eds.), pp. 118–124. New York: American Society of Civil Engineers.

Harr, J. (1995). *A Civil Action*. New York: Vintage Books.

Hass, C.N. and Englebrecht, R.S. (1980). Physiological alterations of vegetative microorganisms resulting from aqueous chlorination. *J. Water Pollut. Control Fed.*, 52(7): 1976–1989.

Haug, R.T. (1980). *Compost Engineering: Principles and Practices*. Lancaster, PA: Technomic.

Haug, R.T. (1986). Composting process design criteria, Part III. *BioCycle*, 27(10): 53–57.

Haug, R.T. and Davis, B. (1981). Composting results in Los Angeles. *BioCycle*, 22(6): 19–24.

Hay, J.C. (1996). Pathogen destruction and biosolids composting. *BioCycle*, 37(6): 67–72.

Herbert, P.D.N. et al. (1989). Ecological and genetic studies on *Dreissmenu polymorpha* (Pallas): a new mollusc in the Great Lakes. *Can. J. Fish. Aquat. Sci.*, 46: 187.

Hering, J.G. and Chiu, V.Q. (1998). The Chemistry of Arsenic: Treatment and Implications of Arsenic Speciation and Occurrence, paper presented at AWWA Inorganic Contaminants Workshop, San Antonio, TX, Feb. 23–24.

Hermanowicz, S.W., Jenkins, D., Merlo, R.P., and Trussell, R.S. (2006). *Effects of Biomass Properties on Submerged Membrane Bioreactor (SMBR) Performance and Solids Processing*, 01-CTS-19UR. Alexandria, VA: Water Environment Federation.

IOA. (1997). *Survey of Water Treatment Plants*. Stanford, CT: International Ozone Association.

Jagger, J. (1967). *Introduction to Research in Ultraviolet Photobiology*. Englewood Cliffs, NJ: Prentice-Hall.

Klerks, P.L and Fraleigh, P.C. (1991). Controlling adult zebra mussels with oxidants. *J. AWWA*, 83(12): 92–100.

Knudson, G.B. (1985). Photoreactivation of UV-irradiated *Legionella pneumophila* and other *Legionella* species. *Appl. Environ. Microbiol.*, 49: 975–980.

Koch, B. et al. (1991). Predicting the formation of DBPs by the simulate distribution system. *J. AWWA*, 83(10): 62–70.

Krasner, S.W. (1989). The occurrence of disinfection byproducts in U.S. drinking water. *J. AWWA*, 81(8): 41–53.

Kucera, J., (2010). *Reverse Osmosis: Industrial Applications and Processes*. New York: Wiley.

Laine, J.M. (1993). Influence of bromide on low-pressure membrane filtration for controlling DBPs in surface waters. *J. AWWA*, 85(6): 87–99.

Lalezary, S. et al. (1986). Oxidation of five earthy-musty taste and odor compounds. *J. AWWA*, 78(3):62.

Linden, K.G., Shin, G.A., Faubert, G., Cairns, W., and Sobsey, M.D. (2002). UV disinfection of *Giardia lamblia* cysts in water. *Environ. Sci. Technol.*, 36, 2519–2522.

Lue-Hing, C., Zenz, D.R., and Kuchenrither, R. (1992). *Municipal Sewage Sludge Management: Processing, Utilization, and Disposal.* Lancaster, PA: Technomic.

Lynch, J.M. and Poole, N.J. (1979). *Microbial Ecology: A Conceptual Approach.* New York: John Wiley & Sons.

Masschelein, W.J. (1992). *Unit Processes in Drinking Water Treatment.* New York: Marcel Dekker.

Matisoff, G. et al. (1996). Toxicity of chlorine dioxide to adult zebra mussels. *J. AWWA*, 88(8): 93–106.

McGhee, T.J. (1991). *Water Supply and Sewerage.* New York: McGraw-Hill.

Metcalf & Eddy. (1991). *Wastewater Engineering: Treatment, Disposal, Reuse*, 3rd ed. New York: McGraw-Hill.

Metcalf & Eddy. (2003). *Wastewater Engineering: Treatment, Disposal, Reuse*, 4th ed. New York: McGraw-Hill.

Middlebrooks, E.J. and Pano, A. (1983). Nitrogen removal in aerated lagoons. *Water Res.*, 17(10): 1369–1378.

Middlebrooks, E.J., Middlebrooks, C.H., Reynolds, J.H., Watters, G.Z., Reed, S.C., and George, D.B. (1982). *Wastewater Stabilization Lagoon Design, Performance and Upgrading.* New York: Macmillan.

Millner, P., Ed. (1995). Bioaerosols and composting. *BioCycle*, 36(1): 48–54.

Montgomery, J.M. (1985). *Water Treatment Principles and Design.* New York: John Wiley & Sons.

Muilenberg, T. (1997). Microfiltration basics: theory and practice. In: *Proceedings of the AWWA Membrane Technology Conference & Exhibition*, New Orleans, LA, Feb. 23–26.

Natvik, O., Dawson, B., Emrick, J., and Murphy, S. (2003). BNR "then" and "now"—a case study: Kalispell advanced wastewater treatment plant. In: *Proceedings of the Water Environment Federation 76th Annual Technical Exhibition & Conference*, Los Angeles, CA, Oct. 11–15.

NEIWPCC. (1988). *Guides for the Design of Wastewater Treatment Works TR-16.* Wilmington, MA: New England Interstate Water Pollution Control Commission.

Nieminski, E.C. et al. (1993). The occurrence of DBPs in Utah drinking waters. *J. AWWA*, 85(9): 98–105.

Oguma, K., Katayama, H., Mitani, H., Morita, S., Hirata, T., and Ohgaki, S. (2001). Determination of pyrimidine dimmers in *Escherichia coli* and *Cryptosporidium parvum* during UV light inactivation, photoreactivation, and dark repair. *Appl. Environ. Microbiol.*, 67: 4630–4637.

Oliver, B.G. and Shindler, D.B. (1980). Trihalomethanes for chlorination of aquatic algae. *Environ. Sci. Tech.*, 14(12): 1502–1505.

Oswald, W.J. (1990a). Advanced integrated wastewater pond systems: supplying water and saving the environment for six billion people. In: *Proceedings of the ASCE Convention, Environmental Engineering Division*, San Francisco, CA, Nov. 5–8.

Oswald, W.J. (1990b). Sistemas Avanzados De Lagunas Integradas Para Tratamiento De Aguas Servidas (SALI). In: *Proceedings of the ASCE Convention, Environmental Engineering Division*, San Francisco, CA, Nov. 5–8.

Oswald, W.J. (1996). *A Syllabus on Advanced Integrated Pond Systems.* Berkeley: University of California.

Outwater, A.B. (1994). *Reuse of Sludge and Minor Wastewater Residuals.* Boca Raton, FL: Lewis Publishers.

Padmanabha, A.P. et al. (1984). Solids processing upgrade challenges Blue Plains. *Water Environ. Technol.*, July, 51–56.

Pano, A. and Middlebrooks, E.J. (1982). Ammonia nitrogen removal in facultative waste water stabilization ponds. *J. WPCF*, 54(4): 2148.

Park, J. (2012). *Biological Nutrient Removal Theories and Design*. Madison, WI: Division of Natural Resources (http://www.dnr.state.wi.us/org/water/wm/ww/biophos/bnr_removal. htm).

Paterson, C. and Curtis, T. (2005). Physical and chemical environments. In: *Pond Treatment Technology* (Shilton, A., Ed.), pp. 49–65. London: IWA Publishing.

Pearson, H. (2005). Microbiology of waste stabilisation ponds. In: *Pond Treatment Technology*. (Shilton, A., Ed.), pp. 14–48. London: IWA Publishing.

Peot, C. (1998). Compost use in wetland restoration: design for success. In: *Proceedings of the 12th Annual Residual and Biosolids Management Conference*. Alexandria, VA: Water Environment Federation.

Pipes, Jr., W.O. (1961). Basic biology of stabilization ponds. *Water Sewage Works*, 108(4): 131–136.

Prendiville, P.W. (1986). Ozonation at the 900 cfs Los Angeles Water Purification Plant. *Ozone Sci. Eng.*, 8: 77.

Rauth, A.M. (1965). The physical state of viral nucleic acid and the sensitivity of viruses to ultraviolet light. *Biophys J.*, 5: 257–273.

Reckhow, D.A. and Singer, P.C. (1985). Mechanisms of organic halide formation during fulvic acid chlorination and implications with respect to prezonation. In: *Water Chlorination: Chemistry, Environmental Impact and Health Effects* (Jolley, R.L. et al., Eds.), Vol. 5, pp. 1229–1257. Chelsea, MI: Lewis Publishers.

Reckhow, D.A. et al. (1986). Ozone as a coagulant aid. In: *Ozonation: Recent Advances and Research Needs, Seminar Proceedings*. Denver, CO: American Water Works Association.

Reckhow, D.A. et al. (1990). Chlorination of humic materials: byproduct formation and chemical interpretations. *Environ. Sci. Technol.*, 24(11): 1655.

Rice, R.G. et al. (1998). Ozone Treatment for Small Water Systems, paper presented at First International Symposium on Safe Drinking Water in Small Systems, NSF International, Arlington, VA, May 10–13.

Richard, M. and Bowman, D. (1991). Troubleshooting the Aerated and Facultative Waste Treatment Lagoon, paper presented at the USEPA Natural/Constructed Wetlands Treatment System Workshop, Denver, CO, Sept. 4–6.

Roberts, R. (1990). Zebra mussel invasion threatens U.S. waters, *Science*, 249: 1370.

Sawyer, C.N., McCarty, P.L., and Parkin, G.F. (1994). *Chemistry for Environmental Engineering*. New York: McGraw Hill.

Scarpino, P.V. et al. (1972). A comparative study of the inactivation of viruses in water by chlorine. *Water Res.*, 6: 959.

Shilton, A., Ed. (2005). *Pond Treatment Technology*. London: IWA Publishing.

Shin, G.A., Linden, K.G., Arrowood, M.J., Faubert, G., and Sosbey, M.D. (2001). DNA repair of UV-irradiated *Cryptosporidium parvum* oocysts and *Giardia lamblia* cysts. In: *Proceedings of the First International Ultraviolet Association Congress*, Washington, DC, June 14–16.

Singer, P.C. and Harrington, G.W. (1989). Correlations between trihalomethanes and total organic halides formed during water treatment. *J. AWWA*, 81: 61–65.

Singleton, P. and Sainsbury, D. (1994). *Dictionary of Microbiology and Molecular Biology*, 2nd ed. New York: John Wiley & Sons.

Simms, J. et al. (2000). Arsenic Removal Studies and the Design of a 20,000 m³ Per Day Plant in U.K., paper presented at AWWA Inorganic Contaminants Workshop, Albuquerque, NM, Feb. 27–29.

Sinclair, R.M. (1964). Clam pests in Tennessee water supplies. *J AWWA*, 56(5): 592.

Singer P.C. (1992). Formation and characterization of disinfection byproducts. In: *First International Conference on the Safety of Water Disinfection: Balancing Chemical and Microbial Risks*. Washington, DC: ILSI.

Singer, P.C., and Chang, S.D. (1989). Correlations between trihalomethanes and total organic halides formed during water treatment. *J. AWWA*, 81(8): 61–65.

Singer, P.C., and Harrington, G.W. (1993). Coagulation of DBP precursors: theoretical and practical considerations. In: *Proceedings, American Water Works Association, Water Quality Technology Conference*, Miami, FL, Nov. 7–11.

Sloan Equipment. (1999). *Aeration Products*. Owings Mills, MD: Sloan Equipment.

Snead, M.C. et al. (1980). *Benefits of Maintaining a Chlorine Residual in Water Supply Systems*, EPA-600/2-80-010. Washington, DC: U.S. Environmental Protection Agency.

Sopper, W.E. (1993). *Municipal Sludge Use in Land Reclamation*. Boca Raton, FL: Lewis Publishers.

Spellman, F.R. (1996). *Stream Ecology and Self-Purification*. Boca Raton, FL: CRC Press.

Spellman, F.R. (1999). *Choosing Disinfection Alternatives for Water/Wastewater Treatment*. Boca Raton, FL: CRC Press.

Spellman, F.R. (2000). *Microbiology for Water and Wastewater Operators*. Boca Raton, FL: CRC Press.

Spellman, F.R. (2007). *The Science of Water*, 2nd ed. Boca Raton, FL: CRC Press.

Stevens, A.A. (1976). Chlorination of organics in drinking water. *J. AWWA*, 8(11): 615.

Subramanian, K.D. et al. (1997). Manganese greensand for removal of arsenic in drinking water. *Water Qual. Res. J. Can.*, 32(3): 551–561.

Suffet, I.H. et al. (1986). Removal of tastes and odors by ozonation. In: *Proceedings, AWWA Annual Conference*, Denver, CO, June 22.

Tchobanoglous, G., Theisen, H., and Vigil, S.A. (1993). *Integrated Solid Waste Management*. New York: McGraw-Hill.

Thibaud, H. et al. (1988). Effects of bromide concentration on the production of chloropicrin during chlorination of surface waters: formation of brominated trihalonitromethanes. *Water Res.*, 22(3): 381.

Toomey, W.A. (1994). Meeting the challenge of yard trimmings diversion. *BioCycle*, 35(5): 55–58.

TWUA. (1988). *Manual of Water Utility Operations*, 8th ed. Austin: Texas Water Utilities Association.

Ullrich, A.H. (1967). Use of wastewater stabilization ponds in two different systems. *J. WPCF*, 39(6): 965–977.

USEPA. (1975). *Process Design Manual for Nitrogen Control*, EPA-625/1-75-007. Cincinnati, OH: U.S. Environmental Protection Agency.

USEPA. (1977a). *Operations Manual for Stabilization Ponds*, EPA-430/9-77-012, NTIS No. PB-279443. Washington, DC: U.S. Environmental Protection Agency.

USEPA. (1977b). *Upgrading Lagoons*, EPA-625/4-73-001, NTIS No. PB 259974. Cincinnati, OH: U.S. Environmental Protection Agency.

USEPA. (1989). *Technical Support Document for Pathogen Reducing in Sewage Sludge*, NTIS No. PB89-136618. Springfield, VA: National Technical Information Service, U.S. Environmental Protection Agency.

USEPA. (1991). *Manual of Individual and Non-Public Works Supply Systems*, EPA-570/9-91-004. Washington, DC: U.S. Environmental Protection Agency.

USEPA. (1993). *Manual: Nitrogen Control*, EPA-625/R-93/010. Cincinnati, OH: U.S. Environmental Protection Agency.

USEPA. (1997a). *Community Water System Survey*, EPA 815-R-97-001a. Washington, DC: U.S. Environmental Protection Agency.

USEPA. (1997b). *Innovative Uses of Compost: Disease Control for Plants and Animals*. Washington, DC: U.S. Environmental Protection Agency.

USEPA. (1998). National drinking water regulations: Interim Enhanced Surface Water Treatment Final Rule. *Fed. Reg.*, 63: 69477.

USEPA. (1999a). *Wastewater Technology Fact Sheet: Ultraviolet Disinfection.* Washington, DC: U.S. Environmental Protection Agency.

USEPA. (1999b). *Wastewater Technology Fact Sheet: Ozone Disinfection.* Washington, DC: U.S. Environmental Protection Agency.

USEPA. (1999c). *Microbial and Disinfection Byproduct Rules Simultaneous Compliance Guidance Manual.* Washington, DC: U.S. Environmental Protection Agency.

USEPA. (1999d). *Lead and Copper Rule Minor Revisions: Fact Sheet,* EPA 815-F-99-010. Washington, DC: U.S. Environmental Protection Agency.

USEPA. (2000a). *Wastewater Technology Fact Sheet: Oxidation Ditches,* EPA-832-F-00-013. Washington, DC: U.S. Environmental Protection Agency.

USEPA. (2000b). *Technologies and Costs for the Removal of Arsenic from Drinking Water,* EPA-815-R-00-028. Washington, DC: US Environmental Protection Agency.

USEPA. (2000c). *Wastewater Technology Fact Sheet: Package Plants.* Washington, DC: U.S. Environmental Protection Agency.

USEPA. (2000d). *Clean Watersheds Needs Survey,* Report to Congress, EPA-832-R-10-002. Washington, DC: U.S. Environmental Protection Agency.

USEPA. (2001). *Memorandum: Development and Adoption of Nutrient Criteria into Water Quality Standards.* Washington, DC: U.S. Environmental Protection Agency.

USEPA. (2006). *UV Disinfection Guidance Manual.* Washington, DC: U.S. Environmental Protection Agency.

USEPA. (2007a). *Wastewater Management Fact Sheet: Membrane Bioreactors.* Washington, DC: U.S. Environmental Protection Agency.

USEPA. (2007b). *Advanced Wastewater Treatment to Achieve a Low Concentration of Phosphorus.* Washington, DC: U.S. Environmental Protection Agency.

USEPA. (2007c). *Biological Nutrient Removal Processes and Costs.* Washington, DC: U.S. Environmental Protection Agency.

USEPA. (2007d). *Fact Sheet: Introduction to Clean Water Act (CWA) Section 303(d) Impaired Waters Lists.* Washington, DC: U.S. Environmental Protection Agency.

USEPA. (2007e). *Innovative Uses of Compost: Disease Control for Plants and Animals,* EPA/530-F-97-044. Washington, DC: U.S. Environmental Protection Agency.

USEPA. (2008). *Municipal Nutrient Removal Technologies Reference Document.* Vol. 2. *Appendices.* Washington, DC: U.S. Environmental Protection Agency.

USEPA. (2011). *Principles of Design and Operations of Wastewater Treatment Pond Systems for Plant Operators, Engineers, and Managers.* Washington, DC: U.S. Environmental Protection Agency.

Van Benschoten, J.E. et al. (1995). Zebra mussel mortality with chlorine. *J. AWWA,* 87(5): 101–108.

Vasconcelos, V.M. and Pereira, E. (2001). Cyanobacteria diversity and toxicity in a wastewater treatment plant (Portugal). *Water Res.,* 35(5): 1354–1357.

Vesilind, P.A. (1980). *Treatment and Disposal of Wastewater Sludges,* 2nd ed. Ann Arbor, MI: Ann Arbor Science Publishers.

Vickers, J.C. et al. (1997). Bench scale evaluation of microfiltration for removal of particles and natural organic matter. In: *Proceedings of Membrane Technology Conference,* New Orleans, LA, Feb. 23–26.

Wallis-Lage, C. et al. (2006). MBR Plants: Larger and More Complicated, paper presented at the Water Reuse Association's 21st Annual Water Reuse Symposium, Hollywood, CA, Sept. 10–13.

Watson, H.E. (1908). A Note on the variation of the rate of disinfection with change in the concentration of the disinfectant. *J. Hyg.,* 8: 538.

WEF. (1985). *Operation of Extended Aeration Package Plants,* Manual of Practice OM-7. Alexandria, VA: Water Environment Federation.

WEF. (1995). *Wastewater Residuals Stabilization*, Manual of Practice FD-9. Alexandria, VA: Water Environment Federation.

WEF. (1998). *Design of Municipal Wastewater Treatment Plants*, Vol. 2, 4th ed., Manual of Practice 8. Alexandria, VA: Water Environment Federation.

Wilbur, C. and Murray, C. (1990). Odor source evaluation. *BioCycle*, 31(3): 68–72.

Witherell, L.E. et al. (1988). Investigation of *Legionella pneumophila* in drinking water. *J. AWWA*, 80(2): 88–93.

White, G.C. (1992). *Handbook of Chlorination and Alternative Disinfectants*. New York: Van Nostrand Reinhold.

Yeager, J.G. and Ward, R.I. (1981). Effects of moisture content on long-term survival and regrowth of bacteria in wastewater sludge. *Appl. Environ. Microbiol.*, 41(5): 1117–1122.

8 RO Applications and Concentrate Disposal

> For the first time in the history of the world, every human being is now subjected to contact with dangerous chemicals, from the moment of conception until death.
>
> **—Rachel Carson, conservationist**

> Please understand that the important thing is not to listen to what I say but to look at the facts, the science, and ask: Is there anything?
>
> **—Frank R. Spellman**

SICK WATER*

The term *sick water* was coined by the United Nations in a 2010 press release addressing the need to recognize that it is time to arrest the global tide of sick water. The gist of the UN report was that transforming waste from a major health and environmental hazard into a clean, safe, and economically attractive resource is emerging as a key challenge in the 21st century. As practitioners of environmental health, we certainly support the UN's view on this important topic.

When we discuss sick water, however, in the context of this text, we need to go a few steps further than the UN in describing the real essence and tragic implications of supposedly potable water that makes people or animals sick or worse. Water that is sick is actually filthy spent water or wastewater—a cocktail of fertilizer runoff and sewage disposal alongside animal, industrial, agricultural, and other wastes. In addition to these listed wastes of concern, other wastes are beginning to garner widespread attention. What are these other wastes? Any waste or product that we dispose of in our waters, that we flush down the toilet, pour down the sink or bathtub drain, or pour down the drain of a worksite deep sink. Consider the following example of pollutants we routinely discharge to our wastewater treatment plants or septic tanks—wastes we don't often consider as waste products but that in reality are.

Each morning a family of four, two adults and two teenagers, wakes up and prepares for the day that lies ahead. Fortunately, this family has three upstairs bathrooms to accommodate everyone's needs, and each day the family's natural wastes, soap suds, cosmetics, hair treatments, vitamins, sunscreen, fragrances, and prescribed medications end up down the various drains. In addition, the overnight deposits of cat and dog waste are routinely picked up and flushed down the toilet. Let's examine a short inventory of what this family of four has disposed of or has applied to themselves during their morning rituals:

---

* From Spellman, F.R., *Handbook of Water and Wastewater Treatment Plant Operations*, 3rd ed., CRC Press, Boca Raton, FL, 2014.

- Toilet-flushed animal wastes
- Prescription and over-the-counter therapeutic drugs
- Veterinary drugs
- Fragrances
- Soap
- Shampoo, conditioner, other hair treatment products
- Body lotion, deodorant, body powder
- Cosmetics
- Sunscreen products
- Diagnostic agents
- Nutraceuticals (e.g., vitamins, medical foods, functional foods)

Even though these bioactive substances have been around for decades, today we group all of them (the exception being animal wastes) under the title of *pharmaceuticals and personal care products*, or PPCPs (see Figure 8.1).

Other sources of PPCPs should also be recognized. Residues from pharmaceutical manufacturing; residues from hospitals, clinics, doctor or veterinary offices, or urgent care facilities; illicit drug disposal (e.g., startled drug user flushing illicit drugs down the toilet and into the wastewater stream); veterinary drug use, especially antibiotics and steroids; and agribusiness are all contributors of PPCPs in the environment.

Returning to our family of four, after having applied, used, or ingested the various substances mentioned earlier, they also add at least traces of these products (PPCPs) to the environment through excretion (the elimination of waste material from the body) and bathing, as well as through disposal of any unwanted medications to sewers and trash. How many of us have found old prescriptions in the family medicine cabinet and disposed of them with a single toilet flush? Many of these medications (e.g., antibiotics) are not normally found in the environment. Earlier we stated that wastewater is a cocktail of fertilizer runoff and sewage disposal with additions of animal, industrial, agricultural, and other wastes. When we add PPCPs to this cocktail we can state that we are simply adding mix to the mix.

This mixed-waste cocktail raises many questions: Does the disposal of antibiotics or other medications into the local wastewater treatment system cause problems for anyone or anything downstream? When we drink locally treated tapwater are we also ingesting flushed-down-the-toilet or rinsed-down-the-drain antibiotics, other medications, illicit drugs, animal excretions, cosmetics, vitamins, personal or household cleaning products, sunscreen products, diagnostic agents, crankcase oil, grease, oil, fats, and veterinary drugs and hormones?

The jury is still out on these questions. We simply do not know what we do not know about the fate of PPCPs or their impact on the environment once they enter our wastewater treatment systems, the water cycle, and eventually our drinking water supply systems. We do know that some PPCPs are easily broken down and processed by the human body or degraded quickly in the environment, but the disposal of certain wastes can be problematic for quite some time. A case in point is the mythical hero Hercules (arguably the world's first environmental engineer), who performed his fifth labor by cleaning up King Augeas' stables. Hercules, faced literally with a

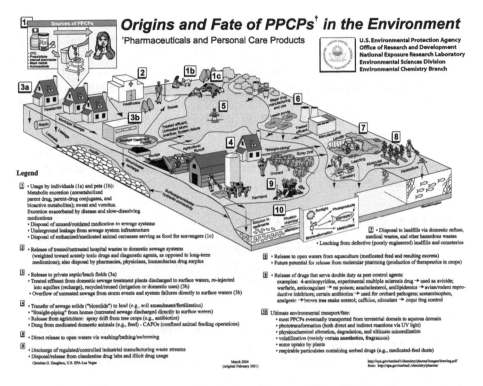

FIGURE 8.1 Origins and fate of PPCPs in the environment. (From USEPA, *Pharmaceuticals and Personal Care Products (PPCPs)*, U.S. Environmental Protection Agency, Washington, DC, 2010, http://www.epa.gov/ppcp/.)

mountain of horse and cattle waste piled high in the stable area, had to devise some method to dispose of the waste. He diverted a couple of rivers to the stable interior, and they carried off all of the animal waste: Out of sight, out of mind. The waste followed the laws of gravity and flowed downstream, becoming someone else's problem. Hercules understood the principal point in pollution control technology, one that is pertinent to this very day: *Dilution is the solution to pollution.*

The fly in this pollution solution ointment is today's modern PPCPs. Although Hercules was able to dispose of animal waste into a running water system where eventually the water's self-purification process cleaned the stream, he did not have to deal with today's personal pharmaceuticals and the hormones that are given to many types of livestock to enhance health and growth.

Studies show that pharmaceuticals are present in our nation's water bodies, and research suggests that certain drugs may cause ecological harm. The USEPA and other research agencies are committed to investigating this topic and developing strategies to help protect the health of both the environment and the public. To date, scientists have found no hard evidence of adverse human health effects from PPCPs in the environment. Some might argue that these PPCPs represent only a small fraction (expressed in parts per trillion, 10^{-12}) of the total volume of water, that we are speaking of a proportion equivalent to 1/20 of a drop of water diluted into an

Olympic-size swimming pool. One student in an environmental health class stated that he did not think the water should be called "sick water," as it was evident to him that water containing so many medications could not be sick. Instead, it might be termed "well water," with the potential to make anyone who drinks it well.

It is important to point out that the term *sick water* can be applied not only to PPCP-contaminated water but also to any filthy, dirty, contaminated, polluted, pathogen-filled drinking water sources. The fact is dirty or sick water means that, worldwide, more people now die from contaminated and polluted water than from all forms of violence, including wars (Corcoran et al., 2010). The United Nations observed that dirty or sick water is a key factor in the rise of deoxygenated dead zones that have been emerging in seas and oceans across the globe.

DRINKING WATER PURIFICATION*

Conventional water and wastewater treatment processes, along with advanced treatment technologies placed within the treatment train for their intended purpose, were presented in the previous chapter. It was also stated earlier that it is important to keep in mind that conventional water treatment along with proper disinfection of water sources used for potable water has served the consumer and public health quite well; it has saved countless numbers of lives. No reasonable person can argue against this fact; however, as explained in this chapter, currently there exist contaminants of concern, emerging contaminants, that conventional water treatment processes do not adequately treat, remove, or neutralize.

This chapter began with a description of *sick water* because it illustrates one of the key points this book is attempting to make. Although we have enough fresh drinking water for everyone on Earth, we have two problems. First is that much of the Earth's clean and safe freshwater is not easily accessible; it is remote and not readily available to people where they reside and where it is needed. For this reason, we have to jump through piping and conduit hoops and loops to transport the water to where it is needed. The second problem was partially indentified in the *sick water* account. First, humans and all members of the animal kingdom pollute anything we touch, breathe, ingest, or urinate or defecate on or in. Nature is usually able to purify all of the environmental mediums (air, water, and soil) that we contaminate, foul, pollute, taint, infect, dirty, or otherwise sully. This is an ongoing process, and with regard to water, our focus here, Nature has a unique self-purifying solution to pollution. As that mythical hero Hercules illustrated, dilution is the solution to pollution, and this can be seen in the self-purifying process that occurs in running water such as streams and rivers that are not overly polluted. Given enough time and distance and if not further polluted by downstream point or nonpoint pollution sources, the water will purify itself. This is Nature's way.

The problem is that, although Nature has her way, there are times she needs help or at least some human respect. With the growing population trend and settlement of humans in areas that were previously unoccupied and considered pristine, drinking

---

* Adapted from Spellman, F.R., *Personal Care Products and Pharmaceuticals in Wastewater and the Environment*, DesTech Publications, Lancaster, PA, 2014.

water has become a limiting factor. Even if a so-called pure mountain-fed stream runs through the pristine wilderness areas where new settlements are established, one must be careful about drinking the water. These pristine streams are not necessarily clean, healthy, or safe to drink. In the wilderness and elsewhere, any location where land runoff occurs can allow land-based contaminants (e.g., fertilizers, feces, pesticides) to be picked up by rainwater, floods, and wind and deposited into streams. Also, in the wilderness and in many rural locations, animals roam here and there, mostly at their own free will. Eventually these animals, domesticated or wild, will urinate, defecate, or die and leave their remains in streams or on land near streams such that the animal urine, feces, and biodegraded remains end up in a river, stream, pond, or lake. When this occurs, the stream typically becomes contaminated with *Giardia lamblia* and other contaminants.

In order to consume water from such a freshwater source and not get sick, the water has to be treated; the contaminants must be removed. This is where conventional water treatment (with disinfection), including filtration and sometimes reverse osmosis (RO) systems, comes into play. Such treatment can remove these contaminants and prevent users of the water from contracting waterborne illnesses. Figure 8.2 shows that water treatment technologies (combined with disinfection) offer a ladder of increasing water quality, and choosing the right level of treatment should be dictated by the end application of the treated water. The emerging contaminants discussed here, the ones currently entering our water sources, require the inclusion of advanced treatment processes, along with RO systems, in the treatment train to ensure our safety.

From reading the *sick water* account earlier, you should now be aware of an emerging worry, an emerging threat, another one of those situations where we do not know what we do not know, but we should know. We simply *need* to know. We certainly do not know everything about everything, but one of the things we *do* know, for example, is that we are quickly approaching a time when we will enter into the fifth generation of people exposed to toxic chemicals from before conception to adulthood. In a few cases, we have identified the hazards of certain chemicals and their compounds and have implemented restrictions. One well-known chemical compound that comes to mind, with regard to its environmental harm and subsequent banning, is dichlorodiphenyltrichloroethane (DDT).

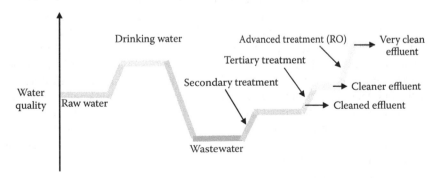

FIGURE 8.2 Treatment technologies available to achieve any desired level of water quality.

The insecticide DDT was first produced in the 1920s and was later developed as a modern synthetic insecticide in the 1940s. It was used extensively and effectively between 1945 and 1965 to control and eradicate insects that were responsible for malaria, typhus, and the other insect-borne human diseases among both military and civilian populations and for insect control in crop and livestock production, institutions, homes, and gardens. DDT was an excellent insecticide because it was very effective at killing a wide variety of insects at low levels. The quick success of DDT as a pesticide and its broad use in the United States and other countries led to the development of resistance by many insect pest species. Moreover, the chemical properties that made this a good pesticide also made it persist in the environment for a long time. This persistence led to accumulation of the pesticide in non-target species, especially raptorial birds (e.g., falcons, eagles). Due to the properties of DDT, the concentration of DDT in birds could be much higher than concentrations in insects or soil. Birds at the top of the food chain (e.g., pelicans, falcons, eagles, grebes) were found to have the highest concentrations of DDT. Although the DDT did not kill the birds outright, it interfered with their calcium metabolism, which led to thin eggshells. As a result, eggs cracked during development, which allowed bacteria to enter and kill the developing embryos. This effect had a great impact on the population levels of these birds. Peregrine falcons and brown pelicans were placed on the endangered species list in the United States, partially due to declining reproductive success of the birds from DDT exposure (Spellman, 2014).

Rachel Carson, that unequaled environmental journalist of profound vision and insight and genius, published *Silent Spring* in 1962, which helped to draw public attention to this problem and to the need for better pesticide controls. This was the very beginning of the environmental movement in the United States and is an excellent example of reporting that identified a problem and warned of many similar problems that could occur unless restrictions were put in place related to chemical pesticide use. Partially as a result of Carson's flagship book, scientists documented the link between DDT and eggshell thinning. This led to the U.S. Department of Agriculture, the federal agency responsible for regulating pesticides before formation of the U.S. Environmental Protection Agency in 1970, taking regulatory actions in the later 1950s and 1960s to prohibit many of the uses of DDT because of mounting evidence of the declining benefits and environmental and toxicological effects of the pesticide.

In 1972, the USEPA issued a cancellation order for DDT based on adverse environmental effects of its use, such as those to wildlife (e.g., raptors), as well as its potential human health risks. Since then, studies have continued, and a causal relationship between DDT exposure and reproductive effects is suspected. Today, DDT is classified as a probable human carcinogen by the U.S. and international authorities. This classification is based on animal studies in which some animals developed liver tumors. DDT is known to be very persistent in the environment, will accumulate in fatty tissues, and can travel long distances in the upper atmosphere. Since the use of DDT was discontinued in the United States, its concentration in the environment and animals has decreased, but because of its persistence, residues of concern from historical use still remain. Moreover, DDT is still used in developing countries because it is inexpensive and highly effective. Other alternatives are too expensive for these other countries to use (USEPA, 2012a).

Let's shift gears for a moment and focus on a brief accounting of the history of the development and use of chemicals. Humankind's development of chemical manufacturing and our use of chemicals can be traced back to the earliest time of recorded history, including the 4000 BCE accounts of glassmaking, brickmaking, and copper smelting practices; the use of coal in 3000 BCE for fuel and to produce asphalt for use in adhesives, waterproofing, and road building; the brewing of beer by the Sumerians in 1750 BCE; Aristotle proposing in 350 BCE the use of distillation to desalinate sea water; sulfur acid being produced by the lead-chamber method in 1746; Charles Goodyear vulcanizing rubber in 1839; Alfred Nobel inventing dynamite in 1866; Alexander Fleming discovering penicillin in 1928; DuPont manufacturing the refrigerant gas Freon®; the drug firm Eli Lilly patenting Prozac® in 1988; and Pfizer introducing Viagra® in 1998. Chemicals continue to be developed, and the motive force driving most chemical development and use is our persistent drive to maintain or achieve the so-called "good life" that many of us desire.

Here are a few fascinating numbers and facts. There are approximately 13,500 chemical manufacturing facilities in the United States owned by more than 9000 companies. There are 84,000 chemicals in use in the United States, with approximately 700 new ones being added each year. Manufacturers generally manufacture chemicals classified into two groups: commodity chemicals and specialty chemicals. Commodity chemical manufacturers produce large quantities of basic and relatively inexpensive compounds in large plants, often built specially to make one chemical. Commodity plants often run continuously, typically shutting down only a few weeks a year for maintenance. Specialty-batch or performance chemical manufacturers produce smaller quantities of more expensive chemicals that are used less frequently on an as-needed basis. Facilities are located all over the country, with many companies located in Texas, Ohio, New Jersey, Illinois, Louisiana, Pennsylvania, and North and South Carolina.

Let's get back to the numbers. In the United States, under the Toxic Substances Control Act (TSCA), five chemicals were banned—only five. This may seem odd because, when the TSCA was passed in 1976, 60,000 chemicals were included in the inventory of existing chemicals. Since that time, the USEPA has only successfully restricted or banned five chemicals and has only required testing on another 200 existing chemicals. An additional 24,000 chemicals have entered the marketplace, so the TSCA inventory now includes more than 84,000 chemicals. The chemical industry is an essential contributor to the U.S. economy, with shipments valued at about $555 billion per year.

If we know several chemicals are dangerous or harmful to us and our environment, why have only five of them been banned? The best answer is that under TSCA it is difficult to ban a chemical that predated the Rule and thus has been grandfathered.

So, which five chemicals has the TSCA banned? If you were a knowledgeable contestant on the television show *Jeopardy*, you might respond with, "What are PCBs, chlorofluorocarbons, dioxin, hexavalent chromium, and asbestos?" An easy question with a straightforward answer, right? Well, not so fast. Technically, you would be incorrect with that answer. Even though the USEPA did initially ban most asbestos-containing products in the United States, in 1991 the rule was vacated and remanded by the Fifth Circuit Court of Appeals. As a result, most of the original bans on the manufacture, importation, processing, or distribution in commerce for most asbestos-containing product categories originally covered in the 1989 final rule were overturned. Only the bans on corrugated paper, rollboard, commercial paper, specialty paper, and flooring felt and any new uses of asbestos remained banned under the 1989 rule. Most asbestos-containing products can still be legally manufactured, imported, processed, and distributed in the United States, even though more than 45,000 Americans have died from asbestos exposure in the past three decades. According to the U.S. Geological Survey, the production and use of asbestos have declined significantly (USEPA, 2014b).

ENDOCRINE DISRUPTORS

A growing body of evidence suggests that humans and wildlife species have suffered adverse health effects after exposure to endocrine-disrupting chemicals (also referred to as *environmental endocrine disruptors*). Environmental endocrine disruptors can be defined as exogenous agents that interfere with the production, release, transport, metabolism binding, action, or elimination of natural hormones in the body responsible for maintaining homeostasis and regulating developmental processes. The definition reflects a growing awareness that the issue of endocrine disruptors in the environment extends considerably beyond that of exogenous estrogens and includes antiandrogens and agents that act on other components of the endocrine system such as the thyroid and pituitary glands (Kavlock et al., 1996). Disrupting the endocrine system can occur in various ways. Some chemicals can mimic a natural hormone, fooling the body into over-responding to the stimulus (e.g., a growth hormone that results in increased muscle mass) or responding at inappropriate times (e.g., producing insulin when it is not needed). Other endocrine-disrupting chemicals can block the effects of a hormone from certain receptors. Still others can directly stimulate or inhibit the endocrine system, causing overproduction or underproduction of hormones. Certain drugs are used to intentionally cause some of these effects, such as birth control pills. In many situations involving environmental chemicals, an endocrine effect may not be desirable.

In recent years, some scientists have proposed that chemicals might inadvertently be disrupting the endocrine system of humans and wildlife. Reported adverse effects include declines in populations, increases in cancers, and reduced reproductive function. To date, these health problems have been identified primarily in domestic or wildlife species with relatively high exposures to organochlorine compounds, including DDT and its metabolites, polychlorinated biphenyls (PCBs), and dioxides, or to naturally occurring plant estrogens (phytoestrogens). However, the relationship of human diseases of the endocrine system and exposure to environmental contaminants is poorly understood and scientifically controversial.

Although domestic and wildlife species have demonstrated adverse health consequences from exposure to elements in the environment that interact with the endocrine system, it is not known if similar affects are occurring in the general human population, but again there is evidence of adverse effects in populations with relatively high exposures. Several reports of declines in the quality and decrease in the quantity of sperm production in humans over the last five decades and the reported increase in incidences of certain cancers (breast, prostate, testicular) that may have an endocrine-related basis

> **DID YOU KNOW?**
>
> During the fourth century BCE, Hippocrates noted that Queen Anne's lace prevented pregnancies. The Greeks used pomegranate as a contraceptive, and modern research confirms strong estrogen activity. Fennel was also used in the ancient world to prevent pregnancy and precipitate abortions.

have led to speculation about environmental etiologies (Kavlock et al., 1996). There is increasing concern about the impact of the environment on public health, including reproductive ability, and controversy has arisen from some reviews claiming that the quality of human semen has declined (Carlson et al., 1992). However, little notice has been paid to these warnings, possibly because they have been based on data on selected groups of men recruited from infertility clinics, from among semen donors, or from candidates for vasectomy. Furthermore, the sampling of publications used for review was not systematic, thus implying a risk of bias. Because a decline in semen quality may have serious implications for human reproductive health, it is of great importance to elucidate whether the reported decrease in sperm count reflects a biological phenomenon or, rather, is due to methodological errors.

Data on semen quality collected systematically from reports published worldwide indicate clearly that sperm density declined appreciably from 1938 to 1990, although we cannot conclude whether or not this decline is continuing today. Concomitantly, the incidence of some genitourinary abnormalities including testicular cancer and possibly also maldescent (faulty descent of the testicle into the scrotum) and hypospadias (abnormally placed urinary meatus) has increased. Such remarkable changes in semen quality and the occurrence of genitourinary abnormalities over a relatively short period are more probably due to environmental rather than genetic factors. Some common prenatal influences could be responsible both for the decline in sperm density and for the increase in cancer of the testis, hypospadias, and cryptorchidism (one or both testicles fail to move to scrotum). Whether estrogens or compounds with estrogen-like activity or other environmental or endogenous factors damage testicular function remains to be determined (Carlson et al., 1992). Even though we do not know what we do not know about endocrine disruptors, it is known that the normal functions of all organ systems are regulated by endocrine factors, and small disturbances in endocrine function, especially during certain stages of the life cycle such as development, pregnancy, and lactation, can lead to profound and lasting effects. The critical issue is whether sufficiently high levels of endocrine-disrupting chemicals exist in the ambient environment to exert adverse health effects on the general population.

Current methodologies for assessing, measuring, and demonstrating human and wildlife health effects (e.g., the generation of data in accordance with testing guideline) are in their infancy. The USEPA has developed testing guidelines and the Endocrine Disruption Screening Program, which is mandated to use validated methods for screening the testing chemicals to identify potential endocrine disruptors, to determine adverse effects and dose–response, to assess risk, and ultimately to manage risk under current laws.

PHARMACEUTICALS AND PERSONAL CARE PRODUCTS*

Pharmaceuticals and personal care products were first referred to as PPCPs only a few years ago, but these bioactive chemicals (substances that have an effect on living tissue) have been around for decades. Their effect on the environment is now recognized as an important area of research. PPCPs include the following:

- Prescription and over-the-counter therapeutic drugs
- Veterinary drugs
- Fragrances
- Cosmetics
- Sunscreen products
- Diagnostic agents
- Nutraceuticals (e.g., vitamins)

Sources of PPCPs include the following:

- Human activity
- Residues from pharmaceutical manufacturing
- Residues from hospitals
- Illicit drugs
- Veterinary drug use, especially antibiotics and steroids
- Agribusiness

The significance of individuals directly contributing to the combined load of chemicals in the environment has gone largely unrecognized, but the presence of PPCPs in the environment illustrates the immediate connection of the actions and activities of individuals with their environment. Individuals add PPCPs to the environment through excretion (the elimination of waste material from the body) and bathing, as well as the disposal of unwanted medications to sewers and trash. Some PPCPs are easily broken down and processed by the human body or degrade quickly in the environment, but others are not easily broken down and processed, so they enter domestic sewers. Excretion of biologically unused and unprocessed drugs depends on

- *Individual drug composition*—Certain excipients, or inert ingredients, can minimize absorption and therefore maximize excretion.
- *Ability of individual bodies to break down drugs*—This ability depends on age, sex, health, and individual idiosyncrasies.

Because they dissolve easily and do not evaporate at normal temperatures or pressure, PPCPs make their way into the soil and into aquatic environments via sewage, treated sewage biosolids (sludge), and irrigation with reclaimed water.

---

* Adapted from Daughton, C.G. and Ternes, T.A., Pharmaceuticals and personal care products in the environment: agents of subtle change?, *Environ. Health Perspect.*, 107(Suppl. 6), 907–938, 1999; Daughton, C.G., *Drugs and the Environment: Stewardship and Sustainability*, U.S. Environmental Protection Agency, Las Vegas, NV, 2010.

For the purposes of this text and the following discussion, pharmaceutical, veterinary, and illicit drugs and the ingredients in cosmetics, food supplements, and other personal care products, together with their respective metabolites and transformation products, are collectively referred to as PPCPs and more appropriately called *micropollutants*. PPCPs are commonly infused into the environment via sewage treatment facilities, outhouses, septic tanks, cesspools, concentrated animal feeding operations (CAFOs), human and animal excretion into the environment (water and soil), and wet weather runoff (e.g., stormwater runoff). In many instances, untreated sewage is discharged into receiving waters by flood overload events, domestic straight-piping, bypassing due to interceptor and pumping failures, or sewage waters lacking municipal treatment.

Note that, even with wastewater treatment, many of the micropollutants in the sewage waste stream remain in the effluent that is discharged into the receiving waters, because many treatment processes are not designed to remove low concentrations of PPCP micropollutants. The big unknown is whether the combined low concentrations from each of the numerous PPCPs and their transformation products have any significance with respect to ecologic function, while recognizing that immediate effects could escape detection if they are subtle and that long-term cumulative consequences could be insidious. Another question is whether the pharmaceuticals remaining in water used for domestic purposes poses long-term risks for human health after lifetime ingestion via potable waters multiple times a day of very low, subtherapeutic doses of numerous pharmaceuticals; however, this issue is not addressed in this text.

The problem is further complicated by the fact that, although the concentration of individual drugs in the aquatic environment could be low (subparts per billion or subnanomolar, often referred to as micropollutants), the presence of numerous drugs sharing a specific mode of action could lead to significant effects through additive exposures. It is also significant that, until very recently, drugs, unlike pesticides, have not been subject to the same scrutiny regarding possible adverse environmental effects. They have therefore enjoyed several decades of unrestricted discharge to the environment, mainly via wastewater treatment works. This is surprising, especially because certain pharmaceuticals are designed to modulate endocrine and immune systems and cellular signal transduction. As such, they have obvious potential to act as endocrine disruptors in the environment, which is especially true for aquatic organisms, whose exposures may be of a more chronic nature because PPCPs are constantly infused into the environment wherever humans live or visit, whereas pesticide fluxes are more sporadic and have greater spatial heterogeneity. At the present time, it is quite apparent that little information exists from which to construct comprehensive risk assessments for the vast majority of PPCPs having the potential to enter the environment.

Although little is known of the occurrence and effects of pharmaceuticals in the environment, more data exist for antibiotics than for any other therapeutic class. This is the result of their extensive use in both human therapy and animal husbandry, their more easily deterred effects end points (e.g., via microbial and immunoassays), and their greater chances of introduction into the environment, not just by wastewater treatment plants but also by runoff and groundwater contamination, especially from confined animal feeding operations. The literature on antibiotics is much more developed because of the obvious issues of direct effects on native microbiota (and consequent alteration of microbial community structure) and development of resistance in potential human pathogens. Because of the considerably larger literature on antibiotics, this text only touches on this issue; for the same reason, it only touches on steroidal drugs purposefully designed to modulate endocrine systems.

PHARMACEUTICALS IN THE ENVIRONMENT

The fact that pharmaceuticals have been entering the environment from a number of different routes and possibly causing untoward effects in biota has been noted in the scientific literature for several decades, but until recently its significance has gone largely unnoticed. This is due in large part to the international regulation of drugs by human health agencies, which usually have limited expertise in environmental issues. In the past, drugs were rarely viewed as potential environmental pollutants, and there was seldom serious consideration given to their fates once they were excreted from the user. Then again, until the 1990s, any concerted efforts to look for drugs in the environment would have met with limited success because the chemical analysis tools required to identify the presence of drugs with low detection limits (i.e., nanograms per liter or parts per trillion) amid countless other substances, native and anthropogenic alike, were not commonly available. Other obstacles that still exist to a large degree are that many pharmaceuticals and cosmetic ingredients and their metabolites are not available in the widely used environmentally oriented mass spectral libraries. They are available in such specialty libraries such as Pfleger (Maurer et al., 2011), which are not frequently used by environmental chemists. Analytical reference standards, when available, are often difficult to acquire and are quite costly. The majority of drugs are also highly water soluble. This precludes the application of straightforward, conventional sample clean-up/preconcentration methods, coupled with direct gas chromatographic separation, that have been used for years for conventional pollutants, which tend to be less polar and more volatile.

Drugs in the environment did not capture the attention of the scientific or popular press until recently. Some early overviews and reviews were presented by Halling-Sorenson et al. (1998), Raloff (1998), Roembke et al. (1996), Ternes et et al. (1998, 1999), and Velagaleti (1998); more recently, by USEPA's Christian G. Daughton (2010). The evidence supports the case that PPCPs refractory to degradation and transformation do indeed have the potential to reach the environment. What is not known, however, is whether these chemicals and their transformation products can elicit physiologic effects on biota at the low concentrations at which they are observed to occur. Another unknown is the actual quantity of each of the numerous commercial drugs that is ingested or disposed of. With respect to determining the potential extent of the problem, this contrasts sharply with pesticides, for which usage is much better documented and controlled.

When discussing disposal or wastage of pharmaceuticals, one thing seems certain—namely, we must understand the terminology currently used in discussing drug disposal or drug wastage. Daughton (2010) pointed out that discussions of drug disposal are complicated enough, but sometimes it is not even clear as to what is meant by the various terms used to describe drugs that are subject to disposal. Terms used in the literature include *unused, unwanted, unneeded, expired, wasted*, and *leftover*. The distinctions between these can be subtle or ambiguous. For instance, "unused" and "expired" are not good descriptors, as they represent only subsets of the total spectrum of medications that can require disposal. "Unused" omits those medications requiring disposal but which have indeed already been used (such as used medical devices). Just because a medication container or package has been opened does not necessarily mean that it has been used. "Unused" can also mean to patients that they are literally no longer using the medication (for its intended purpose), despite the fact that many patients continue using medications on a self-medicating basis—that is, administering the medication for a condition or duration not originally intended, one of the many forms of non-compliance. The term "expired" omits the preponderance of drugs that are discarded before expiry, often soon after they are dispensed. The term "leftover" is sufficiently expansive, as it includes all medications no longer begin used for the original prescribed condition or intended use—or even unintended purpose. Another term often used to refer to unused consumer pharmaceuticals is *home-generated pharmaceuticals* (or *home-generated pharmaceutical waste*), but this too is not a rigorous term, as many drugs for consumer use are not kept in the home but are dispersed in countless locations throughout society (Ruhoy and Daughton, 2008).

A major obstacle in any discussion of drug wastage is what exactly is meant by "wastage." A definition of wastage is notoriously difficult, especially because the topic involves countless variables and perspectives. A simple definition for drug waste is medications dispensed to—or purchased by—a consumer that are never used for the original intended purpose. But, on closer examination, this is not as straightforward as it might first appear. A better term might be *leftover* medications, as this avoids any inference of whether the medications were actually wasted (that is, served no purpose) but does not infer a reason for why the medications accumulated unused or unwanted. Would a medication intended for emergency contingency purposes (and now expired) be considered wasted? After all, such medications served their purpose of being available for possible emergencies. How about medications intended for unscheduled consumption as the situation arises or as needed? These scenarios show that it would not be possible to completely eliminate leftover medications—only to reduce them to a necessary minimum. It could be argued whether or not the basic premise that medications experience undue wastage is even valid. No one really knows how much drug waste occurs in commerce (at the consumer level or in the healthcare setting) in terms of either the total quantity or the cost.

Many statements regarding drug wastage are based on rates of patient compliance, which is an enormously complex and controversial topic by itself. Noncompliance rates, however, include not only the frequency with which drugs go unused but also the frequency with which prescriptions are *not* filled or with which they are consumed incorrectly. Neither of the latter contributes to any need for disposal. Failure to fill a prescription may even reduce the need for disposal, so noncompliance does not necessarily lead to leftover drugs. Few make this distinction in the literature.

A striking difference between pharmaceuticals and pesticides with respect to environmental release is that pharmaceuticals have the potential for ubiquitous direct release into the environment worldwide—anywhere that humans live or visit. Even areas considered relatively pristine (e.g., national parks) are subject to pharmaceutical exposures, especially given that some parks have very large, aging sewage treatment systems, some of which discharge into park surface waters and some which overflow during wet weather events and infrastructure failures. Other possible sources include disposal of unwanted illicit drugs and synthesis byproducts into domestic sewage systems by clandestine drug operations. The disposal of raw products and intermediates (e.g., ephedrine) via toilets is not uncommon in illegal laboratories. Also, in contrast to pesticides, pharmaceuticals in any stage of clinical testing (not yet approved for dispensing by the Food and Drug Administration) are subject to release into the environment, although their overall concentrations would be very low.

Some drugs are excreted essentially unaltered in their free form (e.g., methotrexate, platinum antineoplastics), often with the help of active cellular multidrug transporters for moderately lipophilic drugs. Others are metabolized to various extents, which is partly a function of the individual patient and the circadian timing of the dose (the P450 microsomal oxidase system is a major route of formation of more polar, more easily excreted metabolites). Still others are converted to more soluble forms by the formation of conjugates (with sugars or peptides). The subsequent transformation products—metabolites and conjugates from eukaryotic and prokaryotic metabolism and from physicochemical alteration—add to the already complex picture of thousands of highly bioactive chemicals. The FDA refers to all metabolites and physicochemical transformation products, such as those that range from the dissociated parent compound to photolysis products, for a given drug as *structurally related substances* (SRSs), which can have greater or lesser physiological activity than the parent drug.

As in mammals, the metabolic disposition of lipophilic xenobiotics, such as numerous drugs, in vertebrate aquatic species is largely governed by what is referred to as Phase I and Phase II reactions (James, 1986); less is known about invertebrate metabolism. Phase I makes use of monooxygenases (e.g., cytochrome P450), reductases, and hydrolases (for esters and epoxides) to add reactive functional groups to the molecule. Phase II uses covalent conjugation (glucuronidation) to make the molecule hydrophilic and more excretable. These reactions are catalyzed by glycosyltransferases and sulfotransferases (for hydroxyaromatics and carboxy groups), glutathione S-transferases (for electrophilic functional groups such as halogens, nitro groups, or unsaturated/conjugated sites), acetyltransferases (for primary amines or hydrazines), and aminoacyltransferases (for forming peptides from carboxy groups using free amino acids). This metabolic strategy creates metabolites successively more polar than the parent compound, thereby enhancing excretion. Considerable interspecies and intraspecies diversity, however, can be observed in actual metabolic potentials. Detection of exposure of fish to many drugs can be facilitated through the analysis of bile.

The introduction of drugs into the environment is partly a function of the quantity of drugs manufactured, the dosage frequency and amount, the excretion efficiency of the parent compound and metabolites, propensity of the drug to sorb to solids, and the metabolic transformation capability of subsequent sewage treatment (or landfill) microorganisms. Publicly owned treatment works (POTWs) receive influent from domestic, municipal, and industrial (including pharmaceutical manufacture) sewage systems. The processed liquid effluents from primary and secondary treatments are then discharged to surface waters and the residual solids (biosolids) to landfills/ farms. Land disposal, including manure from treated animals at CAFOs, creates the potential for introduction into groundwaters or surface waters (via wet weather run-off). Theoretically, PPCPs in sewage biosolids applied to crop lands could be taken up by plants.

Compounds surviving the various phases of metabolism and other degradative or sequestering actions (i.e., display environmental persistence) can then pose an exposure risk for organisms in the environment. Even the less/nontoxic conjugates (glucuronides) can later be converted back to the original bioactive compounds via enzymatic (β-glucuronidases) or chemical hydrolysis (e.g., acetylsalicylic acid can be hydrolyzed to the free salicylic acid).

Some degradation products can even be more bioactive than the parent compound. Therefore, conjugates can essentially act as storage reservoirs from which the free drugs can later be released into the environment.

PPCPs and Wastewater Treatment Plants

Treatment facilities, primarily POTWs or wastewater treatment plants (WWTPs), which also include privately owned works, play a key role in the introduction of pharmaceuticals into the environment. (See Rogers, 1996, for a review of the fate of synthetic chemicals in wastewater treatment plants.) WWTPs were designed to handle human waste of mainly natural origin, primarily via the acclimated degradative action of microorganisms (the efficiency of metabolism of a given drug can increase with duration of treatment because of enzyme induction and cellular adaptation) and the coagulation/flocculation of suspended solids; sometimes, tertiary treatment (e.g., chemical or ultraviolet oxidation) is used. Most anthropogenic chemicals introduced along with this normal waste suffer unknown fates. Two primary mechanisms remove substances from the incoming waste stream: (1) microbial degradation to lower molecular weight products, leading sometimes to complete mineralization—CO_2 and H_2O; and (2) sorption to filterable solids which are later removed with the biosolids.

Although the microbiota of wastewater treatment systems may have been exposed to many PPCPs for a number of years, two factors work against the effective microbial removal of these substances for WWTPs. First, the concentrations of most drugs are probably so low that the lower limits for enzyme affinities may not be met. For example, the daily loadings of PPCPs into WWTPs are largely a function of the serviced human population, the dosages/duration of medications consumed, and the metabolic or excretory half-lives, all of which are large variables. As an example, the daily load of a subset of pharmaceuticals to a particular POTW near Frankfurt, Germany, ranged from tens to hundreds of grams, with approximate individual removal efficiencies varying widely from 10 to 100% but trending to around 60% (Ternes, 1998). This particular POTW serviced about a third of a million people at a flow rate of roughly 60,000 m³/day. Despite the number of studies on treatment efficiencies, a widespread investigation is still lacking for the differences in removal efficiencies for distinct types of WWTPs as well as for individual treatment techniques. The extent to which a particular plant uses primary, secondary, and tertiary technologies will greatly influence removal efficiencies; the technologies employed vary widely among cities. The biodegradative fate of most compounds in WWTPs is governed by non-growth-limiting (enzyme-saturating) substrate concentrations (copiotrophic metabolism, which means they thrive in nutrient-rich environments). In contrast, PPCPs are present in WWTPs at concentrations at enzyme-subsaturating levels, which necessitates oligotrophic metabolism (nutrient-poor environments). These micropollutants might be handled by only a small subset of specialist oligotrophic organisms whose occurrence is probably more prevalent in native environments (e.g., lakes) characterized by low-carbon fluxes (e.g., sediments and associated pore waters, where desorption mass transfer is limiting) than in WWTPs.

Many new drugs are introduced to the market each year; some of these drugs are from entirely new classes never seen before by the microbiota of a WWTP. Each of these presents a new challenge to biodegradation. A worst-case scenario may not be unusual—the concentration of a drug leaving a WWTP in the effluent could essentially be the same as that entering. Only a several-fold to multiple order of magnitude dilution when the effluent is mixed into the receiving water, assuming a sufficiently high natural flow, serves to reduce the concentration; obviously, smaller streams have increased potential for having higher concentrations of any PPCP that has been introduced. In general, most pharmaceuticals resist extensive microbial degradation (e.g., mineralization); although some parent drugs often show poor solubility in water, leading to preferential sorption of suspended particles, they can thereby sorb to colloids and therefore be discharged in the aqueous effluent (Velagaleti, 1998). Metabolites, including breakdown products and conjugates, will partition mainly to the aqueous effluent. Some published data demonstrate that many parent drugs do make their way into the environment.

In a 2004–2009 study, scientists found and reported that pharmaceutical manufacturing facilities can be a significant source of pharmaceuticals to the environment. Effluents from two wastewater treatment plants that receive discharge from pharmaceutical manufacturing facilities (PMFs) had 10 to 1000 times higher concentrations of pharmaceuticals than effluents from four WWTPs across the nation that do not receive PMF discharge. The effluents from these two WWTPs are discharged to streams where the measured pharmaceuticals were traced downstream, and as far as 30 kilometers (18 miles) from one plant's outfall. This was the first study to assess PMFs as a potential source of pharmaceuticals in the environment. The PMFs investigated are pharmaceutical formulation facilities, where ingredients are combined to form final drug products and products are packaged for distribution. Although pharmaceuticals have been measured in many streams and aquifers across the nation, levels are generally lower than 1 part per billion (ppb); however, concerns persist in 23 other plants that higher levels may occur in environmental settings where wastewaters are released to the environment.

In this study, 35 to 38 effluent samples were collected from each of three WWTPs in New York State and one effluent sample was collected from each of 23 strategically selected WWTPs across the nation. The samples were analyzed for seven target pharmaceuticals, including opioids and muscle relaxants, some of which had not been previously studied in the environment. Pharmaceutical concentrations in effluents from two of the three WWTPs in New York State, both of which receive more than 20% of their discharge from PMFs, were compared to the measurements made at the third plant in New York State and at other plants across the nation not receiving discharge from PMFs. Maximum pharmaceuticals concentrations in effluent samples from the 24 WWTPs not receiving discharges from PMFs rarely (about 1%) exceeded 1 part per billion. By contrast, maximum concentrations in effluents from the two WWTPs receiving PMF discharge were as high as 3888 ppb of metaxalone (a muscle relaxant), 1700 ppb of oxycodone (an opioid prescribed for pain relief), greater than 400 ppb of methadone (an opioid prescribed for pain relief and drug withdrawal), 160 ppb of butalbital (a barbiturate), and greater than 40 ppb of both phendimetrazine (a stimulant prescribed for obesity) and carisoprodol (a muscle relaxant).

The pharmaceuticals investigated in this study were identified using a forensic approach that identified pharmaceuticals present in samples and subsequently developed methods to quantify these pharmaceuticals at a wide range of concentrations. Additional pharmaceuticals which may be formulated at these sites, also were identified as present in the effluents of these two WWTPs. Ongoing studies are documenting the levels at which these additional pharmaceuticals occur in the environment.

The efficiency of removal of pharmaceuticals by WWTPs is largely unknown. To date the most extensive study of treatment efficiency, Ternes (1998) reported removal from German WWTPs of 14 drugs representing five broad physiologic categories. Removal of the parent compound (keep in mind that possible subsequent metabolites were not accounted for) ranged from 7% (carbamazepine, an antiepileptic) to 96% (propranolol, a beta blocker); most removal efficiencies averaged about 60%. Fenofibrate, acetaminophen, salicylic acid, o-hydroxyhippuric acid, and gentistic acid (acetylsalicylic acid metabolites) could not be detected in effluent; salicylic acid was found in the influent at concentrations up to 54 µg/L. It is important to understand that, absent the stoichiometric accounting of metabolic products, one cannot distinguish between the three major fates of a substance: (1) degradation to lower molecular weight compounds, (2) physical sequestration by solids (and subsequent removal as sludge), and (3) conjugates that can later be hydrolyzed to yield the parent compound (e.g., clofibric and fenofibric acid conjugates). Therefore, by simply following the disappearance (removal) of a substance, one cannot conclude that it was structurally altered or destroyed, as it may simply reside in another state or form. Identifying metabolic products is difficult not only because of the number of metabolites (sometimes several per parent compound) but also because standard reference materials are difficult to obtain commercially and can be costly.

Despite high removal rates in WWTPs for some drugs, upsets in the homeostasis of a treatment plant can result in higher than normal discharges. For example, Ternes (1998) found that wet-weather runoff dramatically reduces the removal rates for certain drugs, such as several nonsteroidal antiinflammatory drugs (NSAIDs) and lipid regulators, in a facility located close to Frankfurt. During the increased period of influent flow, the removal rate dropped to below 5% from over 60% previously; several days were required for the removal rates to recover. Clearly, even for drugs efficiently removed, the operational state of the WWTP can have a dramatic effect on the removal efficiencies. Other transients that could affect removal include transitions between seasons and sporadic plug-flow influx of toxicants from various sources. Overflows from WWTP failure or overcapacity events (e.g., floods, excessive water use) lead to the direct, untreated introduction of sewage into the environment. In efforts to improve tributary conditions by increasing stream flow, some cities (e.g., Portland, Oregon) have considered increasing the percentage of annual overflow events (Learn, 1999). The highest concentration in a WWTP effluent reported by Ternes (1998) was for bezafibrate (4.6 µg/L); the highest concentration in surface water was also for bezafibrate (3.1 µg/L).

PPCPs in Drinking Water

From 1990 to 1995, few pharmaceuticals were identified in domestic drinking water, probably because of the dearth of monitoring efforts and because the required detection limits were too low for the current routine analytical technology. In Germany, however, clofibric acid concentrations up to 165 ng/L (Stan et al., 1994) and 270 ng/L (Heberer et al., 1998) have been measured in tapwater; the presumed source was from recharged groundwaters that had been contaminated by sewage. Stumpf et al. (1996) and Ternes et al. (1998) found several pharmaceuticals in German drinking water in the lower nanograms-per-liter range, with a maximum of 70 ng/L for clofibric acid. Additionally, these investigators found that diclofenac, bezafibrate, phenazone, and carbamazepine were sometimes present. In the majority of the samples analyzed, however, no drugs were observed. These investigations indicate that contamination of drinking water does not appear to be a general problem. Depending on the water source for drinking/water production, however, certain facilities can experience contamination, especially if the source is polluted groundwater and if polishing technology does not remove the PPCP (Heberer et al., 1998; Stumpf et al., 1999). A major unaddressed issue regarding human health is the long-term effects of ingesting via potable waters very low, subtherapeutic doses of numerous pharmaceuticals multiple times a day for many decades. This concern especially relates to infants, fetuses, and people suffering from certain enzyme deficiencies (which can even be food induced, such as microsomal oxidase inhibition by grapefruit juice).

WASTEWATER REUSE: TOILET TO TAP*

When presenting my well-worn college classroom lecture on the basics of water or one of my public speeches on drinking water to various groups, I typically begin my statement with the following:

> All water on Earth is recycled, reclaimed, or reused. We have the same amount of water on Earth today as we did when water first appeared. We are drinking the same water today that all of our predecessors drank, including the water consumed by Neanderthals, cave dwellers, Cleopatra, Caesar, da Vinci, Napoleon, and all humans and animals who ever existed on Earth. The water taken from our kitchen sink taps is the same water that has been in our lakes, streams, oceans, ponds, swimming pools, industrial factories, our gardens and lawns ... and the same water flushed down our toilets.

---

* Adapted from USEPA, *Guidelines for Water Reuse*, U.S. Environmental Protection Agency, Washington, DC, 2012.

It is this last statement, about drinking toilet water that gets the greatest response from my audience. I have found that when you tell people that they drink the same water they just flushed down their toilets, their faces take on agonized looks of disgust, or they act like they are about to be sick, or they give me an evil look suggesting that a physical attack against my person might be forthcoming. These responses are not unexpected; indeed, they are typical. To attain a better understanding of the material that follows, it is necessary to first define the terminology used in treating municipal wastewater (toilet and household wastewater) for reuse as drinking water.

WATER REUSE TERMINOLOGY

The terminology associated with treating municipal wastewater and reusing it varies within the United States and globally. For example, some states and countries use the term *reclaimed water* while others use the term *recycled water*, although the terms are synonymous. Similarly, the terms *water recycling* and *water reuse* have the same meaning. In this book, the terms *reclaimed water* and *water reuse* are used. If you have ever walked the internal passageways of a typical >1-MGD wastewater treatment plant, you might have been amazed to see numerous galleries of various-sized pipe and piping systems running helter-skelter in every direction throughout the plant site. How do the plant operators know which pipes are which? How do they know what process material, chemical, type of air, etc., each pipe conveys? When the piping throughout the plant is color coded and stenciled in capital letters of an appropriate size to indicate the material being conveyed within the pipes, the plant operator (or anyone else who can read and is not colorblind and knows the color code) has no problem identifying the various pipes. Definitions of terms used in this text and generally accepted treatment plant pipe color identification codes and markings are provided below.

> *Agricultural reuse (food crops)*—The use of reclaimed water to irrigate food crops that are intended for human consumption.
> *Agricultural reuse (processed food crops and non-food crops)*—The use of reclaimed water to irrigate crops that are either processed before human consumption or not consumed by humans.
> *De facto reuse (defined as in practice but not necessarily as ordained by law)*—A common situation where reuse of treated wastewater is, in fact, practiced but is not officially recognized or permitted as a reuse project (e.g., a drinking water supply intake located downstream from a WWTP discharge point; see Figure 8.3). Most water treatment public administrators will not openly advertise (for obvious reasons) that the toilet water you flush today is the tapwater you will use and consume tomorrow, but, if asked directly, they will usually admit that this is the case—at least, this has been the author's experience. *De facto reuse* is similar to *indirect potable reuse* (IPR), whereby a drinking water source (surface or groundwater) is augmented with reclaimed water followed by an environmental buffer that precedes drinking water treatment (see Figure 8.3).

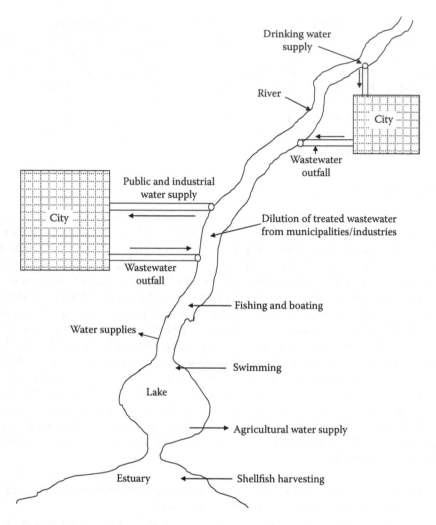

FIGURE 8.3   *De facto* reuse.

Direct potable reuse (DPR)—The introduction of reclaimed water (with or without retention in an engineered storage buffer) directly into a drinking water treatment plant, either collocated or remote from the advanced wastewater treatment system.

Environmental reuse—The use of reclaimed water to create, enhance, sustain, or augment water bodies including wetlands, aquatic habitats, or stream flow.

Graywater—Wastewater generated from wash-hand basins, showers, and baths which can be recycled onsite for uses such as toilet flushing, landscape irrigation, and constructed wetlands. Graywater differs from the discharge of toilets which is designated sewage or *blackwater* to indicate it contains human waste.

Groundwater recharge–non-potable reuse—The use of reclaimed water to recharge aquifers that are not used as a potable water source.

Impoundments (restricted)—The use of reclaimed water in an impoundment where body contact is restricted.

Impoundments (unrestricted)—The use of reclaimed water in an impoundment in which no limitations are imposed on body-contact water recreation activities.

Industrial reuse—The use of reclaimed water in industrial applications and facilities, power production, and extraction of fossil fuels.

Non-potable reuse—All water reuse applications that do not involve potable reuse.

Pipe size and letter size

Pipe Size	Letter Size
Up to 1-1/2 inch	1/2 inch
2 to 6 inches	1-1/4 inches
8 inches and up	2-1/2 inches

Piping color and identification codes

Flow	Color	Abbreviation
Non-potable water	Light green	NPW
Potable water	Dark green	PW
Sanitary sewer force main	Gray	SSFM
Raw wastewater influent	Gray	RWI
Water reclamation	Purple	RWM

Potable reuse—Planned augmentation of a drinking water supply with reclaimed water.

Reclaimed water—Municipal wastewater that has been treated to meet specific water quality criteria with the intent of being used for a range of purposes. The term *recycled water* is synonymous with *reclaimed water.*

Urban reuse (restricted)—The use of reclaimed water for non-potable applications in municipal settings where public access is controlled or restricted by physical or institutional barriers, such as fencing, advisory signage, or temporal access restriction.

Urban reuse (unrestricted)—The use of reclaimed water for non-potable applications in municipal settings where public access is not restricted.

Water reclamation—The act of treating municipal wastewater to make it acceptable for reuse.

Water reuse—The use of treated municipal wastewater (reclaimed water).

Wastewater—Used water discharged from homes, business, industry, and agricultural facilities.

Natural Water Cycle

The water cycle describes how water moves through the environment and identifies the links among groundwater, surface water, and the atmosphere (see Figure 8.4). Water is taken from the Earth's surface to the atmosphere by evaporation from the surface of lakes, rivers, streams, and oceans. This evaporation process occurs when the sun heats water. The heat of the sun energizes surface molecules, allowing them to

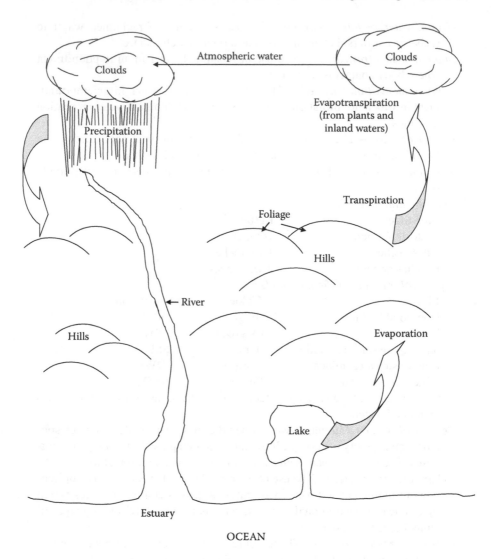

FIGURE 8.4 Natural water cycle.

break free of the attractive force binding them together. They then evaporate and rise as invisible vapor in the atmosphere. Water vapor is also emitted from plant leaves by a process called *transpiration*. Every day, an actively growing plant transpires five to ten times as much water as it can hold at once. As water vapor rises, it cools and eventually condenses, usually on tiny particles of dust in the air. When it condenses, it becomes a liquid again or turns directly into a solid (ice, hail, or snow). These water particles then collect and form clouds. The atmospheric water formed in clouds eventually falls to the ground as precipitation. The precipitation can contain contaminants from air pollution. The precipitation may fall directly onto surface waters, be intercepted by plants or structures, or fall onto the ground. Most precipitation falls in coastal areas or in high elevations. Some of the water that falls in high elevations becomes runoff water, the water that runs over the ground (sometimes collecting nutrients from the soil) to lower elevations to form streams, lakes, and fertile valleys. The water we see is known as *surface water.* Surface water can be broken down into five categories: oceans, lakes, rivers and streams, estuaries, and wetlands.

Because the amount of rain and snow remains almost constant, but population and usage per person are both increasing rapidly, water is in short supply. In the United States alone, water usage is four times greater today than it was in 1900. In the home, this increased use is directly related to increases in the number of bathrooms, garbage disposals, home laundries, and lawn sprinklers. In industry, usage has increased 13 times since 1900.

Over 170,000 small-scale suppliers provide drinking water to approximately 200 million Americans by at least 60,000 community water supply systems, as well as to nonresidential locations, such as schools, factories, and campgrounds. The rest of Americans are served by private wells. The majority of the drinking water used in the United States is supplied from groundwater. Untreated water drawn from groundwater and surface waters and used as a drinking water supply can contain contaminants that pose a threat to human health.

Note: Individual American households use approximately 146,000 gallons of freshwater annually and that Americans drink 1 billion glasses of tap water each day.

Obviously, with a limited amount of drinking water available for use, water that is available must be reused or we will be faced with an inadequate supply to meet the needs of all users. Water use/reuse is complicated by water pollution. Pollution is relative and is difficult to define; for example, floods and animals (dead or alive) are polluters, but their effects are local and tend to be temporary. Today, water is polluted in many ways, and pollution exists in many forms. Pollution may be apparent as excess aquatic weeds, oil slicks, a decline in sport fish populations, or an increase in carp, sludge worms, and other forms of life that readily tolerate pollution. Maintaining water quality is important because water pollution is detrimental not only to health but also to recreation, commercial fishing, aesthetics, and private, industrial, and municipal water supplies.

At this point, the reader might be asking: With all the recent publicity about pollution and the enactment of new environmental regulations, hasn't water quality in the United States improved recently? Answer: The pace at which fishable and swimmable waters have been achieved under the Clean Water Act (CWA) might lead one to believe so. The 1994 *National Water Quality Inventory Report to Congress*, however, indicated that 63% of the nation's lakes, rivers, and estuaries met designated uses, which was only a slight increase over that reported in 1992.

The main culprit is *nonpoint source (NPS) pollution*, which is the leading cause of impairment for rivers, lakes, and estuaries. Impaired sources are those that do not fully support designated uses, such as fish suitable for consumption, drinking water supply, groundwater recharge, aquatic life support, or recreation. The five leading sources of water quality impairment in rivers are agriculture, municipal wastewater treatment plants, habitat and hydrologic modification, resource extraction, and urban runoff and storm sewers (Fortner and Schechter, 1996).

The health of rivers and streams is directly linked to the integrity of habitat along the river corridor and in adjacent wetlands. Stream quality will deteriorate if activities damage vegetation along riverbanks and in nearby wetlands. Trees, shrubs, and grasses filter pollutants from runoff and reduce soil erosion. Removal of vegetation also eliminates shade that moderates stream temperature. Stream temperature, in turn, affects the availability of dissolved oxygen in the water column for fish and other aquatic organisms. Lakes, reservoirs, and ponds may receive water-carrying pollutants from rivers and streams, melting snow, runoff, or groundwater. Lakes may also receive pollution directly from the air.

Thus, in attempting to answer the original question—Has water quality in the United States improved recently?—the best answer probably is that we are holding our own in controlling water pollution, but we need to make more progress. This understates an important point; that is, when it comes to water quality, we need to make more progress on a continuing basis.

Urban Water Cycle

An artificially generated water cycle or the urban water cycle consists of (1) source (surface or groundwater), (2) water treatment and distribution, (3) use and reuse, and (4) wastewater treatment and disposition, as well as the connection of the cycle to the surrounding hydrological basins (see Figure 8.5).

MOTIVATION FOR REUSE

The ability to reuse water, regardless of whether the intent is to augment water supplies or manage nutrients in treated effluent, has possible benefits that are also the key motivators for implementing reuse programs. These benefits include improved agricultural production; reduced energy consumption associated with production, treatment, and distribution of water; and significant environmental benefits, such as reduced nutrient loads to receiving waters due to reuse of the treated wastewater.

Urbanization and Water Scarcity

The current world population of ~7 billion is expected to reach 9.5 billion by 2050 (USCB, 2013). More people, of course, means increased need for more potable water. In addition to the increasing need to meet potable water supply demand and other urban demands (e.g., landscape irrigation, commercial, and industry needs), increased agricultural demands due to greater incorporation of animal and dairy products into the diet has also increased demand on water for food production (Pimentel and Pimentel, 2003). These increases in population and a dependency on high-water-demand agriculture are coupled with increasing urbanization, and all of these factors and others are producing land use changes that exacerbate water supply challenges. Likewise, sea level rise and increasing intensity and variability of local climate patterns are predicted to alter hydrologic and ecosystem dynamics and composition (Bates et al., 2008).

Reuse projects must factor in climate predictions, both for demand projections and for ecological impacts. Municipal wastewater generation in the United States averages approximately 75 gallons per capita daily (gpcd), or 284 liters per capita daily (Lpcd) and is relatively constant throughout the year. Where collection systems

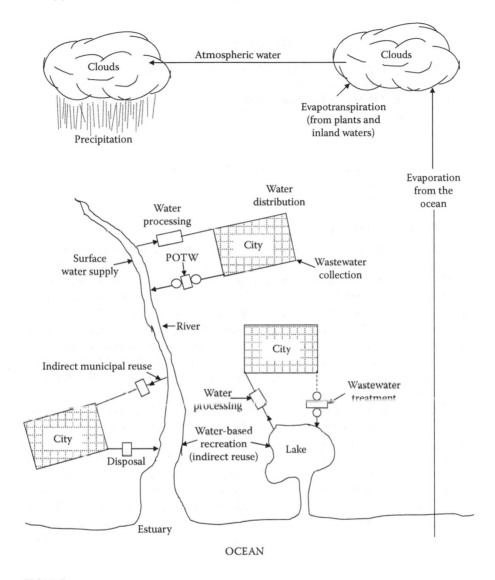

FIGURE 8.5 Urban water cycle.

are in poor condition, the wastewater generation rate may be considerably higher or lower due to infiltration/inflow or exfiltration, respectively. Given losses at various points in the overall system and potential downstream water rights, the actual available water would most likely be about 50% of the potential value, but the resulting impact on the available water supply would still be impressive.

As urban areas continue to grow, pressure on local water supplies will continue to increase. Already, groundwater aquifers used by over half of the world population are being overdrafted (Brown, 2011). As a result, it is no longer advisable to use water once and dispose of it; it is important to identify ways to reuse water. Reuse

DID YOU KNOW?

The energy required to deliver 1 million gallons of clean water from a lake or river is 1400 kilowatt-hours; from groundwater, 1800 kilowatt-hours; from wastewater, 2350 to 3300 kilowatt-hours; and from seawater, 9780 to 16,500 kilowatt-hours (Webber, 2008).

will continue to increase as the world's population becomes increasingly urbanized and concentrated near coastlines, where local freshwater supplies are limited or are available only with large capital expenditure (Creel, 2003).

Water–Energy Connection

"Water is needed to generate energy. Energy is needed to deliver water. Both resources are limiting the other—and both may be running short" (Webber, 2008, p. 2). This is an important statement because most people have no idea (not even a clue) about the water–energy connection. The fact is energy efficiency and sustainability are key drivers of water reuse, which is why water reuse is so integral to sustainable water management. The water–energy connection recognizes that water and energy are mutually dependent—energy production requires large volumes of water, and water infrastructure requires large amounts of energy (NCSL, 2014). Water reuse is a critical factor in slowing the compound loop of increased water and energy use witnessed in the water–energy connection.

Water reuse is integral to sustainable water management because it allows water to remain in the environment and to be preserved for future uses while meeting the water requirements of the present. Water and energy are interconnected, and sustainable management of either resource requires consideration of the other. Water reuse reduces energy use by eliminating additional portable water treatment and associated water conveyance because reclaimed water typically offsets potable water use and is used locally. The energy required for capturing, treating, and distributing water and the water required to produce energy are inextricably linked. Water reuse can achieve two benefits: offsetting water demand and providing water for energy production. Thermoelectric energy generation currently uses about half of the water resources consumed in the United States and is a major potential user of reclaimed water (Kenny et al., 2009).

DID YOU KNOW?

Assuming plants draw and dump water, the amount of water required to generate 1 megawatt-hour of electricity using gas/steam combined cycle is 7400 to 20,000 gallons; for coal and oil, 21,000 to 50,000 gallons; and for nuclear, 25,000 to 60,000 gallons (Webber, 2008).

DID YOU KNOW?

Running the hot water for 5 minutes uses about the same amount of energy as burning a 60-watt light bulb for 14 hours (USEPA, 2014a).

Environmental Protection

Water scarcity and water supply demands in arid and semi-arid regions drive reuse as an alternate water supply; however, there are still many water reuse programs in the United States that have been initiated in response to rigorous and costly require- ments to remove nutrients (mainly nitrogen and phosphorus) from effluent discharge to surface waters. Environmental concerns over negative impacts from increasing nutrient discharges to coastal waters are resulting in mandatory reductions in the number of ocean discharges in Florida and California. By eliminating effluent dis- charges for all or even a portion of the year through water reuse, a municipality may be able to avoid or reduce the need for costly nutrient removal treatment pro- cesses or maintain wasteload allocations while expanding capacity. Avoiding costly advanced wastewater treatment facilities was the key driver for St. Petersburg, Florida, to initiate reclaimed water distribution to residential, municipal, commer- cial, and industrial demands when the state legislature enacted the Wilson–Grizzle Act in 1972, significantly restricting nutrient discharges into Tampa Bay. Today, St. Petersburg serves more than 10,250 residential connections in addition to parks, schools, golf courses, and commercial/industrial applications, including 13 cooling towers. Another example is King County, Washington, which is implementing reuse to reduce the discharge of nutrients into Puget Sound to address the health of this marine water (USEPA, 2012b).

WASTEWATER QUALITY

Based on the author's many years of experience in wastewater treatment plant opera- tions, including advanced wastewater treatment with biological nutrient removal (BNR), rapid and slow sand filtration, membrane, and RO systems, it can be said without reservation that the water treated at the 14 different plants was cleaner than the water bodies the plants outfalled to. Every test conducted on the treated water showed that the effluent was not only clean but also safe to drink.

The doubting Thomases might ask how could the yuck factor possibly be removed from the toilet-flush-to-drinking-water scenario. Consider the drinking water supplies for the cities of Philadelphia, Nashville, Cincinnati, and New Orleans. Philadelphia draws its drinking water from the Delaware River, Nashville draws its drinking water from the Cumberland River, Cincinnati draws its drinking water from the Ohio River, and New Orleans draws its water from the Mississippi River. According to the Environmental Working Group's report on the 50 most polluted rivers in the country, the four rivers mentioned here are ranked as follows:

- The Mississippi River ranks number one; it is the most polluted river in the country.
- The Ohio River is the third most polluted river in the country.
- The Delaware River is the eighth most polluted river in the country.
- The Cumberland River is not ranked among the top 50 most polluted rivers but historically has been listed as a very polluted water body; it must be pointed out, however, that its pollution levels have been lowered in recent years.

This list will probably surprise many people who are aghast at the thought of drinking water from such polluted rivers. So, how is the water from such polluted water sources made safe to drink? Earlier we talked about conventional water treatment and ancillary processes used to treat source waters; it is the unit processes within these treatment systems that make the source water potable and safe for consumption. We are talking about state-of-the-art water treatment, including rapid system filtration and, in some cases, granular activated carbon (GAC) treatment. Moreover, after the clean and safe potable water is used it is disposed of via state-of-art wastewater treatment and then outfalled (as clean, safe, drinkable water) into the polluted rivers where the water originated. Again, properly treated wastewater is many times over (beyond the nth degree) of better quality than the receiving water it is outfalled to. The bottom line is that whether we call it *de facto* (or unplanned) water reuse or something else, the fact is we are using and consuming the water we flush down our toilets and pour down our drains ... and, quite frankly, it tastes pretty good.

REPLACING THE YUCK FACTOR

To date, no regulations or criteria have been developed or proposed in the United States specially for direct potable reuse (DPR, or toilet to tap). Past regulatory evaluations of this practice generally have been deemed unacceptable due to a lack of definitive information related to public health protection. Still, the *de facto* reuse of treated wastewater effluent as a water supply is common in many of the nation's water systems, with some drinking water treatment plants using water for which a large fraction has originated as wastewater effluent from upstream communities, especially under low-flow conditions (Spellman, 2007). Considering that unplanned reuse is already widely practiced, DPR may be a reasonable option based on significant advances in treatment technology and monitoring methodology in the last decade and health effects data from indirect potable reuse (IPR) projects and DPR demonstration facilities, such as the water quality and treatment performance data generated at operational IPR projects such as Montebello Forebay in Los Angeles County (WRRF, 2011).

Water reuse treatment technology consists of a portfolio of treatment options that are capable of mitigating microbial and chemical contaminants in reclaimed water. The options include engineered treatment and natural processes. The holdup on widespread implementation of toilet-to-tap reuse is the lack of guidance for design and operation of natural processes; it is the biggest deterrent to their expanded use in engineered reuse systems.

It is important to point out that included in engineered reuse systems is the application of RO membranes to the treatment of municipal wastewater; this practice has been highly successful. Reverse osmosis can remove dissolved solids that cannot be removed by biological or other conventional municipal treatment processes. In addition, RO membranes can also lower organics, color, and nitrate levels; however, extensive pretreatment and periodic cleaning are usually necessary to maintain acceptable membrane water fluxes. Early studies showed that high removals of total dissolved solids (TDS) and moderate removals of organics could be achieved (Cruver, 1976; Fang and Chian, 1976; Lim and Johnston, 1976). Tsuge and Mori (1977) showed that tubular membranes (with a substantial pretreatment system) could remove both inorganics and organics from municipal secondary effluent and produce water meeting drink water standards. The day is nearing when a pipe-to-pipe connection, wastewater treatment plant to household tap, will be generally accepted without second thoughts.

DESALINATION

Desalination of the sea is not the answer to our water problems. It is survival technology, a life support system, an admission of the extent of our failure.

—John Archer, author

Another well known application of RO systems is desalination (removal of salt from feedwater) of seawater and brackish water. Having served as an engineering enlisted person and engineering officer for many years aboard U.S. Naval ships with extended at-sea tours, the author came to know evaporators and flash distillers intimately. These simple engineering devices were used in conjunction with ship-generated steam within a vacuum-vessel to convert seawater to potable water. I have to say the water tasted normal to all; I never heard anyone say that it had a salty taste.

What qualifies as a salty taste? To answer this question we need to look at the parameters for saline water:

- *Freshwater*—Less than 1000 ppm
- *Slightly saline water*—From 1000 to 3000 ppm
- *Moderately saline water*—From 3000 to 10,000 ppm
- *Highly saline water*—From 10,000 to 35,000 ppm

Converting seawater, which contains about 35,000 ppm of salt, to potable water for sailors at sea is a necessity, of course. In addition, when you consider that more than 99% of the Earth's water is saltwater, that many nations border the seas, and that approximately 60% of the Earth's population lives along ocean coastlines, it only

DID YOU KNOW?

Reverse osmosis is used by 44% of today's desalination plants.

DID YOU KNOW?

- It is estimated that some 30% of the world's irrigated areas suffer from salinity problems, and remediation is seen to be very costly.
- In 2002, there were about 12,500 desalination plants around the world in 120 countries. They produce some 14 million cubic meters per day of freshwater, which is less than 1% of total world consumption.
- The most important users of desalinated water are in the Middle East (mainly Saudi Arabia, Kuwait, United Arab Emirates, Qatar, and Bahrain), which uses about 70% of worldwide capacity, and in North Africa (mainly Libya and Algeria), which uses about 6% of worldwide capacity.
- Among industrialized countries, the United States is one of the most important users of desalinated water, especially in California and parts of Florida. The cost of desalination has kept desalination from being used more often (USGS, 2014).

seems reasonable (and practical to many) for these nations to look at the seemingly endless oceans as a source of potable water. This certainly is the view of regions in the Middle East, where desalting seawater and saline groundwater is normal operating procedure. In fact, the Middle East boasts about two thirds of the world's desalting capacity (Wachinski, 2013).

Distillation dominated the desalination of seawater from ancient times until about 1970. In ancient times, many civilizations used this process on their ships (as we do today in a slightly different manner, of course) to convert seawater into drinking water. Today, desalination plants are used to convert seawater to drinking water on ships and in many arid regions of the world, as well as to treat water in other areas that is fouled by natural and unnatural contaminants. Distillation is perhaps the one water treatment technology that most completely reduces the widest range of drinking water contaminants.

Since 1970, advances and improvements in RO system technology have resulted in a substantial increase in its application. In 1988, 1742 RO plants represented approximately 49% of the total 3527 desalination plants in the world. RO accounts for approximately 23% of the world's desalination capacity (AWWA, 1996).

MISCELLANEOUS RO APPLICATIONS

In addition to purifying drinking water, wastewater purification, and desalination of seawater for potable use, RO systems are used in other applications. For example, RO systems have applications in the food industry, maple syrup production, hydrogen production, reef aquarium upkeep, and window cleaning. Each of these applications is briefly described below.

DID YOU KNOW?

The characteristic flavor and color of maple syrup develop during heating in the evaporator.

FOOD INDUSTRY

Not only is RO a more economical process for concentrating food liquids (such as fruit juices) than conventional heat-treating processes and one that is extensively used in dairy to produce whey protein powders and in the wine-making business, but it is also used, along with thermal processing, irradiation, hydrostatic processing, ohmic processing, and ultrafiltration, to improve the quality of food stuffs.

MAPLE SYRUP PRODUCTION

In the 1940s, some maple syrup producers started using RO to remove about 75 to 90% of the water from sap before the sap is boiled down to syrup. RO typically concentrates maple sap from 2% to 8–12% sugar prior to heating in an evaporator. RO also reduces the amount of energy required to process the syrup and reduces the exposure of the syrup to high temperatures. This can substantially reduce the cost of producing maple syrup by reducing the amount of water that must be removed by heat-driven evaporation and thus the amount of time and evaporator fuel required to process sap into syrup. In the RO system used in processing maple syrup, it is important to monitor the membranes to ensure that they are not contaminated and degraded by microbes.

HYDROGEN PRODUCTION

In small-scale hydrogen production, RO can be used to prevent the formation of minerals on the surface of electrodes.

REEF AQUARIUMS

Because tapwater can contain excessive chlorine, chloramines, copper, nitrites, nitrates, silicates, and phosphates that are detrimental to the sensitive organisms in a reef environment, RO is used to mix an artificial seawater, which is better suited for use in reef aquariums.

WATER-FED POLE WINDOW WASHING

Removing total dissolved solids using RO-produced pure water has been shown to be an effective method for cleaning windows, buildings, cars, truck trailers, vinyl awnings, and other items. City water is full of different kinds of dissolved solids held in suspension in the water. When the water is cycled through an RO system the solids are removed and the water is purified. The problem is water does not like to be purified; it attracts and gloms up dirt like a magnet. This natural dirt attraction tendency of pure water is what makes it so effective in cleaning various surfaces. Using extended pure-water-fed poles makes window washing easier and more effective by removing the dirt and leaving a clean surface.

MEMBRANE CONCENTRATE DISPOSAL

To this point in the book we have discussed the basics of reverse osmosis operating systems, their various applications, their benefits, and their usages at the present time. It is important to point out, however, that along with the good there is the not so good; that is, RO systems have their advantages but they also have a few disadvantages. The disadvantage discussed here is the major one—that is, concentrate disposal. Where is the concentrate waste stream to be disposed of? Earlier we discussed flash distillation (evaporators) aboard ships and how the ships convert seawater to quality potable water. We did not mention, however, that ships at sea have a huge advantage over all other RO system operations in that at sea the concentrate waste stream is easily disposed of. Consisting of a concentrated brine, the shipboard concentrate is simply pumped overboard into the salty sea.

MASS BALANCE

To gain a better understanding of membrane disposal issues and techniques we begin with a discussion of mass balance. The simplest way to express the fundamental engineering principle of mass balance is to say, "Everything has to go somewhere." More precisely, the *law of conservation of mass* says that when chemical reactions take place matter is neither created nor destroyed. What this important concept allows us to do is track materials (concentrates)—that is, pollutants, microorganisms, chemicals, and other materials from one place to another. The concept of mass balance plays an important role in reverse osmosis system operations (especially in desalination) where we assume a balance exists between the material entering and leaving the RO system: "What comes in must equal what goes out." The concept is very helpful in evaluating biological systems, sampling and testing procedures, and many other unit processes within any treatment or processing system.

All desalination processes have two outgoing process streams—the product water, which is lower in salt than the feed water, and a concentrated stream that contains the salts removed from the product water (see Figure 8.6). Even distillation has a "bottoms" solution that contains salt from the vaporized water. As shown in Figure 8.6, the higher concentrated stream is called the *concentrate*. The nature of the concentrate stream depends on the salinity of the feed water, the amount of product water recovered, and the purity of the product water. To determine the volume and concentration of the two outgoing streams, a mass balance is constructed. The recovery rate of water, the rejection rate of salt, and the input flow and concentration are necessary to solve equations for the flow and concentration of the product and

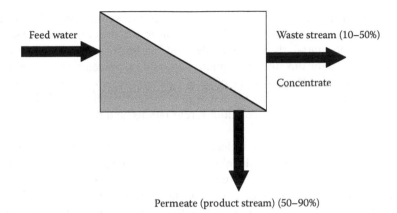

Feed water

Waste stream (10–50%)

Concentrate

Permeate (product stream) (50–90%)

FIGURE 8.6   Reverse osmosis desalination streams.

concentrate. RO concentrate is disposed of by several methods, including surface water discharge, sewer discharge, deep well injection, evaporation ponds, spray irrigation, and zero liquid discharge.

SURFACE WATER AND SEWER DISPOSAL[*]

Disposal of concentrate to surface water and sewers are the two most widely used disposal options for both desalting membrane processes. Post-1992 data provide the following statistics (USDOI, 2006):

Disposal Option	Percent of Desalting Plants
Surface water disposal	45%
Disposal to sewer	42%
Total	87%

This disposal option, although not always available, is the simplest option in terms of equipment involved and is frequently the lowest cost option. As will be seen, however, the design of an outfall structure for surface water disposal can be complex.

Disposal to surface water involves conveyance of the concentrate or backwash to the site of disposal and an outfall structure that typically involves a diffuser and outlet ports or valve mounted on the diffuser pipe. Factors involved in the outfall design are discussed in this section, and cost factors are presented. However, due to the large number of cost factors and the large variability in design conditions associated with surface water disposal, a relatively simple cost model cannot be developed. Disposal to surface waters requires a National Pollutant Discharge Elimination System (NPDES) permit.

---

[*] Adapted from USDOI, *Membrane Concentrate Disposal: Practices and Regulations*, U.S. Department of the Interior, Washington, DC, 2006.

Disposal to the sewer involves conveyance to the sewer site and typically a negotiated fee to be paid to the WWTP. Because the negotiated fees can range from zero to substantial, there is no model that can be presented. No disposal permits are required for this disposal option. Disposal of concentrate or backwash to the sewer, however, affects WWTP effluent that requires an NPDES permit. With regard to design considerations for disposal to surface water, a brief discussion of ambient conditions, discharge conditions, regulations, and the outfall structure are discussed below.

Because receiving waters can include rivers, lakes, estuaries, canals, oceans, and other bodies of water, the range of ambient conditions can vary greatly. *Ambient conditions* include the geometry of the receiving water bottom, and the receiving water salinity, density, and velocity. Receiving water salinity, density, and velocity may vary with water depth, distance from the discharge point, and time of day and year.

Discharge conditions include the discharge geometry and the discharge flow conditions. The discharge geometry can vary from the end of the pipe to a lengthy multiple-port diffuser. The discharge can be at the water surface or submerged. The submerged outfall can be buried (except for ports) or not. Much of the historical outfall design work deals with discharges from WWTPs. These discharges can be very large—up to several hundred million gallons per day in flow. In ocean outfalls and in many inland outfalls, these discharges are of lower salinity than the receiving water, and the discharge has positive buoyancy. The less dense effluent rises in the more dense receiving water after it is discharged.

The volume of flow of membrane concentrates is on the lower side of the range of WWTP effluent volumes, extending up to perhaps 15 MGD at present. Membrane concentrate, as opposed to WWTP effluent, tends to be of higher salinity than most receiving waters, resulting in a condition of negative buoyancy where the effluent sinks after it is discharged. This raises concerns about the potential impact of the concentrate on the benthic community at the receiving water bottom. Any possible effect on the benthic community is a function of the local ecosystem, the composition of the discharge, and the degree of dilution present at the point of contact. The chance of an adverse impact is reduced by increasing the amount of dilution at the point of bottom contact through diffuser design.

With regard to concentrate discharge regulations it is important to note that receiving waters can differ substantially in their volume, flow, depth, temperature, composition, and degree of variability in these parameters. The effect of discharge of a concentrate or backwash to a receiving water can vary widely depending on these factors. The regulation of effluent disposal to receiving water involves several considerations, including the end-of-pipe characteristics of the concentrate or backwash. Comparison is made between receiving water quality standards (dependent on the classification of the receiving water) and the water quality of the effluent to determine disposal feasibility. In addition, in states such as Florida, the effluent must also pass tests where test species, chosen based on the receiving water characteristics, are exposed to various dilutions of the effluent. Because the nature of the concentrate or backwash is different than that of the receiving water, there is a region near the discharge area where mixing and subsequent dilution of the concentrate or backwash occurs.

Where conditions cannot be met at the end of the discharge pipe, a mixing zone may be granted by the regulatory agency. The mixing zone is an administrative construct that defines a limited area or volume of the receiving water where this initial dilution of the discharge is allowed to occur. The definition of an allowable mixing zone is based on receiving water modeling. The regulations require that certain conditions be met at the edge of the mixing zone in terms of concentration and toxicity.

Once the mixing zone conditions are met, then the outfall structure can be properly designed and installed. Actually, the purpose of the outfall structure is to ensure that mixing conditions can be met and that discharge of the effluent, in general, will not produce any damaging effect on the receiving water, its lifeforms, wildlife, and the surrounding area.

In a highly turbulent and moving receiving water with large volume relative to the effluent discharge, simple discharge from the end of a pipe may be sufficient to ensure rapid dilution and mixing of the effluent. For most situations, however, the mixing can be improved substantially through the use of a carefully designed outfall structure. Such a design may be necessary to meet regulatory constraints. The most typical outfall structure for this purpose consists of a pipe of limited length mounted perpendicular to the end of the delivery pipe. This pipe, called a *diffuser*, has one or more discharge ports along its length.

Disposal to the Sewer

Where possible, this means of disposal is simple and usually cost effective. Disposal to a sewer does not require a permit but does require permission from the wastewater treatment plant. The impact of both the flow volume and composition of the concentrate will be considered by the WWTP, as it will affect their capacity buffer and their NPDES permit. The high volume of some concentrates prohibits their discharge to the local WWTP. In other cases, concerns are focused on the increased TDS level of the WWTP effluent that results from the concentrate discharge. The possibility of disposal to a sewer is highly site dependent. In addition to the factors mentioned, the possibility is influenced by the distance between the two facilities, by whether the two facilities are owned by the same entry, and by future capacity increases anticipated. Where disposal to a sewer is allowed, the WWTP may be required to pay fees based on volume or composition.

Deep Well Disposal

Injection wells are a disposal option in which liquid wastes are injected into porous subsurface rock formations. Depths of the wells typically range from 1000 to 8000 feet. The rock formation receiving the waste must possess the natural ability to contain and isolate it. Paramount in the design and operation of an injection well is the ability to prevent movement of wastes into or between underground sources of drinking water. Historically, this disposal option has been referred to as *deep well injection* or *disposal to waste disposal wells*. Because of the very slow fluid movement in the injection zone, injection wells may be considered a storage method rather than a disposal method; the wastes remain there indefinitely if the injection program has been properly planned and carried out.

Because of their ability to isolate hazardous wastes from the environment, injection wells have evolved as the predominant form of hazardous waste disposal in the United States. According to a 1984 study (Gordon, 1984), almost 60% of all hazardous waste disposed of in 1981, or approximately 10 billion gallons, was injected into deep wells. By contrast, only 35% of this waste was disposed of in surface impoundments and less than 5% in landfills. The study also found that a still smaller volume of hazardous waste, under 500 million gallons, was incinerated in 1981. Although RO concentrate is not classified as hazardous, injection wells are widely used for concentrate disposal in the state of Florida.

A study prepared for the Underground Injection Practices Council showed that relatively few injection well malfunctions have resulted in contamination of water supplies (Strycker and Collins, 1987). However, other studies have documented instances of injection well failure resulting in contamination of drinking water supplies and groundwater resources (Gordon, 1984).

Injection of hazardous waste can be considered safe if the waste never migrates out of the injection zone; however, there are at least five ways a water may migrate and contaminate potable groundwater (Strycker and Collins, 1987):

- Wastes may escape through the well bore into an underground source of drinking water because of insufficient casing or failure of the injection well casing due to corrosion or excessive injection pressure.
- Wastes may escape vertically outside of the well casing from the injection zone into an underground source of drinking water (USDW) aquifer.
- Wastes may escape vertically from the injection zone through confining beds that are inadequate because of high primary permeability, solution channels, joints, faults, or induced fractures.
- Wastes may escape vertically from the injection zone through nearby wells that are improperly cemented or plugged or that have inadequate or leaky casing.
- Wastes may contaminate groundwater directly by lateral travel of the injected wastewater from a region of saline water to a region of freshwater in the same aquifer.

EVAPORATION POND DISPOSAL

Solar evaporation, a well-established method for removing water from a concentrate solution, has been used for centuries to recover salt (sodium chloride) from seawater. There are also installations that are used for the recovery of sodium chloride and other chemicals from strong brines, such as the Great Salt Lake and the Dead Sea, and for the disposal of brines resulting from oil well operation (Office of Saline Water, 1970).

Evaporation ponds for membrane concentrate disposal are most appropriate for smaller volume flows and for regions having a relatively warm, dry climate with high evaporation rates, level terrain, and low land costs. These criteria apply predominantly in the western half of the United States—in particular, the southwestern portion. Advantages associated with evaporation ponds include the following:

- They are relatively easy and straightforward to construct.
- Properly constructed evaporation ponds are low maintenance and require little operator attention compared to mechanical equipment.
- Except for pumps to convey the wastewater to the pond, no mechanical equipment is required.
- For smaller volume flows, evaporation ponds are frequently the least costly means of disposal, especially in areas with high evaporation rates and low land costs.

Despite the inherent advances of evaporation ponds, they are not without some disadvantages that can limit their application:

- They can require large tracts of land if they are located where the evaporation rate is low or the disposal rate is high.
- Most states require impervious liners of clay or synthetic membranes such as polyvinylchloride (PVC) or Hypalon®. This requirement substantially increases the costs of evaporation ponds.
- Seepage from poorly constructed evaporation ponds can contaminate underlying potable water aquifers.
- There is little economy of scale (i.e., no cost reduction resulting from increased production) for this land-intensive disposal option. Consequently, disposal costs can be large for all but small-sized membrane plants.

In addition to the potential for contamination of groundwater, evaporation ponds have been criticized because they do not recover the water evaporated from the pond. However, the water evaporated is not "lost"; rather, it remains in the atmosphere for about 10 days and then returns to the surface of the Earth as rain or snow. This hydrologic cycle of evaporation and condensation is essential to life on land and is largely responsible for weather and climate.

With regard to evaporation pond design considerations, sizing of the ponds, determination of the evaporation rate, and pond depth are important parameters. Evaporation ponds function by transferring liquid water in the pond to water vapor in the atmosphere about the pond. The rate at which an evaporation pond can transfer this water governs the size of the pond. Selection of pond size requires determination of both the surface area and the depth needed. The surface area required is dependent primarily on the evaporation rate. The pond must have adequate depth for surge capacity and water storage, storage capacity for precipitated salts, and freeboard for precipitation (rainfall) and wave action.

Proper sizing of an evaporation pond depends on accurate calculation of the annual evaporation rate. Evaporation from a freshwater body, such as a lake, is dependent on local climatological conditions, which are very site specific. To develop accurate evaporation data throughout the United States, meteorological stations have been established at which special pans simulate evaporation from large bodies of water such as lakes, reservoirs, and evaporation ponds. The pans are fabricated to standard dimensions and are situated to be as representative of a natural body of water as possible. A standard evaporation pan is referred to as a Class A pan. The standardized dimensions of the pans and the consistent methods for collecting the evaporation data allow comparatively and reasonably accurate data to be developed for the United States. The data collection must cover several years to be reasonably accurate and representative of site-specific variations in climatic conditions. Published evaporation rate databases typically cover a 10-year or greater period and are expressed in inches per year.

The pan evaporation data from each site can be compiled into a map of pan evaporation rates. Because of the small heat capacity of evaporation ponds, they tend to heat and cool more rapidly than adjacent lakes and to evaporate at a higher rate than an adjacent natural pond of water. In general, experience has shown the evaporation rate from large bodies of water to be approximately 70% of that measured in a Class A pan (Bureau of Reclamation, 1969). This percentage is referred to as the *Class A pan coefficient* and must be applied to measured pan evaporation to arrive at actual lake evaporation. Over the years, site-specific Class A pan coefficients have been developed for the entire Untied States. Multiplying the pan evaporation rate by the pan coefficient results in a mean annual lake evaporation rate for a specific area.

Maps depicting annual average precipitation across the United States also are available. Subtracting the mean annual evaporation from the mean annual precipitation gives the net lake surface evaporation in inches per year. This is the amount of water that will evaporate from a freshwater pond (or the amount the surface level will drop) over a year if no water other than natural precipitation enters the pond. All of these maps assume an impervious pond that allows no seepage. Note that, in some parts of the country, the results of this calculation give a negative number, and in other parts of the country it is a positive number. A negative number indicates a net loss of water from a pond over a year, or a drop in the pond surface level. A positive number indicates more precipitation than evaporation at a particular site. A freshwater pond at one of these sites would actually gain water over a year, even if no water other than natural precipitation were added. Thus, such a site would not be a candidate for an evaporation pond. It is important to realize that data of this type are representative only of the particular sites of the individual meteorological stations, which may be separated by many miles. Climatic data specific to the exact site should be obtained if at all possible before actual construction of an evaporation pond.

The evaporation data described above are for freshwater pond evaporation; however, brine density has a marked effect on the rate of solar evaporation. Most procedures for calculating evaporation rate indicate that evaporation is directly proportional to vapor pressure. Salinity reduces evaporation primarily because the vapor pressure of the saline water is lower than that of freshwater and because dissolved salts lower the free energy of the water molecules. Cohesive forces acting between the dissolved ions and the water molecules may also be responsible for inhibiting evaporation, making it more difficult for the water to escape as vapor (Miller, 1989).

DID YOU KNOW?

Reverse osmosis concentrate streams are not easily disposed of in inland areas, as surface water and sanitary sewer discharges would not be allowed, and deep well injection may not be feasible depending on geologic features.

The lower vapor pressure and lower evaporation rate of saline water result in a lower energy loss and, thus, a higher equilibrium temperature than that of freshwater under the same exposure conditions. The increase in temperature of the saline water would tend to increase evaporation, but the water is less efficient in converting radiant energy into latent heat due to the exchange of sensible heat and long-wave radiation with the atmosphere. The net result is that, with the same input of energy, the evaporation rate of saline water is lower than that of freshwater.

For water saturated with sodium chloride salt (26.4%), the solar evaporation rate is generally about 70% of the rate for freshwater (Office of Saline Water, 1971). Studies have shown that the evaporation rate from the Great Salt Lake, which has a TDS level of between 240,000 and 280,000 mg/L, is about 80 to 82% of the rate for freshwater. Other studies indicate that evaporation rates of 2%, 5%, 10%, and 20% sodium chloride solutions are 97%, 98%, 93%, and 78%, respectively, of the rates of freshwater (Reclamation,1969). These ratios have been determined from both experiment and theory. However, there is no simple relationship between salinity and evaporation, for there are always complex interactions among site-specific variables such as air temperature, wind velocity, relative humidity, barometric pressure, water surface temperature, heat exchange rate with the atmosphere, incident solar absorption and reflection, thermal currents in the pond, and depth of the pond. As a result, these ratios should be used only as guidelines and with discretion. It is important to recognize that salinity can significantly reduce the evaporation rate and to allow for this effect when sizing the evaporation pond's surface area. In lieu of site-specific data, an evaporation ratio of 0.70 is a reasonable allowance for long-term evaporation reduction. This ratio is also considered to be an appropriate factor for evaporation ponds that are expected to reach salt saturation over their anticipated service life.

Pond depth is an important parameter in determining the pond evaporation rate. Studies indicated that pond depths ranging from 1 to 18 inches are optimal for maximizing evaporation rate. However, similar studies indicate only a 4% reduction in the evaporation rate as pond depth is increased from 1 to 40 inches (Reclamation, 1969). Very shallow evaporation ponds are subject to drying and cracking of the liners and are not functional in long-term service for concentrate disposal. From a practical operating standpoint, an evaporation pond must not only evaporate wastewater but also provide

- Surge capacity or contingency water storage
- Storage capacity for precipitated salts
- Freeboard for precipitation and wave action

For an evaporation pond to be a viable disposal alternative for membrane concentrate, it must be able to accept concentrate at all times and under all conditions so as not to restrict operation of the desalination plant. The pond must be able to accommodate variations in the weather and upsets in the desalination plant. The desalination plant cannot be shut down because the evaporation pond level is rising faster than anticipated.

To allow for unpredictable circumstances, it is important that design contingencies be applied to the calculated pond area and depth. Experience from the design of industrial evaporation ponds has shown that discharges are largest during the first

---

DID YOU KNOW?

To gain an understanding of what is meant by incident solar absorption, the following definitions are provided:

Incident ray—A ray of light that strikes (impinges upon) a surface. The angle between this ray and the perpendicular or normal to the surface is the angle of incidence.

Reflected ray—A ray that has rebounded from a surface.

Angle of incidence—The angle between the incident ray and a normal line.

Angle of reflection—The angle between the reflected ray and the normal line.

Angle of refraction—The angle between the refracted ray and the normal line.

Index of refraction—The ratio speed of light (*c*) in a vacuum to its speed (*v*) in a given material; it is always greater than 1.

---

year of plant operation, are reduced during the second year, and are relatively constant thereafter. A long-term, 20% contingency may be applied to the surface areas of the pond or its capacity to continuously evaporate water. The additional contingencies above the 20% (up to 50%) during the first and second years of operation are applied to the depth holding capacity of the pond.

Freeboard for precipitation should be estimated on the basis of precipitation intensity and duration for the specific site. There may also be local codes governing freeboard requirements. In lieu of site-specific data, an allowance of 6 inches for precipitation is generally adequate where evaporation ponds are most likely to be located in the United States. Freeboard for wave action can be estimated as follows (Office of Saline Water, 1970):

$$H_w = 0.047 \times W \times \sqrt{F} \qquad (8.1)$$

where
H_w = Wave height (ft).
W = Wind velocity (mph).
F = Fetch, or straight-line distance the wind can blow without obstruction (miles).

DID YOU KNOW?

Current concentrate disposal of membrane concentrate using evaporation ponds accounts for 5% of total disposal practices.

The run-up of waves on the face of the dike approaches the velocity head of the waves and can be approximated as 1.5 times the wave height (H_w). H_w is the free-board allowance for wave action and typically ranges from 2 to 4 feet. The minimum recommended combined freeboard (for precipitation and wave action) is 2 feet. This minimum applies primarily to small ponds.

Over the life of the pond (which should be sized for the same duration as the projected life of the desalination facility), the water will likely reach saturation and precipitate salts. The type and quantity of salts are highly variable and very site specific. Allowance in pond depth for precipitate salts can be made using Figure 8.7, which provides an estimate for the depth of precipitate produced as a function of

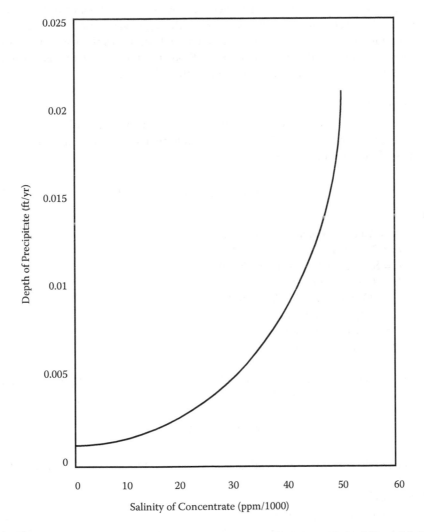

FIGURE 8.7   Rate of precipitation in an evaporation pond. (Adapted from Office of Saline Water, *Disposal of Brine by Solar Evaporation: Field Experiments*, U.S. Department of the Interior, Washington, DC, 1970.)

the salinity of the wastewater discharged to the pond (Office of Saline Water, 1970). For a given salinity, Figure 8.7 provides an estimate of precipitate produced (in feet per year) for each foot of wastewater discharged to the pond. Multiplying the annual deposition depth times the depth of water discharged to the pond each year and then by the life of the pond will result in the necessary allowance for the life of the pond.

In order to provide the non-engineer or non-scientist with an idea of how (for illustrative reasons only) the evaporation pond evaporation rate is mathematically determined, the following example is provided.

Estimating the Evaporation Pond Evaporation Rate

In lake, reservoir, and pond management, knowledge of evaporative processes is important for understanding how water losses through evaporation are determined. Evaporation increases the storage requirement and decreases the yield of lakes and reservoirs. Several models and empirical methods used for calculating lake and reservoir evaporative processes are described in the following text.

Water Budget Model

The water budget model for lake evaporation is used to make estimations of lake evaporation in some areas. It depends on an accurate measurement of the inflow and outflow of the lake and is expressed as

$$\Delta S = P + R + G_I - G_O - E - T - O \tag{8.2}$$

where

ΔS = Change in lake storage (mm).
P = Precipitation (mm).
R = Surface runoff or inflow (mm).
G_I = Groundwater inflow (mm).
G_O = Groundwater outflow (mm).
E = Evaporation (mm).
T = Transpiration (mm).
O = Surface water release (mm).

If a lake has little vegetation and negligible groundwater inflow and outflow, lake evaporation can be estimated by

$$E = P + R - O \pm \Delta S \tag{8.3}$$

Much of the following information is adapted from Mosner and Aulenbach (2003).

Energy Budget Model

According to Rosenberry et al. (1993), the energy budget model is recognized as the most accurate method for determining lake evaporation, although it is also the most costly and time-consuming method. The evaporation rate, E_{EB}, is given by (Lee and Swancar, 1996):

$$E_{EB} \text{ (cm/day)} = \frac{Q_s - Q_r + Q_a + Q_{ar} - Q_{bs} + Q_v - Q_x}{L(1 + BR) + T_0}$$ (8.4)

where

E_{EB} = Evaporation (cm/day).

Q_s = Incident shortwave radiation (cal/cm²/day).

Q_r = Reflected shortwave radiation (cal/cm²/day).

Q_a = Incident longwave radiation from atmosphere (cal/cm²/day).

Q_{ar} = Reflected longwave radiation (cal/cm²/day).

Q_{bs} = Longwave radiation emitted by lake (cal/cm²/day).

Q_v = Net energy advected by streamflow, groundwater, and precipitation (cal/cm²/day).

Q_x = Change in heat stored in water body (cal/cm²/day).

L = Latent heat of vaporization (cal/g).

BR = Bowen ratio (dimensionless).

T_0 = Water surface temperature (°C).

Priestly–Taylor Equation

The Priestly–Taylor equation is used to calculate potential evapotranspiration (Winter et al., 1995), which is a measure of the maximum possible water loss from an area under a specified set of weather conditions or evaporation as a function of latent heat of vaporization and heat flux in a water body:

$$PET \text{ (cm/day)} = \alpha(s/s + \gamma)[(Q_n - Q_x)/L]$$ (8.5)

where

PET = Potential evapotranspiration (cm/day).

α = 1.26 (dimensionless Priestly–Taylor empirically derived constant).

$(s/s + \gamma)$ = Parameters derived from the slope of the saturated vapor pressure–temperature curve at the mean air temperature; γ is the psychrometric constant, and s is the slope of the saturated vapor pressure gradient (dimensionless).

Q_n = Net radiation (cal/cm²/day).

Q_x = Change in heat stored in water body (cal/cm²/day).

L = Latent heat of vaporization (cal/g).

Penman Equation

The Penman equation for estimating potential evapotranspiration, E_0, can be written as (Winter et al., 1995):

$$E_0 = \frac{(\Delta/\gamma)H_e + E_a}{(\Delta/\gamma) + 1}$$ (8.6)

where

Δ = Slope of the saturation absolute humidity curve at air temperature.

γ = Psychrometric constant.

H_e = Evaporation equivalent of the net radiation.
E_a = Aerodynamic expression for evaporation.

DeBruin–Keijman Equation

The DeBruin–Keijman equation determines evaporation rates as a function of the moisture content of the air above the water body, the heat stored in the still water body, and the psychrometric constant, which is a function of atmospheric pressure and latent heat of vaporization (Winter et al., 1995):

$$PET \text{ (cm/day)} = [(SVP/0.95SVP) + 0.63\gamma)] \times (Q_n - Q_x) \tag{8.7}$$

where SVP is the saturated vapor pressure at mean air temperature (millibars/K), and all other terms have been defined previously.

Papadakis Equation

The Papadakis equation does not account for the heat flux that occurs in the still water body to determine evaporation (Winter et al., 1995). Instead, the equation depends on the difference in the saturated vapor pressure above the water body at maximum and minimum air temperatures, and evaporation is defined as

$$PET \text{ (cm/day)} = 0.5625[E_0 max - (E_0 min - 2)] \tag{8.8}$$

where all terms have been defined previously.

SPRAY IRRIGATION DISPOSAL

Land application methods include irrigation systems, rapid infiltration, and overland flow systems (Crites et al., 2000). These methods, and in particular irrigation, were originally used to take advantage of sewage effluent as a nutrient or fertilizer source as well as to reuse the water. Membrane concentrate has been used for land application in the spray irrigation mode. Using the concentrate in lieu of fresh irrigation water helps conserve natural resources, and in areas where water conservation is of great importance, spray irrigation is especially attractive. Because of the higher TDS concentration of RO concentrate, unless it is diluted (recall that dilution is the solution to pollution), the concentrate is less likely to be used for spray irrigation purposes.

Concentrate can be applied to cropland or vegetation by sprinkling or surface techniques for water conservation by exchange when lawns, parks, or golf course are irrigated and for preservation and enlargement of green belts and open spaces. Where

---

DID YOU KNOW?

The removal of nutrients is one advantage spray irrigation has compared to conventional disposal methods such as instream discharge.

the nutrient concentration of the wastewater for irrigation is of little value, hydraulic loading can be maximized to the extent possible, and system costs can be minimized. Crops such as water-tolerant grasses with low potential for economic return but with high salinity tolerance are generally chosen for this type of requirement.

Fundamental considerations in land application systems include knowledge of wastewater characteristics, vegetation, and public health requirements for successful design and operation. Environmental regulations at each site must be closely examined to determine if spray irrigation is feasible. Contamination of the groundwater and runoff into surface water are key concerns. Also, the quality of the concentrate—its salinity, toxicity, and the soil permeability—must be acceptable.

The principal objective in spray irrigation systems for concentrate discharge is ultimate disposal of the applied wastewater. With this objective, the hydraulic loading is usually limited by the infiltration capacity of the soil. If the site has a relatively impermeable subsurface layer or a high groundwater table, underdrains can be installed to increase the allowable loading. Grasses are usually selected for the vegetation because of their high nutrient requirements and water tolerance.

Other conditions must be met before concentrate irrigation can be considered as a practical disposal option. First, there must be a need for irrigation water in the vicinity of the membrane plant. If the need exists, a contract between the operating plant and the irrigation user would be required. Second, a backup disposal or storage method must be available during periods of heavy rainfall. Third, monitor wells must be drilled before an operating permit is obtained (Conlon, 1989).

With regard to design factors the following considerations are applicable to spray irrigation of concentrate for ultimate disposal:

- Salt, trace metals, and salinity
- Site selection
- Preapplication treatment
- Hydraulic loading rates
- Land requirements
- Vegetation selection
- Distribution techniques
- Surface runoff control

Salt, Trace Metals, and Salinity

Three factors that affect an irrigation source's long-term influence on soil permeability are the sodium content relative to calcium and magnesium, the carbonate and bicarbonate content, and the total salt concentration of the irrigation water. Sodium salts remain in the soil and may adversely affect its structure. High sodium concentrations in clay-bearing soils disperse soil particles and decrease soil permeability, thus reducing the rate at which water moves into the soil and educing aeration. If the soil permeability, or infiltration rate, is greatly reduced, then the vegetation on the irrigation site cannot survive. The hardness level (calcium and magnesium) will form insoluble precipitates with carbonates when the water is concentrated. This buildup of solids can eventually block the migration of water through the soil.

The U.S. Department of Agriculture's Salinity Laboratory developed a sodium adsorption ratio (SAR) to determine the sodium limit. It is defined as follows:

$$SAR = Na/[Ca + Mg)/2]^{1/2} \qquad (8.9)$$

where
   Na = Sodium (milliequivalent per liter, meq/L).
   Ca = Calcium (meq/L).
   Mg = Magnesium (meq/L).

High SAR values (>9) may adversely affect the permeability of fine-textured soils and can sometimes be toxic to plants.

Trace elements are essential for plant growth; however, at higher levels, some become toxic to both plants and microorganisms. The retention capacity for most metals in most soils is generally high, especially for pH above 7. Under low pH conditions, some metals can leach out of soils and may adversely affect the surface waters in the area.

Salinity is the most important parameter in determining the impact of the concentrate on the soil. High concentrations of salts whose accumulation is potentially harmful will be continually added to the soil with irrigation water. The rate of salt accumulation depends on the quantity applied and the rate at which it is removed from the soil by leaching. The salt levels in many brackish reverse osmosis concentrates can be between 5000 and 10,000 parts per million, a range that normally rules out spray irrigation.

In addition to the effects of total salinity on vegetation and soil, individual ions can cause a reduction in plant growth. Toxicity occurs when a specific ion is taken up and accumulated by the vegetation, ultimately resulting in damage to it. The ions of most concern in wastewater effluent irrigation are sodium, chloride, and boron. Other heavy metals can be very harmful, even if present only in small quantities. These include copper, iron, barium, lead, and manganese. These all have strict environmental regulations in many states.

In addition to the influence on the soil, the effect of the salt concentrations on the groundwater must be considered. The possible impact on groundwater sources may be a difficult obstacle where soil saturation is high and the water table is close to the surface. The chance of increasing background TDS levels of the groundwater is high with the concentrate. Due to this consideration, spray irrigation requires a runoff control system. An underdrain or piping distribution system may have to be installed under the full areas of irrigation to collect excess seepage through the soil and protect the groundwater sources. If high salinity concentrate is being used, scaling of

DID YOU KNOW?

Soluble salts in a water solution will conduct an electric current; thus, changes in electrical conductivity (EC) can be used to measure the water's salt content in electrical resistance units (deciSiemens per meter, or dS/m).

TABLE 8.1

Site Selection Factors and Criteria

Factor	Criterion
Soil	
Type	Loamy soils are preferred, but most soils from sands to clays are acceptable.
Drainability	Well-drained soil is preferred.
Depth	Uniformly 5 to 6 feet or more throughout sites is preferred.
Groundwater	
Depth to groundwater	A minimum of 5 feet is preferred.
Groundwater control	Control may be necessary to ensure renovation if the water table is less than 10 feet from the surface.
Groundwater movement	Velocity and direction of movement must be determined.
Slopes	Slopes of up to 20% are acceptable with or without terracing.
Underground formations	Formations should be mapped and analyzed with respect to interference with groundwater or percolating water movements.
Isolation	Moderate isolation from public is preferred; the degree of isolation depends on wastewater characteristics, method of application, and crop.
Distance from source of wastewater	An appropriate distance is a matter of economics.

the underdrain may become a problem. The piping perforations used to collect the water can be easily scaled because the openings are generally small. Vulnerability to scaling must be carefully evaluated before a project is undertaken.

Site Selection

Site selection factors and criteria for effluent irrigation are presented in Table 8.1. A moderately permeable soil capable of infiltration up to 2 inches per day on an intermittent basis is preferable. The total amount of land required for land application is highly variable but primarily depends on application rates.

Preapplication Treatment

Factors that should be considered when assessing the need for preapplication treatment include whether the concentrate is mixed with additional wastewaters before application, the type of vegetation grown, the degree of contact with the wastewater by the public, and the method of application. In four Florida sites, concentrate is aerated before discharge, because each plant discharges to a retention pond or ponds before irrigation. Aeration by increasing dissolved oxygen prevents stagnation and algae growth in the ponds and also supports fish populations. The ponds are required for flow equalization and mixing. Typically, concentrate is blended with biologically treated wastewater.

Hydraulic Loading Rates

Determining the hydraulic loading rate is the most critical step in designing a spray irrigation system. The loading rate is used to calculate the required irrigation area and is a function of precipitation, evapotranspiration, and percolation. The following equation represents the general water balance for hydraulic loading based on a monthly time period and assuming zero runoff:

$$HLR = ET + PER - PPT \qquad (8.10)$$

where
  HLR = Hydraulic loading rate.
  ET = Evapotranspiration.
  PER = Percolation.
  PPT = Precipitation.

In most cases, surface runoff from fields irrigated with wastewater is not allowed without a permit or, at least, must be controlled; it is usually controlled just so that a permit does not have to be obtained.

Seasonal variations in each of these values would be taken into account by evaluating the water balance for each month as well as the annual balance. For precipitation, the wettest year in 10 is suggested as reasonable in most cases. Evapotranspiration will also vary from month to month, but the total for the year should be relatively constant. Percolation includes that portion of the water that, after infiltration into the soil, flows through the root zone and eventually becomes part of the groundwater. The percolation rate used in the calculation should be determined on the basis of a number of factors, including soil characteristics underlying geologic conditions, groundwater conditions, and the length of the drying period required for satisfactory vegetation growth. The principal factor is the permeability of hydraulic conductivity of the least permeable layer in the soil profile.

Resting periods, standard in most irrigation techniques, allow the water to drain from the top few inches of soil. Aerobic conditions are thus restored, and air penetrates the soil. Resting periods may range from a portion of each day to 14 days and depend on the vegetation, the number of individual plots in the rotation cycle, and the availability of backup storage capacity.

To properly calculate an annual hydraulic loading rate, it is necessary to obtain monthly evapotranspiration, precipitation, and percolation rates. The annual hydraulic loading rate represents the sum of the monthly loading rates. Recommended loading rates range from 2 to 20 feet per year (Goigel, 1991).

Land Requirements

When a hydraulic loading rate has been determined, the required irrigation area can be calculated using the following equation:

$$A = Q \times K_1/ALR \qquad (8.11)$$

where
  A = Irrigation area (acre).
  Q = Concentrate flow (gpd).
  K_1 = 0.00112 d $\times$ ft^3 $\times$ acres/(hr $\times$ gal $\times$ ft^2).
  ALR = Annual hydraulic loading rate (ft/yr).

The total land area required for spray irrigation includes allowances for buffer zones and storage and, if necessary, land for emergencies or future expansion.

For loadings of constituents such as nitrogen, which may be of interest to golf course managers who need fertilizer for the grasses, the field area requirement is calculated as follows:

$$\text{Field area (acres)} = 3040 \times C \times Q/L_c \qquad (8.12)$$

where

    C = Concentration of constituent (mg/L).

    Q = Flow rate (MGD).

    L_c = Loading rate of constituent (lb/acre-yr).

Vegetation Selection

The important aspects of vegetation for irrigation systems are water needs and tolerances, sensitivity to wastewater constituents, public health regulations, and vegetation management considerations. The vegetation selection depends highly on the location of the irrigation site and natural conditions such as temperature, precipitation, and topsoil condition. Automated watering alone cannot always ensure vegetation propagation. Vegetation selection is the responsibility of the property owners. Woodland irrigation for growing trees is being conducted in some areas. The principal limitations on this use of wastewater include low water tolerances of certain trees and the necessity to use fixed sprinklers, which are expensive. Membrane concentrate disposal will generally be to landscape vegetation. Such application (e.g., to highway median and border strips, airport strips, golf courses, parks and recreational areas, wildlife areas) has several advantages. Problems associated with crops for consumption are avoided, and the irrigated land is already owned, so land acquisition costs are saved.

Distribution Techniques

Many different distribution techniques are available for engineered wastewater effluent applications. For irrigation, two main groups, sprinkling and surface application, are used. Sprinkling systems used for spray irrigation are of two types—fixed and moving. Fixed systems, often called solid set systems, may be either on the ground surface or buried. Both types usually consist of impact sprinklers mounted on risers that are spaced along lateral pipelines, which are, in turn, connected to main pipelines. These systems are adaptable to a wide variety of terrains and may be used for irrigation of either cultivated land or woodlands. Portable aluminum pipe is normally used for aboveground systems. This pipe has the advantage of relatively low capital cost but is easily damaged, has a short expected life because of corrosion, and must be removed during cultivation and harvesting operations. Pipe used for buried systems may be buried as deep as 1.5 feet below the ground surface. Buried systems usually have the greatest capital cost; however, they are probably the most dependable and are well suited to automated control. There are a number of different moving sprinkle systems, including center-pivot, side-roll, wheel-move, rotating-boom, and winch-propelled systems.

Surface Runoff Control

Surface runoff control depends mainly on the proximity of surface water. If run-off drains to a surface water, an NPDES permit may be required. This situation should be avoided if possible due to the complication of quantifying overland runoff. Berms can be built around the irrigation field to prevent runoff. Another alternative, although expensive, is a surrounding collection system. It is best to use precautions and backup systems to ensure that overwatering and subsequent runoff do not occur in the first place.

ZERO LIQUID DISCHARGE DISPOSAL

In this approach, evaporation is used to further concentrate the membrane concentrate. For the extreme limit of processing concentrate to dry salts, the method becomes a zero discharge option. Evaporation requires major capital investment, and the high energy consumption together with the final salt or brine disposal can result in significant disposal costs. Because of this, disposal of municipal membrane concentrate by mechanical evaporation would typically be considered as a last resort—that is, when no other disposal option is feasible. Cost aside, however, zero liquid discharge does offer some advantages:

- It may avoid a lengthy and tedious permitting process.
- It may gain quick community acceptance.
- It can be located virtually anywhere.
- It represents a positive extreme in recycling by efficiently using the water source.

When this thermal process is used following an RO system, for example, it produces additional product water by recovering high-purity distillate from the concentrate wastewater stream. The distillate can be used to help meet the system product water volume requirement. This reduces the size of the membrane system and, thus, the size of the membrane concentrate to be treated by the thermal process. In addition, because the product purity of the thermal process is so high (TDS in the range of 10 mg/L), some of the product water volume reduction of the system may be met by blending the thermal product with untreated source water. The usual concerns and considerations of using untreated water for blending must be addressed. The end result may be a system where the system product requirement is met by three streams: (1) membrane product, (2) thermal process product, and (3) bypass water.

Single- and Multiple-Effect Evaporators

Using steam as the energy source, it takes about 1000 British thermal units (Btu) to evaporate a pound of water. In a single-effect evaporator, heat released by the condensing steam is transferred across a heat exchange surface to an aqueous solution boiling at a temperature lower than that of the condensing stream. The solution absorbs heat, and part of the solution water vaporizes, causing the remaining solution to become richer in solution. The water vapor flows to a barometric or surface condenser, where it condenses as its latent heat is released to cooling water at a lower temperature. The finite temperature differences among the steam, the boiling liquid, and the condenser are the driving forces required for the heat transfer surface area to be less than infinite. Practically all of the heat removed from the condensing stream (which had been generated initially by burning fuel) is rejected to cooling water and is often dissipated to the environment without being of further use.

The water vapor that flows to the condenser in a single-effect evaporator is at a lower temperature and pressure than the heating stream but has almost as much enthalpy. Instead of releasing the latent heat to cooling water, the water vapor may be used as heating steam in another evaporator effect operating at a lower temperature and pressure than the first effect.

Additional effects may be added in a similar manner, each generating additional vapor, which may be used to heat a lower temperature effect. The vapor generated in the lowest temperature effect finally is condensed by releasing its latent heat to cooling water in a condenser. The economy of a single- or multiple-effect evaporator may be expressed as the ratio of kilograms of total evaporation to kilograms of heating steam. As effects are added, the economy increases, representing more efficient energy utilization. Eventually, added effects result in marginal added benefits, and the number of effects is thus limited by both practical and economic considerations. Multiple effect evaporators increase the efficiency (economy) but add capital costs in additional evaporator bodies.

More specifically, the number of effects, and thus the economy achieved, is limited by the total temperature difference between the saturation temperature of the heating steam (or other heat source) and the temperature of the cooling water (or other heat sink). The available temperature difference may also be constrained by the temperature sensitivity of the solution to be evaporated. The total temperature difference, less any losses, becomes allocated among effects in proportion to their resistance to heat transfer, the effects being thermal resistances in series.

The heat transfer surface area for each effect is inversely proportional to the net temperature difference available for that effect. Increasing the number of effects reduces the temperature difference and evaporation duty per effect, which increases the total area of the evaporator in rough proportion to the number of effects. The temperature difference available to each effect is reduced by boiling point elevation and by the decrease in vapor saturation temperature due to pressure drop. The boiling point elevation of a solution is the increase in boiling point of the solution compared to the boiling point of pure water at the same pressure; it depends on the nature of the solute and increases with increasing solute concentration. In a multiple-effect evaporator, the boiling point elevation and vapor pressure drop losses for all the effects must be summed and subtracted from the overall temperature difference between the heat source and sink to determine the net driving force available for heat transfer.

Vapor Compression Evaporator Systems (Brine Concentrators)

A vapor compression evaporator system, or brine concentrator, is similar to a conventional single-effect evaporator, except that the vapor released from the boiling solution is compressed in a compressor. Compression raises the pressure and saturation temperature of the vapor so that it may be returned to the evaporator steam chest to be used as heating steam. The latent heat of the vapor is used to evaporate more water instead of being rejected to cooling water. The compressor adds energy to the vapor to raise its saturation temperature above the boiling temperature of the solution by whatever net temperature difference is desired. The compressor is not completely efficient, as it is subject to small losses due to mechanical friction and larger losses due to non-isentropic compression. However, the additional energy required because of non-isentropic compression is not lost from the evaporator system; instead, it serves to superheat the compressed vapor. The compression energy added to the vapor is of the same magnitude as energy required to raise feed to the boiling point and make up for radiation and venting losses. By exchanging heat between the condensed vapors (distillate) and the product with the feed, it is usually possible to operate with little or no makeup heat in addition to the energy necessary to drive the compressor. The compressor power is proportional to the increase in saturation temperature produced by the compressor. The evaporator design must trade off between compressor power consumption and heat transfer surface area. Using the vapor compression approach to evaporate water requires only about 100 Btu to evaporate a pound of water. Thus, one evaporator body driven by mechanical vapor compression is equivalent to 10 effects, or a 10-body system driven by steam.

Although most brine concentrators have been used to process cooling water, concentrators have also been used to concentrate reject from RO plants. Approximately 90% of these concentrators operate with a seeded slurry process that allows the reject to be concentrated as much as 40 to 1 without scaling problems developing in the evaporator. Brine concentrators also produce a distilled product water that can be used for high-purity purposes or for blending with other water supplies. Because of the ability to achieve such high levels of concentration, brine concentrators can reduce or eliminate the need for alternative disposal methods such as deep well injection or solar evaporation ponds. When operated in conjunction with crystallizers or spray dryers, brine concentrators can achieve zero liquid discharge of RO concentrate under all climatic conditions.

Individual brine concentrator units range in capacity from approximately 10 to 700 gpm of feedwater flow. Units below 150 gpm of capacity are usually skid mounted, and larger units are field fabricated. A majority of operating brine concentrators are single-effect, vertical-tube, falling-film evaporators that use a calcium sulfate-seeded slurry process. Energy input to the brine concentrator can be provided by an electric-driven vapor compressor or by process steam from a host industrial facility. Steam-driven systems can be configured with multiple effects to minimize energy consumption. Product water quality is normally less than 10 mg/L TDS. Brine reject from the concentrator typically ranges between 2 and 10% of the feedwater flow, with TDS concentrations as high as 250,000 mg/L.

DID YOU KNOW?

In the British system of units, the unit of heat is the British thermal unit, or Btu. One Btu is the amount of heat required to raise 1 pound of water 1°F at normal atmospheric pressure (1 atm).

DID YOU KNOW?

Solids in water occur either in solution or in suspension and are distinguished by passing the water sample through a glass-fiber filter. The suspended solids are retained on top of the filter, and the dissolved solids pass through the filter with the water. When the filtered portion of the water sample is placed in a small dish and then evaporated, the solids in the water that remain as residue in the evaporating dish are the total dissolved solids (TDS). Dissolved solids may be organic or inorganic. Water may come into contact with these substances within the soil, on surfaces, and in the atmosphere. The organic dissolved constituents of water are from the decay products of vegetation, from organic chemicals, and from organic gases. Removing these dissolved minerals, gases, and organic constituents is desirable, because they may cause physiological effects and produce aesthetically displeasing color, taste, and odors.

Because of the corrosive nature of many wastewater brines, brine concentrators are usually constructed of high-quality materials, including titanium evaporator tubes and stainless-steel vessels suitable for 30-year evaporator life. For conditions of high chloride concentrations or other more corrosive environments, brine concentrators can be constructed of materials such as AL-6XN®, Incoloy® 825, or other exotic metals to meet performance and reliability requirements.

Figure 8.8 shows a schematic diagram of a typical single-effect, vertical-tube brine concentrator. RO concentrate enters a tank where the pH is adjusted to prepare for deaeration. The concentrate then passes through a heat exchanger and enters a deaerator, where noncondensable gases are removed. From the deaerator, the concentrate enters the evaporator sump, where it mixes with the brine slurry. The slurry is constantly recirculated from the sump to a floodbox at the top of the evaporator tube bundle. Water from the floodbox flows through brine distributors and moves as a thin film down the interior walls of the evaporator tubes.

Some of the brine evaporates and flows through mist eliminators before entering the vapor compressor, where additional heat is added. Vapor from the compressor then flows to the outside of the evaporator tubes, where its heat is transferred to the cooler brine falling inside the tubes. As the compressed vapor gives up heat, it condenses as product water and is collected and pumped through the feedwater heat exchanger, where it transfers its heat to the incoming feedwater.

The slurry process prevents scaling of the evaporator tubes. Calcium sulfate and silica precipitates build on calcium sulfate seed crystals in the recirculation brine instead of scaling on heat transfer surfaces. With the seeded slurry system, concentrations of up to 30% total solids can be reached in the recirculation water without scaling.

Brine concentrator technology was developed in the early 1970s to help thermal power stations achieve zero discharge of wastewater. At present, approximately 75 brine concentrators are in operation in the United States and overseas. Of these, about a dozen are being used to concentrate reject streams (RO concentrate) from industrial RO plants. The operating experiences of these plants have shown that

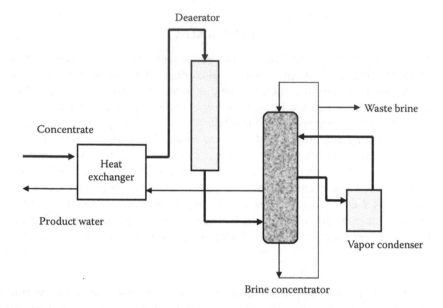

FIGURE 8.8 Schematic diagram of brine concentrator processor flow (pumps not shown). (Adapted from sales literature for Resources Conservation Co., Bellevue, WA.)

using brine concentrator evaporators for concentration of RO concentrate is a viable application and that the systems are highly reliable. Many operating systems have achieved on-stream operating availabilities greater than 90% over an extended period of years.

The specific design features and performance of brine concentrator systems are usually developed in conjunction with the equipment suppliers, based on the flow, chemistry, and economic factors involved in each case. The suppliers use proprietary methods to determine concentration factors to minimize brine concentrator blowdown rates while controlling scaling in the evaporator tubes. Process water recovery typically is limited by the formation of a double salt that is a combination of sodium and calcium sulfate; thus, recovery is dependent on the site-specific feedwater quality but is usually in the range of 90 to 98%.

Brine concentrators can be applied to a majority of RO concentrate streams. For such streams that are already saturated in calcium sulfate, brine concentrators operate without calcium sulfate addition. If the concentrations of calcium sulfate in the concentrate stream are insufficient, calcium sulfate is added as required to support the seeded slurry process.

Blowdown from brine concentrators is high in dissolved and suspended solids and saturated in calcium sulfate. Disposal can be handled in several ways. In areas where evaporation ponds are feasible and cost effective, brine concentrator blowdown can be settled in a decant basin and then pumped to an evaporation pond. Settled solids then are removed by a front-end loader, clamshell, or other device and transported to a land disposal facility. Blowdown can also be sent directly to a disposal pond, where the solids periodically can be removed, or to a pond constructed deep enough so that

> **DID YOU KNOW?**
>
> The method of evaporation selected is based on the characteristics of the RO membrane concentrate and the type of energy source to be used.

solids removal will not be required during the design life of the facility. In areas with negative net evaporation rates, or with expensive construction requirements for evaporation ponds, brine concentrator blowdown can be concentrated to a wet cake or dry powder using crystallizers or spray dryers.

Crystallizers

Crystallizer technology has been used for many years to concentrate feed streams in industrial processes. More recently, as the need to concentrate wastewaters has increased, this technology has been applied to reject from desalination processes, such as brine concentrate evaporators, to reduce wastewater to a transportable solid. Crystallizer technology is especially applicable in areas where solar evaporation pond construction costs are high, solar evaporation rates are negative, or deep well disposal is costly, geologically not feasible, or not permitted.

Crystallizers used for wastewater disposal range in capacity from about 2 to 50 gpm. These units have vertical cylindrical vessels with heat input from vapor compressors or an available steam supply. For small systems ranging from 2 to 6 gpm, steam-driven crystallizers are more economical. Steam can be supplied by a package boiler or a process source, if one is available. For larger systems, electrically driven vapor compressors are normally used to supply heat for evaporation.

Figure 8.9 shows a schematic of a forced-circulation vapor compression crystallizer. Wastewater, in the form of brine concentrator blowdown or from another source, is fed to the sump of the crystallizer. The incoming wastewater joins the circulating brine and is pumped to a shell-and-tube heat exchanger, where it is heated by vapor from the vapor compressor. Because the tubes in the heat exchanger are submerged, the brine is under pressure and will not boil. This arrangement prevents scaling in the tubes. The recirculating brine enters the crystallizer vapor body at an angle and swirls in a vortex. A small amount of the brine evaporates. As water is evaporated from the brine, crystals form. Most of the brine is recirculated to the heater. A small stream from the recirculating loop is sent to a centrifuge or filter to separate remaining water from the crystals. The vapor is compressed in a vapor compressor. Vapor from the compressor heats the recirculating brine as it condenses on the shell side of the heat exchanger. Condensate is collected and may be recycled to other processes requiring high-quality water. The crystallizer system produces a wet solid that can readily be transported for land disposal.

Typically, the crystallizer requires a purge stream of about 2% of the feed to the crystallizer. This is necessary to prevent extremely soluble species (such as calcium chloride) from building up in the vapor body and to prevent production of dry cake solids. The suggested disposal of this stream is to a small evaporation pond. The crystallizer produces considerable solids that can be disposed of to commercial

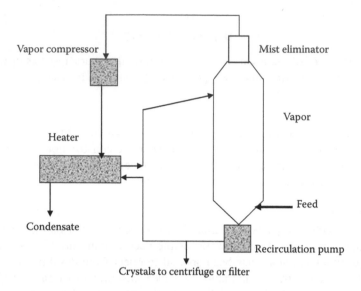

FIGURE 8.9 Schematic diagram of forced-circulation, vapor compression crystallizer process flow. (Adapted from sales literature for Resources Conservation Co., Bellevue, WA.)

landfill. The first crystallizers applied to power plant wastewater disposal experienced problems related to materials selection and process stability, but subsequent design changes and operating experience have produced reliable technology. For RO concentrate disposal, crystallizers would normally be operated with a brine concentrator evaporator to reduce brine concentrator blowdown to a transportable solid. Crystallizers can be used to concentrate RO reject directly, but their capital costs and energy usage are much higher than for a brine concentrator of equivalent capacity.

Spray Dryers

Spray dryers provide an alternative to crystallizers for concentration of wastewater brines to dryness. Spray dryers are generally more cost effective for smaller feed flows of less than 10 gpm. Figure 8.10 shows a schematic of a spray dryer. The system includes a feed tank, vertical spray drying chamber, and dried brine separator (bag filter) to collect dried solids. Concentrate from the desalination plant is routed to the feed tank, where it is recirculated and mixed to keep solids in suspension. From the feed tank, brine is pumped to the top of the drying chamber, where it is distributed into the chamber through a centrifugal brine atomizer. The atomizer consists of a shaft and a rotating disc that protrudes into the hot, gas stream. Air, heated by a gas, oil, or electric-powered heater, is also introduced at the top of the drying chamber. Hot air is pulled into the chamber and through the bag filter by the suction of an exhaust fan. The bag filter separates dry powder from the drying chamber from the hot air stream. Powder in the drying chamber is collected in a hopper, and the air exits to the atmosphere. Dry powder is discharged from the hopper to a pneumatic conveyor that transports it to a storage silo for transfer to a disposal site.

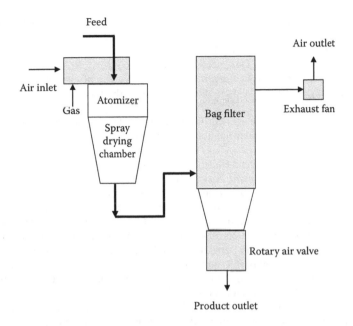

FIGURE 8.10   Schematic diagram of a typical spray dryer. (Adapted from sales literature for Resources Conservation Co., Bellevue, WA.)

Spray dryer technology for wastewater concentration was developed in the early 1980s. Like crystallizers, spray dryers offer an alternative to evaporation ponds, percolation ponds, and deep well disposal for RO concentrate disposal. For such applications, spray dryers are usually operated in conjunction with brine concentrator evaporators for feedwater flows up to 10 gpm. If the RO concentrate stream ranges from 1 to 10 gpm, spray dryers can be cost effective when applied directly to the steam, thus eliminating the brine concentrator evaporator.

REFERENCES AND RECOMMENDED READING

Adams, N.R. (1995). Organizational and activational effects of phytoestrogens on the reproductive tract of the ewe. *Proc. Soc. Exp. Biol. Med.*, 208: 87–91.

Adlercreutz, C.H., Goldin, B.R., Gorback, S.L., Hockerstedt, K.A., Watanabe, S., Hamalainen, E.K. et al. (1995). Soybean phytoestrogens intake and cancer risk. *J. Nutr.*, 125(Suppl. 3): 7578–7708.

Adlercreutz, H., Honjo, H., Higashi, A., Fotsis, T., Hamalainen, E., Haseqa, T. et al. (1991). Urinary excretion of lignans and isoflavonoid phytoestrogens in Japanese men and women consuming a traditional Japanese diet. *Am. J. Clin. Nutr.*, 54: 1093–1100.

Adlercreutz, H., Fotsis, T., Bannwart, C., Wahala, K., Makela, T., Brunow, G. et al. (1986). Determination of urinary lignans and phytoestrogen metabolites, potential antiestrogens and anticarcinogens in urine of women on various habitual diets. *J. Steroid Biochem.*, 25: 791–797.

Aguilar, A. and Raga, J.A. (1983). The striped dolphin epizootic in the Mediterranean Sea. *Ambio*, 22: 524–528.

Ahlborg, U.G., Becking, G., Birnbaum, L. et al. (1994). Toxic equivalency factors for dioxin-like PCBs. Report on a WHO–ECEH and IPCS consultation, December 1993. *Chemosphere*, 28(6): 1049–1067.

Alavanja, M.D., Samanic, C., Dosemeci, M., Lubin, J., Tarone, R., Lynch, C.F., Knott, C., Thomas, K., Hoppin, J.A., Barker, J. et al. (2003). Use of agricultural pesticides and prostate cancer risk in the Agricultural Health Study cohort. *Am. J. Epidemiol.*, 157: 800–814.

Albertazzi, P., Pansini, F., Bottazi, M., Bonaccorsi, G., De Aloysio, D., and Morton, M.S. (1999). Dietary soy supplementation and phytoestrogen levels. *Obstet. Gynecol.*, 94(2): 229–231.

Alexeeff, G.V., Kilgore, W.W., and Li, M.Y. (1990). Ethylene dibromide: toxicology and risk assessment. *Environ. Contam. Toxicol.*, 112: 49–122.

Anway, M. and Skinner, M. (2008). Transgenerational effects of the endocrine disruptor vinclozolin on the prostate transcriptome and adult onset disease. *Prostate*, 68: 515–529.

Anway, M., Cupp, A.S., Uzumeu, M., and Skinner, M.K. (2005). Epigenetic transgenerational actions of endocrine disruptors and male fertility. *Science*, 308: 1466–1469.

Arai, Y., Mori, T., Suzuki, Y., and Bern, H.A. (1983). Long-term effects of perinatal exposure to sex steroids and diethylstilbestrol on the reproductive system of male mammals. In: *International Review of Cytology* (Bourne, G.H. and Danielli, J.F., Eds.), Vol. 81, pp. 235–268. New York: Academic Press.

Arai, Y., Uehara, M., Sato, Y., Kimura, M.I., Eboshida, A., Adlercreutz, H. et al. (2000). Comparison of isoflavones among dietary intake, plasma concentration and urinary excretion for accurate estimation of phytoestrogen intake. *J. Epidemiol.*, 10: 127–135.

Arkoosh, M.R. (1989). Development of Immunological Memory in Rainbow Trout (*Oncorhynchus mykiss*) and Aflatoxin Modulation of the Response, PhD dissertation, Oregon State University, Corvallis.

Arkoosh, M.R. and Kaattari, S. (1987). The effect of early aflatoxin B1 exposure on *in vivo* and *in vitro* antibody response in rainbow trout, *Salmo gairdneri*. *J. Fish Biol.*, 31(Suppl. A): 19–22.

Auger, J., Kunstman, J.M., Czyglik, F., and Jouannet, P. (1995). Decline in semen quality among fertile men in Paris during the past 20 years. *N. Engl. J. Med.*, 332: 281–285.

Austin, H., Keil, J.E., and Cole, P. (1989). A prospective follow-up of cancer mortality in relation to serum DDT. *Am. J. Public Health*, 79: 43–46.

AWWA. (1996). *Water Treatment Membrane Processes*. Denver, CO: American Water Works Association.

Barbash, J.E. and Reinhard, M. (1989). Abiotic dehalogenation of 1,2-dichloroethane and 1,2-dibromethane in aqueous solution containing hydrogen sulfide. *Environ. Sci. Technol.*, 23: 1349–1358.

Barnes, D., Alford-Stevens, A., Birnbaum, L., Kutz, F., Wood, W., and Patton, D. (1991). Toxicity equivalency factors for PCBs? *Qual. Assur.*, 1(1): 70–81.

Bates, B.C., Kundzewicz, Z.W., Wu, S., and Palutikof, J.P. (2008). *Climate Change and Water*. Geneva, Switzerland: Intergovernmental Panel on Climate Change.

Baumann, P.C., Harshbarger, J.C., and Harman, K.J. (1990). Relationship between liver tumors and age in brown bullhead populations from two Lake Erie tributaries. *Sci. Total Environ.*, 94: 71–87.

Beland, P., DeGuise, S., Girard, C., Lagace, A., Martineau, D., Michaud, R., Muir, E.C.G., Norstrom, R.J., Pelletier, E., Ray, S., and Shugart, L.R. (1993). Toxic compounds and health and reproductive effects in St. Lawrence beluga whales. *J. Great Lakes Res.*, 19: 766–775.

Benbrahim-Tallaa, L. and Waalkes, M.P. (2008). Inorganic arsenic and human prostate cancer. *Environ. Health Perspect.*, 116: 158–164.

Benbrahim-Tallaa, L., Liu, J., Webber, M.M., and Waalkes, M.P. (2007). Estrogen signaling and disruption of androgen metabolism in acquired androgen independence during cadmium carcinogenesis in human prostate epithelial cells. *Prostate*, 67: 135–145.

Biava, C.G., Smuckler, E.A., and Whorton, M.D. (1978). The testicular morphology of individuals exposed to dibromochloropropane (DBCP). *Exp. Mol. Pathol.*, 29: 448–458.

Birnbaum, L.S. (1994). The mechanism of dioxin toxicity: relationship to risk assessment. *Environ. Health Perspect.*, 102(Suppl. 9): 157–167.

Birnbaum, L.S. (1999). TEFs: a practical approach to a real-world problem. *Hum. Ecol. Risk Assess.*, 5: 13–24.

Birnbaum, L.S. and DeVito, J.J. (1995). Use of toxic equivalency factors for risk assessment for dioxins and related compounds. *Toxicology*, 105(2–3): 391–401.

Bishop, C.A., Brooks, R.J., Carey, J.H., Ng, P., Norstrom, R.J., and Lean, D.R.S. (1991). The case for a cause–effect linkage between environmental contamination and development in eggs of the common snapping turtle (*Chelydra s. serpentina*) from Ontario, Canada. *J. Toxicol. Environ. Health*, 33: 521–547.

Blomkvist, G., Roos, A., Jensen, S., Bignert, A., and Olsson, M. (1992). Concentrations of DDT and PCB in seals from Swedish and Scottish waters. *Ambio*, 21: 539–545.

Borgmann, U. and Whittle, D.M. (1991). Contaminant concentration trends in Lake Ontario Lake trout (*Salvelinus namaycush*). *J. Great Lakes Res.*, 17: 368–381.

Boss, W.R. and Witschi, E. (1943). The permanent effects of early stilbestrol injections on the sex organs of the herring gull (*Larus argentatus*). *J. Exp. Zool.*, 94: 181–209.

Bowser, D., Frenkel, K., and Zelikoff, J.T. (1994). Effects of *in vitro* nickel exposure on macrophage-mediate immunity in rainbow trout. *Bull. Environ. Contam. Toxicol.*, 52: 367–373.

Breithaupt, H. (2004). A cause without a disease. *EMBO Rep.*, 5(1): 16–18.

Brown, D.P. (1987). Mortality of workers exposed to polychlorinated biphenyl—an update. *Arch. Environ. Health*, 42: 333–339.

Brown, L.R. (2011). The new geopolitics of food. *Foreign Policy*, 186, 54–63.

Buist, A.S. and Vollmer, W.M. (1990). Reflections of the rise in asthma morbidity and mortality. *JAMA*, 264: 1719–1720.

Bureau of Reclamation. (1969). *Disposal of Brine Effluents from Desalting Plants*. Washington, DC: U.S. Department of the Interior.

Burkhard, L.P. and Ankley, G.T. (1989). Identifying toxicants: NETAC's toxicity-based approach. *Environ. Sci. Technol.*, 23: 1438–1443.

Bysshe, S.E. (1982). Bioconcentration factor in aquatic organisms. In: *Handbook of Chemical Property Estimation Methods* (Lyman, W.J., Reehl, W.F., and Rosenblatt, D.H., Eds.), Chap. 5. New York: McGraw-Hill.

California Department of Water Resources. (2011). *SWPAO—Notice to State Water Project Contractors*. Sacramento: State Water Project Analyst's Office (http://www.water.ca.gov/swpao/notices.cfm).

Carlson, E., Giwercam, A., Keiding, N., and Skakkebaek, N.E. (1992). Evidence for decreasing quality of semen during past 50 years. *Br. Med. J.*, 305: 609–612.

Cassidy, A., Brown, J.E., Hawdon, A., Faughnan, M.A., King, L.J., Millward, J. et al. (2006). Factors affecting the bioavailability of soy isoflavones in humans after ingestion of physiologically relevant levels from different soy foods. *J. Nutr.*, 136: 45–51.

CDC. (2005). *Third National Report on Human Exposure to Environmental Chemicals*. Atlanta, GA: Centers for Disease Control and Prevention.

CDC. (2013). *Cancer among Women*. Atlanta, GA: Centers for Disease Control and Prevention (http://www.cdc.gov/cancer/dcpc/data/women.htm).

Chen, Z., Zheng, W., Custer, L.J., Dai, Q., Shu, X., Jin, F. et al. (1999). Usual dietary consumption of soy foods and its correlation with the excretion rate of isoflavonoids in overnight urine samples among Chinese women in Shanghai. *Nutr. Cancer*, 33: 82–87.

Colborn, T. (1991). Epidemiology of Great Lakes bald eagles. *J. Toxicol. Environ. Health*, 33: 395–453.

Colborn, T. and Clement C., Eds. (1992). *Chemically Induced Alterations in Sexual and Functional Development: The Wildlife/Human Connection*. Princeton, NJ: Princeton Scientific Publishing.

Colborn, T., vom Saal, F.S, and Soto, A.M. (1993). Developmental effects of endocrine-disrupting chemicals in wildlife and humans. *Environ. Health Perspect.*, 101(5): 378–384.

Cole, G.A. (1994). *Textbook of Limnology*, 4th ed. Prospect Heights, IL: Waveland Press.

Conlon, W.J. (1989). Disposal of concentrate from membrane process plants. *Waterworld News*, Jan./Feb.: 18–19.

Corcoran, E., Nellemann, C., Baker, E., Bos, R., Osborn, D., and Savelli, H., Eds. (2010). *Sick Water? The Central Role of Wastewater Management in Sustainable Development. A Rapid Response Assessment*. Geneva, Switzerland: United Nations Environment Programme.

Cornwell, T., Cohick, W., and Raskin, I. (2004). Dietary phytoestrogens and health. *Phytochemistry*, 65(8): 995–1016.

Couch, J.A. and Harshbarger, J.C. (1985). Effects of carcinogenic agents on aquatic animals: an environmental and experimental overview. *Environ. Carcinog. Rev.*, 3: 63–105.

Council on Environmental Quality. (1993). *Environmental Quality: 23rd Annual Report of the Council on Environmental Quality*. Washington, DC: U.S. Government Printing Office.

Creel, L. (2003). *Ripple Effects: Population and Coastal Regions*. Washington, DC: Population Reference Bureau.

Crites, R.W., Reed, S.C., and Bastian, R.K. (2000). *Land Treatment Systems for Municipal and Industrial Wastes*. McGraw-Hill, New York.

Cruver, J. (1976). Waste-treatment applications of reverse osmosis. *Trans. ASME*, 246.

Danish Environmental Protection Agency. (1995). *Male Reproductive Health and Environmental Chemicals with Estrogenic Effects*, Miljoprojekt 20. Copenhagen: Ministry of the Environment and Energy.

Daughton, C.G. (2010). *Drugs and the Environment: Stewardship and Sustainability*. Las Vegas, NV: U.S. Environmental Protection Agency.

Daughton, C.G. (2011). *Illicit Drugs in Municipal Sewage: Proposed New Non-Intrusive Tool to Heighten Public Awareness of Societal Use of Illicit/Abused Drugs and Their Potential for Ecological Consequences*. Washington, DC: U.S. Environmental Protection Agency (http://www.epa.gov/esd/bios/daughton/book-conclude.htm).

Davis, D. and Safe, S. (1990). Immunosuppressive activities of polychlorinated biphenyls in C57BL/6N mice: structure–activity relationships as Ah receptor agonists and partial antagonists. *Toxicology*, 63: 97–111.

Davis, E.P. and Bortone, S.A. (1992). Effects of kraft pulpmill effluent on the sexuality of fishes: an environmentally early warning? In: *Chemically Induced Alterations in Sexual and Functional Development: The Wildlife/Human Connection* (Colborn, T. and Clement, C.E., Eds.), pp. 113–127. Princeton, NJ: Princeton Scientific Publishing.

Dean, J.H., Cornacoff, G.F., Rosenthal, G.J., and Luster, M.I. (1994a). Immune system: evaluation of injury. In: *Principles and Methods of Toxicology* (Hayes, A.W., Ed.), pp. 1065–1090. New York: Raven Press.

Dean, J.H., Luster, M.I., Munson, A., and Kimoor, I., Eds. (1994b). *Immunotoxicology and Immunopharmacology*, 2nd ed. New York: Raven Press.

DeGuise, S., Martineau, D., Beland, P., and Fournier, M. (1995). Possible mechanisms of action of environmental contamination on St. Lawrence beluga whales (*Delphinapterus leucas*). *Environ. Health Perspect.*, 103: 73–77.

deSwart, R.L., Ross, P.S., Vedder, L.J., Timmerman, H.H., Heisterkamp, S., Van Loveren, H., Vos, J.G., Reijnders, P.J.H., and Osterhaus, A.D.M.E. (1994). Impairment of immune function in harbor seals (*Phoca vitulina*) feeding on fish from polluted waters. *Ambio*, 23: 155–159.

DeVito, M.J. and Birnbaum, L.S. (1995). Dioxins model chemicals for assessing receptor-mediated toxicity. *Toxicology*, 102: 115–123.

DeWally, E., Dodin, S., Verreault, R., Ayotte, P., Sauvé, L., Morin, J., and Brisson, J. (1994). High organochlorine body burden in women with estrogen receptor-positive breast cancer. *J. Natl. Cancer Inst.*, 86: 232–234.

Dixon, R.A. and Ferreira, D. (2002). Genistein. *Phytochemistry*, 60(3): 205–211.

Doerge, D.R. and Sheehan, D.M. (2002). Goitrogenic and estrogenic activity of soy isoflavones. *Environ. Health Perspect.*, 110(Suppl. 3): 349–353.

Doerge, D.R., Chang, H.C., Churchwell, M.I., and Holder, C.L. (2002). Analysis of soy isoflavone conjugation *in vitro* and in human blood using liquid chromatography–mass spectrometry. *Drug Metab. Dispos.*, 283: 298–307.

Driscoll, S.G. and Taylor, S.H. (1980). Effects of prenatal estrogen on the male urogenital system. *Obstet. Gynecol.*, 56: 537–542.

Dunnick, J.K., Elwell, M.R., Huff, J., and Barrett, J.C. (1995). Chemically induced mammary gland cancer in the National Toxicology Program's carcinogenesis bioassay. *Carcinogenesis*, 16: 173–179.

Eadon, G.L., Kaminsky, J., Silkworth, K. et al. (1986). Calculation of 2,3,7,8-TCDD equivalent. *Environ. Health Perspect.*, 70: 221–227.

Egeland, F.M., Sweeney, M.H., Fingerhut, M.A., Willie, K.K., Schnorr, T.M., and Halperin, W.E. (1994). Total serum testosterone and gonadotropins in works exposed to dioxin. *Am. J. Epidemiol.*, 139(3): 272–281.

Eisenreich, S.J., Looney, B.B., and Thorton, J.D. (1981). Airborne organic contaminants in the Great Lakes ecosystem. *Environ. Sci. Technol.*, 15: 30–38.

Environmental Working Group. (1996). *Dishonorable Discharge: The 50 Most Polluted Rivers in the Country*. Washington, DC: Environmental Working Group (http://www.ewg.org/research/dishonorable-discharge/50-most-polluted-rivers-country).

Erdman, Jr., J.W., Badge, T.M., Lampe, J.W., Setchell, K.D.R., and Messina, M. (2004). Not all soy products are created equal: caution needed in interpretation of research results. *J. Nutr.*, 134: 1229S–1233S.

Ezatti, T.M., Massey, J.T., Waksburg, J., Chu, A., and Maurer, K.R. (1992). *Sample Design: Third National Health and Nutrition Survey*, Publ. No. PHS 92-1387. Washington, DC: U.S. Department of Health and Human Services.

Facemire, C.F., Gross, T.S., and Gillette, Jr., L.H. (1995). Reproductive impairment in the Florida panther: nature or nurture? *Environ. Health Perspect.*, 103: 79–86.

Falck, F., Ricci, A., Wolff, M.S., Godbold, J., and Deckers, P. (1992). Pesticides and polychlorinated biphenyl residues in human breast lipids and their relation to breast cancer. *Arch. Environ. Health*, 47: 143–146.

Fang, H. and Chian, E. (1976). Reverses osmosis separation of polar organic compounds in aqueous solution. *Environ. Sci. Technol.*, 10: 36.

FDER. (1981). Tower Chemical Company Investigation, Lake County, Florida, Project Files, File Reference T-15.E.5. Orlando: Florida Department of Environmental Regulation.

Fortner, B. and Schechter, D. (1996). U.S. water quality shows little improvement over 1992 inventory. *Water Environ. Technol.*, 8(2): 15–16.

Fox, G.A., Gilman, A.P., Peakall, D.B., and Anderka, F.W. (1978). Aberrant behavior of nesting gulls. *J. Wildl. Manage.*, 42: 477–483.

Fox, G.A., Collins, B., Hayakawa, E., Weseloh, D.V., Ludwig, J.P., Kubiak, T.J., and Erdman, T.C. (1991a). Reproductive outcomes in colonial fish-eating birds: a biomarker for developmental toxicants in Great Lakes food chains. *J. Great Lakes Res.*, 17: 158–167.

Fox, G.A., Gilbertson, M., Gilman, A.P., and Kubiak, T.J. (1991b). A rationale for the use of colonial fish-eating birds to monitor the presence of developmental toxicants in Great Lakes fish. *J. Great Lakes Res.*, 17: 151–152.

Friess, E., Schiffelholz, T., Steckler, T., and Steiger, A. (2000). Dehydroepiandrosterone—a neurosteroid. *Eur. J. Clin. Invest.*, 30(Suppl. 3): 46–50.

Fry, D.M. and Toone, C.K. (1981). DDT-induced feminization of gull embryos. *Science*, 213: 922–924.

Fry, D.M., Toone, C.K., Speich, S.M., and Peard, R.J. (1987). Sex ratio skew and breeding patterns of gulls: demographic and toxicological considerations. *Stud. Avian Biol.*, 10: 26–43.

Ganong, W.F. (2005). *Review of Medical Physiology*, 22nd ed. New York: McGraw-Hill.

Gergen, P.J. and Weiss, K.B. (1990). Changing pattern of asthma hospitalization among children: 1979–1987. *JAMA*, 264: 1688–1692.

Gilberson, M., Kubiak, T., Ludwig, J., and Fox, G.A. (1991). Great Lakes embryo mortality, edema, and deformities syndrome (GLEMEDS) in colonial fish-eating birds: similarity to chick-edema disease. *J. Toxicol. Environ. Health*, 33: 455–520.

Gill, W.H., Schumacher, F.B., Bibbo, M., Straus, F.H., and Schoenberg, H.W. (1979). Association of diethylstilbestrol exposure *in utero* with cryptorchidism, testicular hypoplasia and semen abnormalities. *J. Urol.*, 122: 36–39.

Gingell, R., Beatty, P.W., Mitschke, H.R., Page, A.C., Sawin, V.L., Putcha, L., and Kramer, W.G. (1987). Toxicokinetics of 1,2-dibromo-3-chloropropane (DBCP) in the rat. *Toxicol. Appl. Pharmacol.*, 91: 386–394.

Gladen, B.C. and Rogan, W.J. (1995). DDE and shortened duration of lactation in a northern Mexican town. *Am. J. Public Health*, 85: 504–508.

Goigel, J.F. (1991). Regulatory Investigation and Cost Analysis of Concentrate Disposal from Membrane Plants, master's thesis, University of Central Florida, Orlando.

Gonzalez, R., Matsiola, P., Torchy, C., de Kinkelin, P., and Avrameas, S. (1989). Natural anti-TNP antibodies from rainbow trout interfere with viral infection *in vitro*. *Res. Immunol.*, 140: 675–684.

Gordon, W. (1984). *A Citizen's Handbook on Ground Water Protection*. New York: Natural Resources Defense Council.

Grace, P.B., Taylor, J.I., Low, Y., Luben, R.N., Mulligan, A.A., Botting, N.P. et al. (2004). Phytoestrogen concentrations in serum and spot urine as biomarkers for dietary phytoestrogen intake and their relation to breast cancer risk in European prospective investigation of cancer and nutrition—Norfolk. *Cancer Epidemiol. Biomarkers Prev.*, 13(5): 698–708.

Grassman, K.A. (1995). Immunological and Hematological Biomarkers from Contaminants in Fish-Eating Birds of the Great Lakes, Ph.D. dissertation, Virginia Polytechnic University, Blacksburg.

Grassman, K.A. and Scanlon, P.F. (1995). Effects of acute lead ingestion and diet on antibody and T-cell-mediated immunity in Japanese quail. *Arch. Environ. Contam. Toxicol.*, 28: 161–167.

Gray, L.E., Ostby, J., Ferrell, J., Rehnberg, G., Linder, R., Cooper, R. Goldman, J., Slott, V., and Laskey, J. (1989). A dose–response analysis of methoxychlor-induced alterations of reproductive development and function in the rat. *Fund. Appl. Toxicol.*, 12: 92–108.

Grossman, C.J. (1985). Interactions between the gonadal steroids and the immune system. *Science*, 227(4684): 257–261.

Guillette, L.J., Gross, T.S., Masson, G.R., Matter, J.M., Percival, H.F., and Woodward, A.R. (1994). Developmental abnormalities of the gonad and abnormal sex hormone concentrations in juvenile alligators from contaminated and control lakes in Florida. *Environ. Health Perspect.*, 102: 680–688.

Guillette, L.J., Crain, D.A., Rooney, A.A., and Pickford, D.B. (1995a). Organization versus activation: the role of endocrine-disrupting contaminants (EDCs) during embryonic development in wildlife. *Environ. Health Perspect.*, 103(Suppl. 7): 157–164.

Guillette, L.J., Gross, T.S., Gross, D.A., Rooney, A.A., and Percival, H.F. (1995b). Gonadal steroidogenesis *in vitro* from juvenile alligators obtained from contaminated or control lakes. *Environ. Health Perspect.*, 103: 31–36.

Guo, Y.L., Lai, T.J., Ju, S.H., Chen, Y.C., and Hsu, C.C. (1993). Sexual developments and biological findings in Yucheng children. *Organohalogen Compd.*, 14: 235–238.

Guzelian, P.S. (1982). Comparative toxicology of chlordecone (Kepone) in humans and experimental animals. *Ann. Rev. Pharmacol. Toxicol.*, 22: 89–113.

Halling-Sorenson, B., Nors Nielsen, S., Lanzky, P.F., Ingerlev, F., Holten Lutzhof, H.C., and Jergensen, S.E. (1998) Occurrence fate and effects of pharmaceutical substances in the environment—a review. *Chemosphere*, 36(2): 357–393.

Hankinson, O. (1995). The aryl hydrocarbon receptor complex. *Ann. Rev. Pharmacol. Toxicol.*, 35: 307–340.

Harshbarger, J.C. and Clark, J.B. (1990). Epizootiology of neoplasms in boy fish of North America. *Sci. Total Environ.*, 94: 1–32.

Haws, L.C., Su, S.H., Harris, M. et al. (2006). Development of a refined database of mammalian relative potency estimates for dioxin-like compounds. *Toxicol. Sci.*, 89(1): 4–30.

Heberer, T. and Stan, H.-J. (1997). Determination of clofibric acid and *N*-(phenylsulfonyl)-sarcosine in sewage river and drinking water. *Int. J. Environ. Anal. Chem.*, 67: 113–124.

Heberer, T., Schmidt-Baumier, K., and Stand, H.-J. (1998). Occurrence and distribution of organic contaminants in the aquatic system in Berlin. Part I. Drug residues and other polar contaminants in Berlin surface and ground water. *Acta Hydrochim. Hydrobiol.*, 26(5): 272–278.

Heberer, T., Gramer, S., and Stan, H.-J. (1999). Occurrence and distribution of organic contaminants in the aquatic system in Berlin. Part III. Determination of synthetic musks in Berlin surface water applying solid-phase microextraction (SPME) and gas chromatography–mass spectrometry (GC/MS). *Acta Hydrochim. Hydrobiol.*, 27: 150–156.

Hebert, C.E., Norstrom, R.J., Simon, M., Braune, B.M., Weseloh, D.V., and Macdonald, C.R. (1994). Temporal trends and sources of PCDDs and PCDFs in the Great Lakes: herring gull egg monitoring 1981–1991. *Environ. Sci. Technol.*, 281: 1266.

Henderson, A.K., Rosen, D., Miller, G.L., Figgs, L.W., Zahm, S.H., Sieber, S.M., Humphrey, H.B., and Sinks, T. (1995). Breast cancer among women exposed to polybrominated biphenyls. *Epidemiology*, 6(5): 544–546.

Herbst, A.L., Ulfelder, H., and Poskanzer, D.C. (1971). Adenocarcinoma of the vagina. Association of maternal diethylstilbestrol therapy with tumor appearance in young women. *N. Engl. J. Med.*, 284: 878–881.

Hesselberg, R.J. and Gannon, J.E. (1995). Contaminant trends in Great Lakes fish. In: *Our Living Resources: A Report to the Nation on the Distribution, Abundance, and Health of U.S. Plants, Animals, and Ecosystems* (LaRoe, E.T., Farris, G.S., Puckett, C.E., Doran, P.D., and Mat, M.J., Eds.), pp. 242–244. Washington, DC: U.S. Department of the Interior.

Hesselberg, R.J., Hickey, J.P., Nortrup, D.A., and Willford, W.A. (1990). Contaminant residues in the bloater (*Coregonus hoyi*) of Lake Michigan 1969–1986. *J. Great Lakes Res.*, 16: 121–129.

Hodges, L.R. and Lear, B. (1974). Persistence and movement of DBCP in three types of soil. *Soil Sci.*, 118: 127–130.

Hoffman, D.J., Rattner, B.A., Sileo, L., Docherty, D., and Kubiak, T.J. (1987). Embryotoxicity, teratogenicity and arylhydrocarbon hydroxylase activity in Foster's terns on Green Bay, Lake Michigan. *Environ. Res.*, 42: 176–184.

Homo-Delarche, F., Fitzpatrick, F., Christeff, N., Nunez, E.A., Bach, J.F., and Dardenne, M. (1991). Sex steroids, glucocorticoids, stress and autoimmunity. *J. Steroid Biochem.*, 40: 619–637.

Horn-Ross, P.L., Barnes, S., Kirk, M., Coward, L., Parsonnet, J., and Hiatt, R.A. (1997). Urinary phytoestrogen levels in young women from a multiethnic population. *Cancer Epidemiol. Biomarkers Prev.*, 6: 339–345.

Hutchins, A.M., Lampe, J.W., Martini, M.C., Campbell, D.R., and Slavin, J.L. (1995a). Vegetables, fruits and legumes: effect on isoflavonoid phytoestrogen and lignan excretion. *J. Am. Diet. Assoc.*, 95: 769–774.

Hutchins, A.M., Slavin, J.L., and Lampe, J.W. (1995b). Urinary isoflavonoid phytoestrogen and lignin excretion after consumption of fermented and unfermented soy produces. *J. Am. Diet. Assoc.*, 95: 545–551.

James, M.O. (1986). Overview of *in vivo* metabolism of drugs by aquatic species. *Vet. Hum. Toxicol.*, 28(Suppl. 1): 2–8.

Jensen A.A. and Slorach, S.A. (1991). *Chemical Contaminants in Human Milk*. Boca Raton, FL: CRC Press.

Kaattari, S.L., Adkison, M., Shapiro, D., and Arkoosh, A.R. (1994). Mechanisms of immunosuppression by aflatoxin B1. In: *Modulators of Fish Immune Response* (Stolen, J., Ed.), Vol. 1, pp. 151–156. Fair Haven, NJ: SOS Publications.

Kaplan, N.M. (1959). Male pseuodhemaphrodism: report of a case with observations on pathogenesis. *N. Engl. J. Med.*, 261: 641–644.

Karr, S.C. Lampe, J.W., Hutchins, A.M., and Slaving, J.L. (1977). Urinary isoflavonoid excretion in humans is dose dependent at low moderate levels of soy-protein consumption. *Am. J. Clin. Nutr.*, 66: 46–51.

Kavlock, R.J. et al. (1996). Research needs for the risk assessment of health and environmental effects of endocrine disruptors: a report of the U.S. EPA-sponsored workshop. *Environ. Health Perspect.*, 104(Suppl. 4): 715–740.

Kawasaki, M. (1980). Experiences with the test scheme under the chemical control law of Japan: an approach to structure-activity correlations. *Ecotoxicol. Environ. Saf.*, 4: 444–454.

Kenny, J.F., Barber, N.L., Hutson, S.S., Linsey, K.S., Lovelace, J.K., and Maupin, M.A. (2009). *Estimated Use of Water in the United States in 2005*. Washington, DC: U.S. Geological Survey (http://pubs.usgs.gov/circ/1344/).

Kharrazi, M., Potashnik, G., and Goldsmith, J.R. (1980). Reproductive effects of dibromochloropropane. *Isr. J. Med. Sci.*, 16: 403–406.

Kirkman, L.M., Lampe, J.W., Campbell, D., Martini, M., and Salving, J. (1995). Urinary lignan and isoflavonoid excretion in men and women consuming vegetable and soy diets. *Nutr. Cancer*, 24: 1–12.

Klein, K.O. (1998). Isoflavones, soy-based infant formulas, and relevance to endocrine function. *Nutr. Rev.*, 56: 193–204.

Krieger, N., Wolff, M.S., Hiatt, R.A., Rivera, M., Vogelman, J., and Orentreich, N. (1994). Breast cancer and serum organochlorines: a prospective study among white, black, and Asian women. *J. Natl. Cancer Inst.*, 86(9): 589–599.

Krishnan, A.V., Statnis, P., Permuth, S.F., Tokes, L., and Feldman, D. (1993). Bisphenol-A: an estrogenic substance is released form polycarbonate flasks during autoclaving. *Endocrinology*, 132: 2279–2286.

Lahvis, G.P., Wells, R.S., Kuehl, D.W., Stewart, J.L., Rhinehart, H.L., and Via, C.S. (1995). Decreased lymphocyte responses in free-ranging bottlenose dolphins (*Tursiops truncates*) are associated with increased concentrations of PCBs and DDT in peripheral blood. *Environ. Health Perspect.*, 103: 67–72.

Lampe, J.W., Gustafson, D.R., Hutchins, A.M., Martini, M.G., Li, S., Wahala, K. et al. (1999). Urinary isoflavonoid and lignan excretion on a Western diet: relation to soy, vegetable and fruit intake. *Cancer Epidemiol. Biomarkers Prev.*, 8: 699–707.

Learn, S. (1999). Sewage tradeoff: add overflow, help tributaries. *The Oregonian*, April 22.

Leary, F.J., Resseguie, L.J., Kurland, L.T., O'Brien, P.C., Emslander, R.F., and Noller, K.L. (1984). Males exposed *in utero* to diethylstilbestrol. *JAMA*, 252: 2984–2989.

Leatherland, J.F. (1993). Field observations on reproduce and developmental dysfunction in introduced and native salmonids from the Great Lakes. *J. Great Lakes Res.*, 19: 737–751.

Lee, T.M. and Swancar, A. (1996). *Influence of Evaporation, Ground Water, and Uncertainty in the Hydrologic Budget of Lake Lucerne, a Seepage Lake in Polk County, Florida*, USGS Water-Supply Paper 2439. Washington, DC: U.S. Geologic Survey

Lim, M. and Johnston, H. (1976). Reverse osmosis as an advanced treatment process. *J. WPCF*, 48, 1820.

Loose, L.D., Pittman, K.A., Benitz, K.F., and Silkworth, J.B. (1977). Polychlorinated biphenyl and hexchlorobenzene induced humoral immunosuppression. *J. Reticuloendothel. Soc.*, 22: 253–271.

Lu, L.J., Grady, J.J., Marshall, M.V., Ramanujam, V.M., and Anderson, K.E. (1995). Altered time course of urinary daidzein and genistein excretion during chronic soya diet in healthy male subjects. *Nutr. Cancer*, 24: 311–323.

Lund, T.D., Munson, D.J., Haldy, M.E., Setchell, K.D.R., Lepahrt, E.D., and Handa, R.J. (2004). Equol is novel antiandrogen that inhibits prostate growth and hormone feedback. *Biol. Reprod.*, 70: 1188–1195.

Luster, M.I., Germolec, D.R., and Rosenthal, G.J. (1990). Immunotoxicology: review of current status. *Ann. Allergy*, 64: 427–432.

Luster, M.I., Munson, A.E., Thomas, P.T., Holsapple, M.P., Fenters, J.D. et al. (1988). Development of a testing battery to assess chemical-induced immunotoxicity: National Toxicology Program's guidelines for immunotoxicity evaluation in mice. *Fund. Appl. Toxicol.*, 10: 2–19.

Lyman, W.J., Reehl, W.F., and Rosenblatt, D.H., Eds. (1982). *Handbook of Chemical Property Estimation Methods*. New York: McGraw-Hill.

Mac, M.J. and Edsall, C.C. (1991). Environmental contaminants and the reproductive success of lake trout in the Great Lakes: an epidemiological approach. *J. Toxicol. Environ. Health*, 33: 375–394.

Mac, M.J., Schwartz, T.R., Edsall, C.C., and Frank, A.M. (1993). Polychlorinated biphenyls in Great Lakes lake trout and their eggs: relations to survival and congener composition 1979–1988. *J. Great Lakes Res.*, 19: 752–765.

Magee, P.J. and Rowland, I.R. (2004). Phyto-oestrogens, their mechanism of action: current evidence of a role in breast and prostate cancer. *Br. J. Nutr.*, 91: 513–531.

Marcholonis, J.J., Hohman, V.S., Thomas, C., and Schulter, S.F. (1993). Antibody production in sharks and humans: a role for natural antibodies. *Dev. Comp. Immunol.*, 17: 41–53.

Maskarinec, G., Singh, S. Meng, L., and Franke, A.A. (1998). Dietary soy intake and urinary isoflavone excretion among women from a multiethnic population. *Cancer Epidemiol. Biomarkers Prev.*, 7: 613–619.

Maurer, H.H., Pfleger, K., and Weber, A.A. (2011). *Mass Spectral and GC Data of Drugs, Poisons, Pesticides, Pollutants, and Their Metabolites*, 4th ed. Weinheim, Germany VCH.

McArthur, M.L.B., Fox, G.A., Peakall, D.B., and Philogene, B.J.R. (1983). Ecological significance of behavior and hormonal abnormalities in breeding ring doves fed an organochlorine chemical mixture. *Arch. Environ. Contam. Toxicol.*, 12: 343–353.

McKinney, J.D., Chae, K., Gupia, B.N., Moore, J.A., and Goldstein, J.A. (1976). Toxicological assessment of hexchlorobiphenyl isomers and 2,3,7,8-tetrachlorodibenzofuran in chicks. *Toxicol. Appl. Pharmacol.*, 36: 65–80.

Means, G.D., Kilgore, M.W., Mahendroo, M.S., Mendelson, C.R., and Simpson, E.R. (1991). Tissue-specific promoters regulate aromatase cytochrome P450 gene expression human ovary and fetal tissues. *Mol. Endocrinol.*, 5(12): 2005–2013.

Medical Research Council. (1990). *IEH Assessment on Environmental Oestrogens: Consequences to Human Health and Wildlife.* Leicester, U.K.: University of Leicester.

Melancon, M.J. (1995). Bioindicators used in aquatic and terrestrial monitoring. In: *Handbook of Ecotoxicology* (Hoffman D.J., Rattner, B.A., Burton, Jr., G.A., and Cairns, J., Eds.), pp. 220–239. Boca Raton, FL: Lewis Publishers.

Messina, M., McCaskill-Stevens, W., and Lampe, J.W. (2006). Addressing the soy and breast cancer relationship: review, commentary, and workshop proceedings. *J. Natl. Cancer Inst.*, 98(18): 1275–1284.

Meyers, M.S., Stehr, C.M., Olson, A.P., Jonson, L.L., McCain, B.B. et al. (1994). Relationships between toxicopathic hepatic lesions and exposure to chemical contaminants in English sole (*Pleuronectes vetulus*), starry flounder (*Platichtys stellatus*), and white croaker (*Genyonemus lineatus*) from selected marine sites on the Pacific coast, USA. *Environ. Health Perspect.*, 102: 200–215.

Michel, C., Gonzalez, R., Bonjour, E., and Avrameas, S. (1990). A concurrent increasing of natural antibodies and enhancement to furunculosis in rainbow trout. *Ann. Rech. Vet.*, 21: 211–218.

Mickley, M.C. (1996). Environmental considerations for the disposal of desalination concentrates. *Int. Desal. Water Reuse Q.*, 5(4): 56–61.

Mickley, M.C., Hamilton, R., Gallegos, L., and Truesdall, J. (1993). *Membrane Concentrate Disposal.* Denver, CO: American Water Works Association.

Miller, W. (1989). Estimating evaporation from Utah's Great Salt Lake using thermal infrared satellite imagery. *Water Res. Bull.*, 25: 541–542.

Mo, Q., Lu, S.F., and Simon, N.G. (2006). Dehydroepiandrosterone and its metabolites: differential effects on androgen receptor trafficking and transcriptional activity. *J. Steroid Biochem. Mol. Biol.*, 99(1): 50–58.

Mosner, M.S. and Aulenbach, B.T. (2003). *Comparison of Methods Used to Estimate Lake Evaporation for a Water Budget of Lake Seminole, Southwestern Georgia and Northwestern Florida.* Atlanta, GA: U.S. Geological Survey.

Munnecke, D.E. and VanGundy, S.D. (1979). Movement of fumigants in soil, dosage responses, and differential effects. *Annu. Rev. Phytopathol.*, 17: 405–429.

Mussalo-Rauhamaa, H., Hasanen, E., Pyysalo, H., Antervo, K., Kauppila, R., and Pantzar, P. (1990). Occurrence of beta-hexachlorocyclohexane in breast cancer patients. *Cancer*, 66: 2124–2128.

NAMS. (2000). The role of isoflavones in menopausal health. Consensus opinion of the North American Menopause Society. *Menopause*, 7(4): 215–229.

National Research Council. (2006). *Health Risks from Dioxin and Related Compounds: Evaluation of the EPA Reassessment.* Washington, DC: National Academies Press.

National Research Council. (2012). *Water Reuse: Potential for Expanding the Nation's Water Supply Through Reuse of Municipal Wastewater.* Washington, DC: National Academies Press.

NCSL. (2014). *Overview of the Water–Energy Nexus in the U.S.* Washington, DC: National Conference of State Legislatures (http://www.ncsl.org/research/environment-and-natural-resources/overviewofthewaterenergynexusintheus.aspx).

Nedrow, A., Miller, J., Walker, M., Nygren, P., Huffman, L.H., and Nelson, H.D. (2006). Complementary and alternative therapies for the management of menopause-related symptoms. A systematic evidence review. *Arch. Intern. Med.*, 166: 1453–1465.

Nesbitt, P.D., Lam, Y., and Thompson, L.U. (1999). Human metabolism of mammalian lignan precursors in raw and processed flaxseed. *Am. J. Clin. Nutr.*, 69(3): 549–555.

Noller, K.L., Blair, P.B., O'Brien, P.C., Melton III, L., Offord, J.R., Kaufman, R.H., and Colton, T. (1988). Increased occurrence of autoimmune disease among women exposed *in utero* to diethylstilbestrol. *Fertil. Steril.*, 49(6): 1080–1082.

Office of Saline Water. (1970). *Disposal of Brine by Solar Evaporation: Field Experiments.* Washington, DC: U.S. Department of the Interior.

Peakall, D.B. and Fox, G.A. (1987). Toxicological investigations of pollutant-related effects in Great Lakes gulls. *Environ. Health Perspect.*, 77: 187–193.

Pelletier, L., Castedo, M., Bellon, B., and Druet, P. (1994). Mercury and autoimmunity. In: *Immunotoxicology and Immunopharmacology* (Dean, J.H., Luster, M.I., Munsen, A.E., and Kimber, I., Eds.), pp. 539–552. New York: Raven Press.

Penny, R. (1982). The effects of DES on male offspring. *West. J. Med.*, 136: 329–330.

Pimentel, D. and Pimentel, M. (2003). Sustainability of meat-based and plant-based diets and the environment. *Am. J. Clin. Nutr.*, 78(3): 660S–663S.

Purdom, C.E., Hardiman, P.A., Bye, V.J., Eno, N.C., Tyler, C.R., and Sumter, J.P. (1994). Estrogenic effects of effluents form sewage treatment works. *Chem. Ecol.*, 8: 275–285.

Raloff, J. (1998). Drugged waters—does it matter that pharmaceuticals are turning up in water supplies? *Sci. News*, 153: 187–189.

Ratcliffe, J.M., Schrader, S.M, Steenland, K., Clapp, D.E., Turner, T., and Hornung, R.W. (1987). Semen quality in papaya workers with long term exposure to ethylene dibromide. *Br. J. Ind. Med.*, 44: 317–326.

Rodigo, D., Lopez Calva, E.J., and Cannan, A. (2012). *Total Water Management*, EPA 600/R-12/551. Washington, DC: U.S. Environmental Protection Agency.

Roembke, J., Kacker, Th., and Stahlschmidt-Allner, P. (1996). *Studie uber Umweltprobleme im/Zusammernhang mit Arzneimittein* [study about environmental problems in context with drugs]. Berlin: Federal Ministry of Research and Development.

Rogan, W.J., Gladen, B.C., and McKinney, J.D. (1987). Polychlorinated biphenyls (PCBs) and dichlorodiphenyl dichloroethane (DDE) in human milk: effects on growth, morbidity and duration of lactation. *Am. J. Public Health*, 77: 1294–1297.

Rogers, H.R. (1996). Sources, behavior and fate of organic contaminants during sewage treatment and in sewage sludges. *Sci. Total Environ.*, 185: 3–26.

Rogers, R.S. (1998). Sepracor: skating on thin "ice." *Chem. Eng. News*, 30: 11–13.

Rosenberry, D.O., Sturrock, A.M., and Winter, T.C. (1993). Evaluation of the energy budget method of determining evaporation at Williams Lake, Minnesota, using alternative instrumentation and study approaches. *Water Resour. Res.*, 29(8): 2473–2248.

Ross, P.S., deSwart, R.L., Reijnders, P.J.H. et al. (1995). Contaminant-related suppression of delayed-type hypersensitivity and antibody response in harbor seals fed herring form the Baltic Sea. *Environ. Health Perspect.*, 103: 162–167.

Rowland, I, Wiseman, H., Sanders, T.A.B., Adlercreutz, H., and Bower, E.A. (2000). Interindividual variation in metabolism of soy isoflavones and lignans: influence of habitual diet on equol production by the gut microflora. *Nutr. Cancer*, 36: 27–32.

Rowland, I., Faughnan, M., Hoey, L., Wahala, K., Williamson, G., and Cassidy, A. (2003). Bioavailability of phytoestrogens. *Br. J. Nutr.*, 89(Suppl. 1): S45–S58.

Rozman, K.K., Bhatia, J. Calafat, A.M., Chambers, C., Culty, M., Etzel, R.A. et al. (2006a). NTP-CERHR expert panel report on the reproductive and developmental toxicity of soy formula. *Birth Defects Res. B Dev. Reprod. Toxicol.*, 77: 280–397.

Rozman, K.K., Bhatia, J., Calafat, A.M., Chambers, C., Culty, M., Etzel, R.A. et al. (2006b). NTP-CERHR expert panel report on the reproductive and developmental toxicity of genistein. *Birth Defects Res. B Dev. Reprod. Toxicol.*, 77: 485–638.

Ruhoy, I.S. and Daughton, C.G. (2008). Beyond the medicine cabinet: an analysis of where and why medications accumulate. *Environ. Int.*, 34(8): 1157–1169.

Sacks, F.M., Lichtenstein, A., Van Horn, L., Harris, W., Kris-Etherton, P., and Winston, M. (2006). Soy protein, isoflavones, and cardiovascular health. An American Heart Association Science Advisory for Professionals from the Nutrition Committee. *Circulation*, 113: 1034–1044.

Safe, S. (1990). Polychlorinated biphenyls (PCBs), dibenzo-*p*-dioxins (PCDDs), dibenzofu-rans (PCDFs), and related compounds: environmental and mechanistic considerations which support the development of toxic equivalency factors (TEFs). *Crit. Rev. Toxicol.*, 21(1): 51–88.

Schmitt, C.J. and Brumbaugh, W.C. (1990). National Contaminant Biomonitoring Program: concentrations of arsenic, cadmium, copper, lead, mercury, selenium, and zinc in fresh-water fishes of the United States 1976–1984. *Arch. Environ. Contam. Toxicol.*, 19: 731–747.

Schmitt, C.J. and Bunck, C.M. (1995). Persistent environmental contaminants in fish and wild-life. In: *Our Living Resources: A Report to the Nation on the Distribution, Abundance, and Health of U.S. Plant, Animals, and Ecosystems* (LaRoe, E.T., Faris, G.S., Puckett, C.E., Doan, P.D., and Mack J.J., Eds.), pp. 413–416. Washington, DC: National Biological Service.

Schmitt, C.J., Zajicek, J.L., and Peterman, P.L. (1990). National Contaminant Biomonitoring Program: residues of organochlorine chemicals in freshwater fishes of the United States 1976–1984. *Arch. Environ. Contam. Toxicol.*, 19: 748–782.

Schrader, S.M., Turner, T.W., and Ratcliffe, J.M. (1988). The effects of ethylene dibromide on semen quality: a comparison of short-term and chronic exposure. *Reprod. Toxicol.*, 2: 191–198.

Schrank, C.S., Cook, M.E., and Hansen, W.R. (1990). Immune response of mallard ducks treated with immunosuppressive agents: antibody response to erythrocytes and *in vivo* response to phytohemagglutinin. *J. Wildl. Dis.*, 26: 307–315.

Schulman, R.A. and Dean, C. (2007). *Solve It with Supplements*. New York: Rodale.

Schuurs, A.H.W.M. and Veheul, H.A.M. (1990). Effects of gender and sex steroids on the immune response. *J. Steroid Biochem.*, 35: 157–172.

Scott, G.I., Moore, D.W., Chandler, G.T., Key, P.B., Hampton, T.W. et al. (1990). *Agricultural Insecticide Runoff Effects on Estuarine Organisms: Correlating Laboratory and Field Toxicity Tests with Ecotoxicological Biomonitoring*, Final Report to U.S. Environmental Protection Agency, Gulf Breeze, FL, 495 pp.

Scott, T. (1996). *Concise Encyclopedia Biology*. New York: Walter de Gruyter.

Seow, A., Shi, C.Y., Franke, A.A., Hankin, J.H., Lee, H., and Yu, M.C. (1998). Isoflavonoid levels in spot urine are associated with frequency of dietary soy intake in a popula-tion-based sample of middle-aged and older Chinese in Singapore. *Cancer Epidemiol. Biomarkers Prev.*, 7: 135–140.

Setchell, K.D.R. and Cassidy, A. (1999). Dietary isoflavones: biological effects and relevance to human health. *J. Nutr* 129(3): 758S–767S.

Setchell, K.D.R., Zimmer-Nechemias, L., Cai, J., and Heubi, J.E. (1997). Exposure of infants to phyto-oestrogens from soy-based infant formula. *Lancet*, 350(9070): 23–27.

Setchell, K.D.R., Brown N.M., Desai, P., Zimmer-Nechemias, L., Wolfe, B.E., Brasher, W.T. et al. (2001). Bioavailability of pure isoflavones in healthy humans and analysis of com-mercial soy isoflavone supplements. *J. Nutr.*, 131(Suppl.): 1362S–1375S.

Setchell, K.D.R., Brown, N.M., and Lydeking-Olsen, E. (2002). The clinical importance of the metabolite equol—a clue to the effectiveness of soy and its isoflavones. *J. Nutr.*, 132: 3577–3584.

Setchell, K.D.R., Brown, N.M., Desai, P.B., Zimmer-Nechemias, L., Wolfe, B., Jakate, A.S. et al. (2003a). Bioavailability, disposition, and dose–response effects of soy isoflavones when consumed by healthy women at physiologically typical dietary intakes. *J. Nutr.*, 133: 1027–1035.

Setchell, K.D.R., Gaughnan, M.S., Acades, T., Zimmer-Nechemias, L., Brown, N.M., Wolfe, B.E. et al. (2003b). Comparing the pharmacokinetics of daidzein and genistein with the use of ^{13}C-labeled tracers in premenopausal women. *Am. J. Clin. Nutr.*, 77: 411–419.

Sexton, K., Callahan, M.S., Bryan, E.F., Saint, C.G., and Wood, W.P. (1996). Informed decisions about protecting and promoting public health: rationale for a National Human Exposure Assessment Survey. *J. Exposure Anal. Environ. Epidemiol.*, 5(3): 233–236.

Sharpe, R.M. (1993). Declining sperm counts in men: is there an endocrine cause? *J. Endocrinol.*, 136: 357–360.

Sharpe, R.M. and Skakkebaek, NE. (1993). Are oestrogens involved in falling sperm counts and disorders of the male reproductive tract? *Lancet*, 341: 1392–1395.

Sinks, T., Steele, G., Smith, A.B., Rinsky, R., and Watkins, K. (1992). Risk factors associated with excess mortality among polychlorinated biphenyl exposed workers. *Am. J. Epidemiol.*, 136: 389–398.

Sirtori, C.R., Arnold, A., and Johnson, S.K. (2005). Phytoestrogens: end of a tale? *Ann. Med.*, 37: 423–438.

Slavin, J.L., Karr, S.C., Hutchins, A.M., and Lampe, J.W. (1998). Influence of soybean processing, habitual diet, and soy dose on urinary isoflavonoid excretion. *Am. J. Clin. Nutr.*, 68(Suppl. 6): 1492S–1495S.

Sonawane, B.R. (1995). Chemical contaminants in human milk: an overview. *Environ. Health Perspect.*, 103(Suppl. 6): 197–205.

Spellman, F.R. (2007). *Ecology for Non-Ecologists*. Lanham, MD: Government Institutes.

Spellman, F.R. (2014). *Personal Care Products and Pharmaceuticals in Wastewater and the Environment*. Lancaster, PA: DesTech Publications.

Spellman, F.R. and Drinan, J. (2001). *Stream Ecology and Self-Purification*, 2nd ed. Lancaster, PA: Technomic.

Stan, H.J. and Heberer, T. (1997). Pharmaceuticals in the aquatic environment. *Anal. Mag.*, 25(7): 20–23.

Stan, H.J., Heberer, T., and Linkerhagner, M. (1994). Occurrence of clofibric acid in the aquatic system—is the use in human medical care the source of the contamination of surface ground and drinking water? *Vom Wasser*, 83: 57–68.

Steinberg, S.M., Pignatello, J.J., and Sawhney, D.L. (1987). Persistence of 1,2-dibromoethane in soil: entrapment in intraparticle micropores. *Environ. Sci. Technol.*, 21: 1201–1208.

Stelljes, M.E. (2002). *Toxicology for Non-Toxicologists*. Rockville, MD: Government Institutes.

Stevens, J.T., Breckenridge, C.B., Wetzel, L.T., Gillis, H.H., Luempert III, L.G., and Eldridge, J.C. (1994). Hypothesis for mammary tumorigenesis in Sprague–Dawley rates exposed to certain triazine herbicides. *J. Toxicol. Environ. Health*, 43: 139–153.

Stob, M. (1983). Naturally occurring food toxicants: estrogens. In: *Handbook of Naturally Occurring Chemicals* (Recheigl, M., Ed.), pp. 81–102. Boca Raton, FL: CRC Press.

Stow, C.A., Carpenter, S.R., Eby, L.A., Amrhein, J.F., and Hesselberg, R.J. (1995). Evidence that PCBs are approaching stable concentrations in Lake Michigan fishes. *Ecol. Appl.*, 5: 248–260.

Strom, B.L., Schinnar, R., Ziegler, E.E., Barnhart, K.T., Sammel, M.D., Macones, G.A. et al. (2001). Exposure to soy-based formula in infancy and endocrinological and reproductive outcomes in young adulthood. *JAMA*, 286: 807–814.

Strycker, A. and Collins, A.G. (1987). *State-of-the-Art Report: Injection of Hazardous Wastes into Deep Wells*. Ada, OK: Robert S. Kerr Environmental Research Laboratory.

Stumpf, M., Ternes, T.A., Haberer, K., Seel, P., and Baumann, W. (1996). Nachweis von Arzneimittleruckstanden in Klaranlagen und Fliebgewassern [determination of drugs in sewage treatment plants and river water]. *Vom Wasser*, 86: 291–303.

Stumpf, M., Ternes, T.A., Wilken, R.-D., Rodrigues, S.V., and Baumann, W. (1999). Polar drug residues in sewage and natural waters in the state of Rio de Janeiro Brazil. *Sci. Total Environ.*, 225(1): 135–141.

Sumpter, J.P. and Jobling, S. (1995). Vitellogenesis as a biomarker for oestrogenic contamination of the aquatic environment. *Environ. Health Perspect.*, 103(Suppl. 7): 173–178.

278   Reverse Osmosis: A Guide for the Nonengineering Professional

Swenson, B.G., Hallberg, T., Nilsson, A., Schutz, A., and Hagmar, L. (1994). Parameters of immunological competence in subjects with high consumption of fish contaminated with persistent organochlorine compounds. *Int. Arch. Occup. Environ. Health*, 65: 351–358.

Ternes, T.A. (1998). Occurrence of drugs in German sewage treatment plants and rivers. *Waste Res.*, 32(11): 3245–3260.

Ternes, T.A. and Hirsch, R. (2000). Occurrence and behavior of iodinated contrast media in the aquatic environment. *Environ. Sci. Technol.*, 34: 2741–2748.

Ternes, T.A. and Wilkens, R.-D., Eds. (1998). Drugs and hormones as pollutants of the aquatic environment: determination and ecotoxicological impacts. *Sci. Total Environ.*, 225(1–2): 1–176.

Ternes, T.A., Stumpf, M., Schuppert, B., and Haberer, K. (1998). Simultaneous determination of antiseptics and acid drugs in sewage and river. *Vom Wasser*, 90: 295–309.

Ternes, T.A., Hirsch, R., Stumpf, M., Eggert, T., Schuppert, B., and Haberer, K. (1999). *Identification and Screening of Pharmaceuticals, Diagnostics and Antiseptics in the Aquatic Environment*. Wiesbaden, Germany: ESWE—Institut fuer Wasserforschung und Wassertechnologie G.m.b.H.

Torkelson, T.R., Sadek, S.E., and Rowe, V.K. (1961). Toxicologic investigations of 1,2-dibromo-3-chloropropane. *Toxicol. Appl. Pharmacol.*, 3: 545–559.

Tsuge, H. and Mori, K. (1977). Reclamation of municipal sewage by reverse osmosis. *Desalination*, 23, 123.

Turgeon, D.D. and Robertson, A. (1995). Contaminants in coastal fish and mollusks. In: *Our Living Resources: A Report to the Nation on the Distribution, Abundance, and Health of U.S. Plants, Animals, and Ecosystems* (LaRoe, E.T., Farris, G.S., Puckett, C.E., Doran, P.D., and Mac, J.J., Eds.), pp. 408–412. Washington, DC: U.S. Department of the Interior.

Uehara, M., Arai, Y., Watanabe, S., and Adlercreutz, H. (2000a). Comparison of plasma and urinary phytoestrogens in Japanese and Finnish women by time-resolved fluoroimmunoassay. *Biofactors*, 12: 217–225.

Uehara, M., Lapeik, O., Hampl, R., Al-maharik, N., Makela, T., Wahala, K. et al. (2000b). Rapid analysis of phytoestrogens in human by time-resolved fluoroimmunoassay. *J. Steroid Biochem. Mol. Biol.* 72: 273–282.

Unger, M., Olsen, J., and Clausen, J. (1984). Organochlorine compounds in the adipose tissue of deceased persons with and without cancer: a statistical survey of some potential confounders. *Environ. Res.*, 29: 371–376.

U.S. Census Bureau. (2013). *World Population*. Washington, DC: U.S. Census Bureau (http://www.census.gov/population/international/data/worldpop/table_population.php).

USDOI. (2006). *Membrane Concentrate Disposal: Practices and Regulation*, Desalination and Water Purification Research and Development Program Report No. 123, 2nd ed. Washington, DC: U.S. Department of the Interior.

USEPA. (1980). *Investigation Report: Tower Chemical Company and Surrounding Area, Clermont, Florida*. Athens, GA: U.S. Environmental Protection Agency.

USEPA. (1983). *Superfund Cleanup, Tower Chemical Company Site, Clermont, Florida*. Athens, GA: U.S. Environmental Protection Agency.

USEPA. (1985). *Exposure Analysis Modeling System: Reference Manual for EXAMS II*. EPA/00/3-85/038. Athens, GA: U.S. Environmental Protection Agency.

USEPA. (1994). Testing guidelines for developmental and reproduce toxicity. *Fed. Reg.*, 59: 422272.

USEPA. (2003). *Toxic Equivalency Factors (TEFs) for Dioxin and Related Compounds: Exposure and Human Health Reassessment of 2,3,7,8-Tetrachlorodizbenzo-p-Dioxin (TCDD) and Related Compounds*. Part II. *Health Assessment for 2,3,7,8-Tetrachlorodizbenzo-p-Dioxin (TCDD) and Related Compounds*, NCEA-1-0836. Washington, DC: Office of Research and Development, National Center for Environmental Assessment.

USEPA. (2008). *Framework for Application of the Toxicity Equivalence Methodology for Polychlorinated Dioxins, Furans, and Biphenyls in Ecological Risk Assessment*, EPA/100/R-08-004. Washington, DC: U.S. Environmental Protection Agency.

USEPA. (2012a). *Eagles, Other Birds Thrive after EPA's 1972 DDT Ban*. Washington, DC: U.S. Environmental Protection Agency (http://www2.epa.gov/aboutepa/eagles-other-birds-thrive-after-epas-1972-ddt-ban).

USEPA. (2012b). *Energy/Water*. Washington, DC: U.S. Environmental Protection Agency (http://water.epa.gov/action/energywater.cfm).

USEPA. (2013). *DDT—A Brief History and Status*. Washington, DC: U.S. Environmental Protection Agency (http://www2.epa.gov/ingredients-used-pesticide-products/ddt-brief-history-and-status).

USEPA. (2014a). *Water–Energy Connection*. Washington, DC: U.S. Environmental Protection Agency (http://www.epa.gov/region9/waterinfrastructure/waterenergy.html).

USEPA. (2014b). *Chemical Manufacturing*. Washington, DC: U.S. Environmental Protection Agency (http://www.epa.gov/sectors/sectorinfo/sectorprofiles/chemical.html).

USGS. (2014). *Saline Water: Desalination*. Washington, DC: U.S. Geological Survey.

Vafeiadou, K., Hall, W.L., and Williams, C.M. (2006). Does genotype and equol-production status affect response to isoflavones? Data from a pan-European study on the effects of isoflavones on cardiovascular risk markers in post-menopausal women. *Proc. Nutr. Soc.*, 65: 106–115.

Valentin-Blasini, L., Blount, B., Caudill, S., and Needham, L. (2003). Urinary and serum concentrations of seven phytoestrogens in a human reference population subset. *J. Exposure Anal. Environ. Epidemiol.*, 13: 276–282.

Velagaleti, R. (1998). Behavior of pharmaceutical drugs (human and animal health) in the environment. *Drug Inform. J.*, 32: 715–722.

Vonier, P.M., Crain, D.A., McLachlan, J.A., Guillette, Jr., L.J., and Arnold, S.F. (1996). Interaction of environmental chemicals with the estrogen and progesterone receptors from the oviduct of the American alligator. *Environ. Health Perspect.*, 104: 1318–1322.

Wachinski, A.M. (2013). *Membrane Processes for Water Use*. New York: McGraw-Hill.

Walker, N.H, Crockett, P.W., Nyska, A. et al. (2005). Dose-additive carcinogenicity of a defined mixture of "dioxin-like compounds." *Environ. Health Perspect.*, 113(1): 43–48.

Webb, P., Lopez, G.N., Uht, R.M., and Kushner, P.J. (1995). Tamoxifen activation of the estrogen receptor/AP-1 pathway: potential origin for the cell-specific estrogen-like effects of antiestrogens. *Mol. Endocrinol.*, 9: 443–456.

Webb, S.J., Geoghegan, T.E., Prough, R.A., and Miller, M. (2006). The biological action of dehydroepiandrosterone involves multiple receptors. *Drug Metab. Rev.*, 38(1–2): 89–116.

Webber, M. (2008). Catch-22: water vs. energy. *Sci. Am.*, 18: 34–41.

Weiss, K.B. and Wagner, D.K. (1990). Changing patterns of asthma mortality: identifying target populations at high risk. *JAMA*, 264: 1688–1692.

Wetzel, L.T., Leumpert III, L.G., Breckenridge, C.B., Tisdel, M.O., Stevens, J.T., Thakur, A.K., Extrom, P.J., and Eldridge, J.C. (1994). Chronic effects of atrazine on estrus and mammary tumor formation in female Sprague–Dawley and Fischer 344 rats. *J. Toxicol. Environ. Health*, 43: 169–182.

Wetzel, R.G. (1975). *Limnology*. Philadelphia, PA: W.B. Saunders.

White, D.H. and Hoffman, D.J. (1995). Effects of polychlorinated dibenzo-*p*-dioxins and dibenzofurans on nesting wood ducks (*Aix sponsa*) at Bayou Meto, Arkansas. *Environ. Health Perspect.*, 103: 37–39.

White, R., Jobling, S., Hoare, S.A., Sumpter, J.P., and Parker, M.G. (1994). Environmentally persistent alkylphenolic compounds are estrogenic. *Endocrinology*, 135: 175–182.

Whittle, D.M. and Baumann, P.C. (1988). The status of selected organics in the Laurentian Great Lakes: an overview of DDT, PCBs, dioxins, furans, and aromatic hydrocarbons. *Aquat. Toxicol.*, 11: 241–257.

Whorton, M.D., Krauss, R.M., Marshall, S., and Milby, T.H. (1977). Infertility in male pesticide workers. *Lancet*, 2: 1259–1261.

Wilcox, A.J., Baird, D.D., Weinberg, C.R., Hornsby, P.P., and Herbst, A.L. (1995). Fertility in men exposed prenatally to diethylstilbestrol. *N. Engl. J. Med.*, 332: 1411–1415.

Winter, T.C., Rosenberry, D.O., and Sturrock, A.M. (1995). Evaluation of eleven equations for determining evaporation for a small lake in the north central United States, *Water Resour. Res.*, 31(4): 983–993.

Wolff, M.S., Toniolo, P.G., Lee, E.W., Rivera, M., and Dubin, N. (1993). Blood levels of organochlorine residues and risk of breast cancer. *J. Natl. Cancer Inst.*, 85: 648–652.

Woodward, A.R., Percival, H.F., Jennings, M.L., and Moore, C.T. (1993). Low clutch viability of American alligators on Lake Apopka. *Fla. Sci.*, 56: 52–63.

Wren, C.D. (1991). Cause–effect linkages between chemicals and populations of mink (*Mustela vison*) and otter (*Utra canadensis*) in the Great Lakes basin. *J. Toxicol. Environ. Health*, 33: 549–585.

WRRF. (2011). *Development and Application of Tools to Assess and Understand the Relative Risks of Regulated Chemicals in Indirect Potable Reuse Projects—The Montebello Forebay Groundwater Recharge Project. Tools to Assess and Understand the Relative Risks of Indirect Potable Reuse and Aquifer Storage and Recovery Projects*, Vol. 1A, WR-06-018-1A. Alexandria, VA: WateReuse Foundation.

Zelikoff, J.T., Enane, N.A., Bowser, D., Squibb, K., and Frenkel, K. (1991). Development of fish peritoneal macrophages as a model for higher vertebrates in immunotoxicological studies. *Fund. Appl. Toxicol.*, 16: 576–589.

Zelikoff, J.T., Smialowicz, R., Bigazzim, P.E., Goyer, R.A., Lawrence, D.A., and Maibach Gardner, D. (1994). Immunomodulation by metals. *Fund. Appl. Toxicol.*, 22: 1–8.

Zelikoff, J.T., Bowser, D., Squibb, K.S., and Frenkel, K. (1995). Immunotoxicity of low cadmium exposure in fish: alternative animal models for immunotoxicological studies. *J. Toxicol. Environ. Health*, 45: 235–248.

Zheng, W., Dai, Q., Custer, L.J., Shu, X.O., Wen, W.Q., Jin, F. et al. (1999). Urinary excretion of isoflavonoids and the risk of breast cancer. *Cancer Epidemiol. Biomarkers Prev.*, 8(1): 35–40.

Glossary

A

Absorption, chemical: Any process by which one substance penetrates the surface of another substance.

Activated sludge: The solids formed when microorganisms are used to treat wastewater using the activated sludge treatment process: mixing primary effluent with bacteria-laden sludge, which is then agitated and aerated to promote biological treatment. This process speeds the breakdown of organic matter in raw sewage undergoing secondary treatment. Activated sludge includes organisms, accumulated food materials, and waste products from the aerobic decomposition process.

Adsorption: The process by which one substance is attracted to and adheres to the surface of another substance without actually penetrating its internal structure.

Advanced waste treatment: A treatment technology used to produce an extremely high-quality discharge.

Aeration: A physical treatment method that promotes biological degradation of organic matter. The process may be passive (when waste is exposed to air) or active (when a mixing or bubbling device introduces air).

Aerobic: Refers to the condition when free, elemental oxygen is present; also used to describe organisms, biological activity, or treatment processes that require free oxygen.

Aerobic bacteria: A type of bacteria that requires free oxygen to carry out metabolic function.

Alum cake: Dewatered alum sludge.

Alum sludge: Solids removed from the sedimentation of raw water that has undergone coagulation, flocculation, and sedimentation.

Aluminum sulfate (alum; $Al_2(SO_4)_3$): Coagulant added to raw water to form floc for solids removal in water treatment.

Anaerobic: Refers to the condition when no oxygen (free or combined) is available; also used to describe organisms, biological activity, or treatment processes that function in the absence of oxygen.

Anion: The ion in an electrolytic solution that migrates to the anode. It carries a negative charge.

Anoxic: Refers to condition when no free, elemental oxygen is present and the only source of oxygen is combined oxygen such as that found in nitrate compounds; also used to describe biological activity or treatment processes that function only in the presence of combined oxygen.

Applied pressure: Feedwater hydraulic pressure.

Array: Group of pressure vessels installed in parallel and in series with common feedwater, product, and concentrate lines.

Autogenous/autothermic combustion (incinerator): The burning of a wet organic material where the moisture content is at such a level that the heat of combustion of the organic material is sufficient to vaporize the water and maintain combustion. No auxiliary fuel is required except for startup.

Average monthly discharge limitation: The highest allowable discharge over a calendar month.

Average weekly discharge limitation: The highest allowable discharge over a calendar week.

B

Beneficial uses: The many ways water can be used, either directly by people or for their overall benefit.

Biochemical oxygen demand (BOD): The amount of oxygen required by bacteria to stabilize decomposable organic matter under aerobic conditions.

Biochemical oxygen demand (BOD$_5$): The amount of organic matter that can be biologically oxidized under controlled conditions (5 days at 20°C in the dark).

Biofouling: Membrane fouling that is attributable to the deposition and growth of microorganisms on the membrane surface.

Biological treatment: A process that uses living organisms to bring about chemical changes.

Biosolids: Solid organic matter recovered from a sewage treatment process and used especially as fertilizer; usually referred to in the plural.

Biosolids cake: The solid discharged from a dewatering apparatus.

Biosolids concentration: The weight of solids per unit weight of biosolids.

Biosolids moisture content: The weight of water in a biosolids sample divided by the total weight of the sample. Normally determined by drying a biosolids sample and weighing the remaining solids, the total weight of the biosolids sample equals the weight of water plus the weight of the dry solids.

Biosolids quality parameters: The three main USEPA parameters used in gauging biosolids quality are (1) the relevant presence or absence of pathogenic organisms, (2) pollutants, and (3) the degree of attractiveness of the biosolids to vectors.

Brackish water: In general, water having a total dissolved solids (TDS) concentration ranging from about 1000 to 10,000 mg/L.

Brine: The concentrate stream associated with a desalination process.

Bucketing: Simple, effective, but labor-intensive method for cleaning large amounts of debris from a sewer line. Workers load buckets from within the line and haul the solids to the surface for disposal.

Buffer: A substance or solution that resists changes in pH.

Building service: Collection system connection and pipe carrying wastewater flow from the generation point to a main.

C

Cake solids discharge rate: The dry solids cake discharged from a centrifuge.

Carbonaceous biochemical oxygen demand (CBOD$_5$): The amount of biochemical oxygen demand that can be attributed to carbonaceous material.

Cation: A positively charged ion in solution that migrates to the cathode.

Centrate: The effluent or liquid portion of a biosolids removed by or discharged from a centrifuge.

Chemical oxygen demand (COD): The amount of chemically oxidizable materials present in the wastewater.

Chemical treatment: A process that results in the formation of a new substance or substances. The most common chemical water or wastewater treatments include coagulation, disinfection, water softening, and oxidation.

Chlorine: A strong oxidizing agent that has strong disinfecting capability. A yellow-green gas, it is extremely corrosive and is toxic to humans in extremely low concentrations in air.

Chlorine demand: A measure of the amount of chlorine that will combine with impurities and therefore will not be available to act as a disinfectant.

Clarifier: A device designed to permit solids to settle or rise and be separated from the flow; also known as a settling tank or sedimentation basin.

Clean Water Act (CWA): Federal law passed in 1972 (with subsequent amendments) with the objective to restore and maintain the chemical, physical, and biological integrity of the nation's waters. Its long-range goal is to eliminate the discharge of pollutants into navigable waters and to make and keep national waters fishable and swimmable.

Clean zone: Any part of a stream upstream of the point of pollution entry.

Cleanout points: Collection system points that allow access for cleaning equipment and maintenance into the sewer system.

Coagulants: Chemicals that cause small particles to stick together to form larger particles.

Coagulation: A chemical water treatment method that causes very small suspended particles to attract one another and form larger particles. This is accomplished by adding a coagulant that neutralizes the electrostatic charges on the particles that cause them to repel each other. The larger particles are easier to trap, filter, and remove.

Coliform: A type of bacteria used to indicate possible human or animal contamination of water.

Collectors or subcollectors: Collection system pipes that carry wastewater flow to trunk lines.

Combined sewer: A collection system that carries both wastewater and stormwater flows.

Comminution: A process used to shred solids into smaller, less harmful particles.

Community water system: A public water system that serves at least 15 service connections used by year-round residents or regularly serves at least 25 year-round residents.

Composite sample: A combination of individual samples taken in proportion to flow.

Concentrate: The membrane output stream containing water that has not passed through the membrane barrier, as well as concentrated feedwater constituents rejected by the membrane (also known as reject, retentate, brine, or residual stream).

Contact time: The length of time the disinfecting agent and the wastewater remain in contact.

Contaminant: A toxic material found as a residue in or on a substance where it is not wanted.

Cross-connection: Any connection between safe drinking water and a nonpotable water or fluid.

CT value: The product of the residual disinfectant concentration (C, in mg/L) and the corresponding disinfectant contact time (T, in minutes). Minimum CT values are specified by the Surface Water Treatment Rule as a means of ensuring adequate kill or inactivation of pathogenic microorganisms in water.

D

Daily discharge: The discharge of a pollutant measured during a calendar day or any 24-hour period that reasonably represents a calendar day for the purposes of sampling. Limitations expressed as weight are total mass (weight) discharged over the day. Limitations expressed in other units are average measurements of the day.

Daily maximum discharge: The highest allowable values for a daily discharge.

Delayed inflow: Stormwater that may require several days or more to drain through the sewer system. This category can include the discharge of sump pumps from cellar drainage as well as the slowed entry of surface water through manholes in ponded areas.

Demand: The chemical reactions that must be satisfied before a residual or excess chemical will appear.

Desalting: A process that removes salts from feedwater.

Detention time: The theoretical time water remains in a tank at a given flow rate.

Dewatering: The removal or separation of a portion of water present in a sludge or slurry.

Direct flow: Type of inflow having a direct stormwater runoff connection to the sanitary sewer and causing an almost immediate increase in wastewater flows. Possible sources are roof leaders, yard and areaway drains, manhole covers, cross-connections from storm drains and catch basins, and combined sewers.

Direct potable reuse: The piped connection of water recovered from wastewater to a potable water-supply distribution system or a water treatment plant, often implying the blending of reclaimed wastewater.

Direct reuse: The use of reclaimed wastewater that has been transported from a wastewater reclamation point to the water reuse site, without intervening discharge to a natural body of water (e.g., agricultural and landscape irrigation).

Discharge Monitoring Report (DMR): The monthly report required by the treatment plant's National Pollutant Discharge Elimination System (NPDES) discharge permit.

Disinfectant: (1) Any oxidant, including but not limited to chlorine, chlorine dioxide, chloramine, and ozone, added to water in any part of the treatment or distribution process that is intended to kill or inactivate pathogenic microorganisms. (2) A chemical or physical process used to kill pathogenic organisms in water; chlorine is often used to disinfect sewage treatment effluent, water supplies, wells, and swimming pools.

Disinfectant contact time (T in CT calculation): The time (T, in minutes) required for water to move from the point of disinfectant application or the previous point of disinfection residual measurement to a point before or at the point where residual disinfectant concentration (C) is measured. Where only one C is measured, T is the time in minutes that it takes for water to move from the point of disinfectant application to a point before or where residual disinfectant concentration (C) is measured.

Disinfection: Inactivating virtually all recognized pathogenic microorganisms but not necessarily all microbial life (differs from pasteurization or sterilization) by the addition of a substance (e.g., chlorine, ozone, hydrogen peroxide) that destroys or inactivates harmful microorganisms or inhibits their activity. Also, the selective destruction of disease-causing organisms. All organisms are not destroyed during the process. This differentiates disinfection from sterilization, which is the destruction of all organisms.

Disinfection byproducts: Compounds formed by the reaction of a disinfectant such as chlorine with organic material in the water supply.

Dissolved oxygen (DO): Free or elemental oxygen that is dissolved in water.

Dose: The amount of chemical being added, expressed in milligrams per liter (mg/L).

Drinking water standards: Water quality standards measured in terms of suspended solids, unpleasant taste, and microbes harmful to human health. Drinking water standards are included in state water quality rules.

Drinking water supply: Any raw or finished water source that is or may be used as a public water system or as drinking water by one or more individuals.

Drying hearth: A solid surface in an incinerator upon which wet waste materials (or waste matter that may turn to liquid before burning) are placed to dry or to burn with the help of hot combustion gases.

E

Effluent: The flow leaving a tank, channel, or treatment process.

Effluent limitations: Standards developed by the USEPA to define the levels of pollutants that can be discharged into surface waters or any restriction imposed by the regulatory agency on quantities, discharge rates, or concentrations of pollutants that are discharged from point sources into state waters.

Electrolytic recovery: Process that uses ion-selective membranes and an electric field to separate anions and cations in solution; used primarily for the recovery of metals from process streams or wastewaters.

Estuaries: Coastal bodies of water that are partly enclosed.

Evaporation: The process by which water as a liquid changes to water vapor.

F

Facultative: Organisms that can survive and function in the presence or absence of free, elemental oxygen.

Facultative bacteria: A type of anaerobic bacteria that can metabolize its food either aerobically or anaerobically.

Fecal coliform: A type of bacteria found in the bodily discharges of warm-blooded animals; used as an indicator organism.

Federal Water Pollution Control Act of 1972: This Act outlines the objective "to restore and maintain the chemical, physical, and biological integrity of the nation's waters." This Act and subsequent Clean Water Act amendments are the most far-reaching water pollution control legislation ever enacted.

Feed rate: The amount of chemical being added in pounds per day.

Filtrate: The effluent or liquid portion of a biosolids removed by or discharged from a centrifuge.

Filtration: Physical treatment method to remove solid (particulate) matter from water by passing the water through porous media such as sand or a man-made filter.

Flashpoint: The lowest temperature at which evaporation of a substance produces sufficient vapor to form an ignitable mixture with air, near the surface of the liquid.

Floc: Solids that join together to form larger particles that will settle better.

Flocculation: The water treatment process following coagulation that uses gentle stirring to bring suspended particles together to form larger, more settleable clumps called floc.

Flume: A flow-rate measurement device.

Flushing: Line-clearing technique that adds large volumes of water to the sewer at low pressures to move debris through the collection system.

Flux: The rate of water flow across a unit surface area, expressed in gallons per day per square foot or liters per hour per square meter.

Food-to-microorganism ratio (F/M): An activated sludge process control calculation based on the amount of food (BOD_5 or COD) available per pound of mixed liquor volatile suspended solids.

Friction head: The energy required to overcome friction in the piping system. It is expressed in terms of the added system head required.

G

Grab sample: An individual sample collected at a randomly selected time.

Grit: Heavy inorganic solids such as sand, gravel, or metal filings.

Groundwater: The freshwater found under the Earth's surface, usually in aquifers. Groundwater is a major source of drinking water, and concern is growing over areas where leaching agricultural or industrial pollutants or substances from leaking underground storage tanks are contaminating groundwater.

H

Head: The equivalent distance water must be lifted to move from the supply tank or inlet to the discharge. Head can be divided into three components: *static head, friction head*, and *velocity head*.

Horizontal directional drilling (HDD): Technique used to drill or bore a tunnel through the soil and pull or push new pipe in behind the drill head.

Hydraulic line-cleaning: Clearing built-up debris or blockages from sewer lines using hydraulic tools such as balls, kites, pills, pigs, scooters, and bags. They work by partially plugging a flooded upstream main. The movement of the tool itself and the force of the water pressure from the partially blocked line work to loosen and flush away debris.

Hydrologic cycle: Literally, the water–Earth cycle; the movement of water in all three of its physical forms (water, vapor, and ice) through the various environmental mediums (air, water, biota, and soil).

Hygroscopic: A substance that readily absorbs moisture.

I

Incineration: An engineered process using controlled flame combustion to thermally degrade waste material.

Indirect potable reuse: The potable reuse by incorporation of reclaimed wastewater into a raw water supply; the wastewater effluent is discharged to the water source and mixed and assimilated with it, with the intent of reusing the water instead of disposing of it. This type of potable reuse is becoming more common as water resources become less plentiful.

Indirect reuse: The use of wastewater reclaimed indirectly by passing it through a natural body of water or use of groundwater that has been recharged with reclaimed wastewater. This type of potable reuse commonly occurs whenever an upstream water user discharges wastewater effluent into a watercourse that serves as a water supply for a downstream user.

Industrial wastewater: Wastes associated with industrial manufacturing processes.

Infiltration: Water entering the collection system through cracks, joints, or breaks. Infiltration includes steady inflow, direct flow, total inflow, and delayed inflow.

Influent: Water, wastewater, or other liquid flowing into a reservoir, basin, or treatment plant.

Inorganic: Mineral materials such as salt, ferric chloride, iron, sand, and gravel.

Interceptors: Collection system pipes carrying wastewater flow to the treatment plant.

J

Jetting: Line-cleaning technique that cleans and flushes the line in a single operation, using a high-pressure hose and a variety of nozzles to combine the advantages of hydraulic cleaning with mechanical cleaning.

Junction boxes: Collection system constructions that occur when individual lines meet and are connected.

L

Land application: Discharge of wastewater onto the ground for treatment or reuse.

Lift stations: Pump installations designed to pump wastes to a higher point through a force main, when gravity flow does not supply enough force to move the wastewater through the collection system.

Lime sludge: Solids removed from water softening processes.

Line-cleaning: See *hydraulic line-cleaning* and *mechanical line-cleaning.*

M

Mains: Collection system pipes that carry wastewater flow to collection sewers.

Manholes: Collection system entry points that allow access into the sewerage system for inspection, preventive maintenance, and repair.

Maximum contaminant level (MCL): The maximum allowable concentration of a contaminant in drinking water, as established by state and/or federal regulations. Primary MCLs are health related and mandatory; secondary MCLs are related to water quality aesthetic considerations and are highly recommended but not required.

Mean cell residence time (MRCT): The average length of time a mixed liquor suspended solids particle remains in the activated sludge process; may also be known as sludge retention time.

Mechanical line-cleaning: Methods such as rodding or bucketing used to clean stoppages and blockages from sewer lines.

Milligrams per liter (mg/L): An expression of the weight of one substance contained within another. Commonly used to express the weight of a substance within a given weight of water and wastewater, it is sometimes expressed as parts per million (ppm), which is equal to mg/L.

Mixed liquor: The combination of return activated sludge and wastewater in the aeration tank.

Mixed liquor suspended solids (MLSS): The suspended solids concentration of the mixed liquor.

Mixed liquor volatile suspended solids (MLVSS): The concentration of organic matter in the mixed liquor suspended solids.

Moisture content: The amount of water per unit weight of biosolids. The moisture content is expressed as a percentage of the total weight of the wet biosolids. This parameter is equal to 100 minus the percent solids concentration.

N

National Pollutant Discharge Elimination System (NPDES): A requirement of the Clean Water Act that discharges must meet certain requirements prior to discharging waste to any water body. It sets the highest permissible effluent limits, by permit, prior to making any discharge.

Near Coastal Water Initiative: 1985 USEPA initiative designed to manage specific problems not dealt with in other programs for waters near coastlines.

Nitrogenous oxygen demand (NOD): A measure of the amount of oxygen required to biologically oxidize nitrogen compounds under specified conditions of time and temperature.

Nonbiodegradable: A substance that does not break down easily in the environment.

NPDES permit: The National Pollutant Discharge Elimination System permit that authorizes the discharge of treated wastes and specifies the conditions that must be met for discharge.

Nutrients: Substances required to support living organisms; usually refers to nitrogen, phosphorus, iron, and other trace metals.

O

Organic: Materials that consist of carbon, hydrogen, oxygen, sulfur, and nitrogen. Many organics are biologically degradable. All organic compounds can be converted to carbon dioxide and water when subjected to high temperatures.

Osmosis: The natural tendency of water to migrate through semipermeable membranes from the weaker solution to the more concentrated solution, until hydrostatic pressure equalizes the chemical balance.

Oxidation: When a substance gains oxygen or loses hydrogen or electrons in a chemical reaction; one of the chemical treatment methods.

Oxidizer: A substance that oxidizes another substance.

P

Part per million: An alternative (but numerically equivalent) unit used in chemistry is milligrams per liter (mg/L).

Pathogenic: Disease causing; a pathogenic organism is capable of causing illness.

Permeate: The portion of the feed stream that passes through a reverse osmosis membrane.

Physical treatment: Any process that does not produce a new substance; for example, in wastewater treatment, that would include screening, adsorption, aeration, sedimentation, and filtration.

Pipe bursting: Trenchless technology method that destroys the old pipe while pulling the new pipe in behind.

Planned reuse: The deliberate direct or indirect use of reclaimed wastewater without relinquishing control over the water during its delivery.

Point source: Any discernible, defined, and discrete conveyance from which pollutants are or may be discharged.

Pollutant: Any substance introduced into the environment that adversely affects the usefulness of the resource.

Pollution: The presence of matter or energy whose nature, location, or quantity produces undesired environmental effects. Under the Clean Water Act, for example, the term is defined as a man-made or man-induced alteration of the physical, biological, and radiological integrity of water.

Potable water reuse: A direct or indirect augmentation of drinking water with reclaimed wastewater that is highly treated to protect public health.

Precipitation: Atmospheric water that falls to the ground as rain (a liquid) or snow, sleet, or hail (a solid).

Pressure: The force exerted per square unit of surface area. May be expressed as pounds per square inch.

Pretreatment: Any physical, chemical, or mechanical process used before the main water/wastewater treatment processes. It can include screening, presedimentation, and chemical addition. Also, the practice of removing toxic pollutants from wastewaters before they are discharged into a municipal wastewater treatment plant.

Primary disinfection: Refers to the initial killing of *Giardia* cysts, bacteria, and viruses.

Primary Drinking Water Standards: Safe Drinking Water Act regulations regarding drinking water quality that are considered essential for the preservation of public health.

Primary treatment: The first step of treatment at a municipal wastewater treatment plant. It typically involves screening and sedimentation to remove materials that float or settle.

Publicly owned treatment works (POTW): A waste treatment works owned by a state, local government unit, or Indian tribe, usually designed to treat domestic wastewaters.

R

Receiving waters: A river, lake, ocean, stream, or other water source into which wastewater or treated effluent is discharged.

Recharge: The process by which water is added to a zone of saturation, usually by percolation through the soil.

Reclaimed wastewater: Wastewater that, as a result of wastewater reclamation, is suitable for a direct beneficial use or a controlled use that would not otherwise occur.

Recovery zone: The point in a stream where, as the organic wastes decompose, the stream quality begins to return to more normal levels.

Residual: The amount of disinfecting chemical remaining after the demand has been satisfied.

Return activated sludge solids (RASS): The concentration of suspended solids in the sludge flow being returned from the settling tank to the head of the aeration tank.

Reverse osmosis (RO): Process where solutions of differing ion concentration are separated by a semipermeable membrane. Typically, water flows from the chamber with lesser ion concentration into the chamber with the greater ion concentration, resulting in hydrostatic or osmotic pressure. In RO, enough external pressure is applied to overcome this hydrostatic pressure, thus reversing the flow of water. This results in the water on the other side of the membrane becoming depleted in ions and demineralized.

Rodders: Tool used to clear obstructions such as heavy root accumulations or large, soft obstructions from collection system lines to restore flow.

S

Safe Drinking Water Act (SDWA): Federal law passed in 1974 to establish federal standards for drinking water quality, protect underground sources of water, and set up a system of state and federal cooperation to ensure compliance with the law.

Sanitary sewer: Collection system that carries human wastes in wastewater from residences, businesses, and some industries to the treatment facility.

Sanitary wastewater: Wastewater that includes sewage and industrial wastes discharged from residences and from commercial, institutional, and similar facilities.

Scaling: The precipitation of inorganic salts on the feed-concentrate side of the membrane.

Screening: A pretreatment method that uses coarse screens to remove large debris from the water to prevent clogging of pipes or channels to the treatment plant.

Scum: The mixture of floatable solids and water that is removed from the surface of the settling tank.

Secondary disinfection: The maintenance of a disinfectant residual to prevent regrowth of microorganisms in the water distribution system.

Secondary Drinking Water Standards: Regulations developed under the Safe Drinking Water Act that establish maximum levels of substances affecting the aesthetic characteristics (taste, color, or odor) of drinking water.

Secondary treatment: The second step of treatment at a municipal wastewater treatment plant, which uses growing numbers of microorganisms to digest organic matter and reduce the amount of organic waste.

Sedimentation: Physical treatment method reduces the velocity of water in basins so that the suspended material settles out by gravity.

Septic: Refers to wastewater with no dissolved oxygen present; generally characterized by a black color and rotten egg (hydrogen sulfide) odor.

Septic zone: The point in a stream where pollution causes dissolved oxygen levels to sharply drop, affecting stream biota.

Settleability: A process control test used to evaluate the settling characteristics of the activated sludge. Readings taken at 30 to 60 minutes are used to calculate the settled sludge volume (SSV) and the sludge volume index (SVI).

Settled sludge volume: The volume in percent occupied by an activated sludge sample after 30 to 60 minutes of settling. It is normally written as SSV with a subscript to indicate the time of the reading used for calculation (SSV_{60} or SSV_{30}).

Sewage: The waste and wastewater produced by residential and commercial establishments and discharged into sewers.

Silt density index (SDI): A dimensionless value resulting from an empirical test used to measure the level of suspended and colloidal material in water.

Sludge: The mixture of settleable solids and water removed from the bottom of the settling tank.

Sludge loading rate: The weight of wet biosolids fed to the reactor per square foot of reactor bed area per hour ($lb/ft^2/hr$).

Sludge retention time (SRT): See *mean cell residence time.*

Sludge volume index (SVI): A process control calculation that is used to evaluate the settling quality of the activated sludge; it requires the SSV_{30} and mixed liquor suspended solids test results to calculate.

Solids concentration: The weight of solids per unit weight of sludge.

Solids content (percent total solids): The weight of total solids in biosolids per unit total weight of biosolids expressed in percent. Water content plus solids content equals 100%. This includes all chemicals and other solids added to the biosolids.

Solids loading rate (drying beds): The weight of solids on a dry weight basis applied annually per square foot of drying bed area.

Solids recovery (centrifuge): The ratio of cake solids to feed solids for equal sampling times. It can be calculated with suspended solids and flow data, or with only suspended solids data. The centrate solids must be corrected if chemicals are fed to the centrifuge.

Static head: The actual vertical distance from the system inlet to the highest discharge point.

Steady inflow: Water discharged from cellar and foundation drains, cooling water discharges, and drains from springs and swampy areas. This type of inflow is steady and is identified and measured along with infiltration.

Sterilization: The removal of all living organisms.

Storm sewer: A collection system designed to carry only stormwater runoff.

Stormwater: Runoff resulting from rainfall and snowmelt.

Stream self-purification: The innate ability of healthy streams (and their biota) to rid themselves of small amounts of pollution. Successful self-purification depends on the volume of water the stream carries, the amount of pollution, and the speed the stream travels.

Supernatant: In a digester, the amber-colored liquid above the sludge.

Surface water: All water naturally open to the atmosphere, and all springs, wells, or other collectors that are directly influenced by surface water.

T

Tertiary treatment: The third step in wastewater treatment, sometimes employed at municipal wastewater treatment plants. It consists of advanced cleaning, which removes nutrients and most BOD.

Total dynamic head: The total of the static head, friction head, and velocity head.

Total inflow: The sum of the direct inflow at any point in the system, plus any flow discharged from the system upstream through overflows, pumping station bypasses, etc.

Total suspended solids (TSS): Solids present in wastewater.

Transpiration: The process by which plants give off water to the atmosphere.

Trunk lines: Collection system pipes that carry wastewater flow to interceptors.

Turbidity: A measure of the cloudiness of water. Turbidity is caused by the presence of suspended matter; it shelters harmful microorganisms and reduces the effectiveness of disinfecting compounds.

Turbulence: A state of high agitation. In turbulent fluid flow, the velocity of a given particle changes constantly both in magnitude and direction.

U

Urban water cycle: A local subsystem of the water cycle that is created by human water use; also called an integrated water cycle. These artificial cycles involve surface water withdrawal, processing, and distribution; wastewater collection, treatment, and disposal back to surface water by dilution; and natural purification in a river. The cycle is repeated by communities downstream.

V

Velocity: The speed of a liquid moving through a pipe, channel, or tank; may be expressed in feet per second.

Velocity head: The energy required to keep the liquid moving at a given velocity, expressed in terms of the added system head required.

Vents: Collection system ventilation points that ensure that gases that build up within sewer systems from the wastes they carry are safely removed from the system.

W

Waste activated sludge solids (WASS): The concentration of suspended solids in sludge removed from the activated sludge process.

Wastewater: The spent or used water from individual homes, communities, farms, or industries that contains dissolved or suspended matter.

Wastewater collection system: Community sewerage system to collect and transport wastewater (1) from residential, commercial, and industrial customers; and (2) from stormwater runoff through storm sewers. Wastewater is transported through the sanitary sewer or combination system to a treatment facility. Stormwater is transported through a storm sewer system or combined system to a treatment facility or approved discharge point

Wastewater reclamation: The treatment or processing of wastewater to make it reusable.

Wastewater reuse: The use of treated wastewater for beneficial use such as industrial cooling.

Water cycle: See *hydrologic cycle.*

Water softening: A chemical treatment method that uses either chemicals to precipitate or a zeolite to remove metal ions (typically Ca^{2+}, Mg^{2+}, Fe^{3+}) responsible for hard water from drinking water supplies. The waste byproduct is lime sludge.

Waterborne disease: Illness caused by pathogenic organisms in water.

Watershed: The land area that drains into a river, river system, or other body of water.

Weir: A device used to measure wastewater flow.

Wellhead protection: The protection of the surface and subsurface areas surrounding a water well or wellfield supplying a public water system that may be contaminated through human activity.

Z

Zone of recent pollution: The pollution discharge point where the stream becomes turbid.

Zoogleal slime: The biological slime that forms on fixed-film treatment devices. It contains a wide variety of organisms essential to the treatment process.

Index

A

absorption, defined, 161
acidification, 41
acids, as water contaminants, 49
acquired immune deficiency syndrome (AIDS), *Cryptosporidium* and, 71, 75
activated carbon, 40, 41, 95, 181, 236
activated sludge process, 120, 139, 145, 160–172, 191
  control parameters, 169–172
  equipment, 164
  factors affecting, 165
  membrane filtration, and, 180
  operational control levels, 170–171
  overview, 164–166
  seasonal variations, and, 172
  sidestreams, 171
  startup, 172
  terminology, 161–163
adsorption, 91, 93, 98, 108, 153, 182, 183
  carbon, 181
  defined, 161
  ratio, sodium, 254
Advanced Integrated Wastewater Pond System® (AIWPS®), 143
advanced wastewater treatment, 120, 125, 136, 137, 177–184
aerated grit removal systems, 134, 136
aerated ponds, 141, 144
aerated static pile, 192, 195–196
aeration, 92, 93, 94, 97, 98, 102–103, 134, 136, 144, 161, 191, 193, 255
  activated sludge process, 164
  basin pH, 169
  diffused air, 162
  tank, 160, 164, 169, 191
    activated sludge process, 165
aerobic
  bacteria, 58, 145, 146, 147, 151, 161, 193
  basin, 167, 168
  biological stabilization, 153
  decomposition, 120, 142, 159, 161, 187
  defined, 120, 161, 193
  digestion, 191
  oxidation, 194
  ponds, 141–143, 146, 151
  processes, 52, 139, 143, 144, 169
  protozoa, 60
  respiration, 147–148

Aggressive Index (AI), 102
agricultural reuse of water, 227
air binding, 110, 112
airborne pathogens, 188–189
algae, 52, 53, 56, 91, 92, 104, 108, 129, 141, 142, 143, 144, 145–146, 147, 149, 150, 151, 153, 154, 163, 255
  bacteria, and, 146
  chlorine demand, and 115
  copper sulfate, and, 49
  overabundance of nutrients, and, 177, 183
  types of, 146
algal bloom, 53, 146, 177, 183
alkaline stabilization, 196–197
alkalinity, 12, 49, 50, 127, 150
  activated sludge process, 169
  aerobic digestion, and, 191
  anaerobic digestion, and, 192
  biological nitrification, and, 181
  carbon dioxide, and, 47–48
  coagulation, and, 105
  corrosion, and, 102
  defined, 128
  influent, 191
  pH, and, 149
aluminum sulfate (alum), 105
ambient conditions, concentrate discharge and, 242
ammonia, 139, 148, 153, 168, 177, 180, 181
ammonium, 148
amoebae, 59, 61, 129
amoebic dysentery, 61, 62
amoebic hepatitis, 62
amplifying host, 68
anaerobic, 193
  bacteria, 53, 58, 145, 191, 193
  decomposition, 92, 142, 151, 171, 187
  defined, 53, 120, 193
  digestion, 137, 138, 191–192
  ponds, 140, 141, 142, 143–144, 145, 146, 152
  protozoa, 62
  tank, 166
animal wastes, 68, 126
anions, 6, 13, 18
anoxic
  activated sludge system, 181
  defined, 120
  pond, 144
  tank, 166–168
anoxygenic photosynthesis, 147